M. M. Green, H. A. Wittcoff

Organic Chemistry Principles and Industrial Practice

WILEY-VCH GmbH & Co. KGaA

Prof. Dr. Mark M. Green
Polytechnic University
Brooklyn
NY 11201
USA

Prof. Dr. Harold Wittcoff
Nexant-ChemSystems
44 S. Broadway
White Plains, N.Y. 10601 USA

Library of Congress Card No.: applied for

British Library Cataloguing-in-Publication Data
A catalogue record for this book is available from the British Library.

**Bibliographic information published
by Die Deutsche Bibliothek**
Die Deutsche Bibliothek lists this publication
in the Deutsche Nationalbibliografie; detailed
bibliographic data is available in the Internet at
<http://dnb.ddb.de>

© 2003 WILEY-VCH Verlag GmbH & Co. KGaA,
Weinheim

Printed in the Federal Republic of Germany
Printed on acid-free paper

Typesetting K+V Fotosatz GmbH, Beerfelden
Printing betz-druck gmbH, Darmstadt
Bookbinding J. Schäffer GmbH & Co. KG, Grünstadt

ISBN 3-527-30289-1

M. M. Green, H. A. Wittcoff

Organic Chemistry Principles and Industrial Practice

Dedications

M. M. G. Dedicated to the memory of my father, Irving Green, who worked so hard to make so much more out of life than he was given, and with love and gratitude to my family.

H. A. W. This book is dedicated to my wife Dorothy; and to our childern Mike and Adelle, Ralph and Virginia, and Ted and Cindy; and to our grandchildren Ron and Julie, Mark and Denise and to Wendy, Michelle, and Bessie.

"No, a thousand times no; there does not exist a category of science to which one can give the name applied science. There are science and the applications of science, bound together as the fruit to the tree which bears it."

Louis Pasteur

Contents

Preface

This book is intended for students of organic chemistry who already know something about the subject. It will serve as a text for a quarter or term course to be used in conjunction with any sophomore level organic chemistry text. We have used it successfully in this manner and found that the book works well, using lectures to elaborate on the principles encountered. The book should also prove interesting to those chemistry students whose knowledge of the subject is more advanced. Teachers of chemistry at all levels may find helpful information in this book about the relationships between organic chemical principles and the application of these principles in the chemical industry. We hope that the discovery of these relationships between principles and practice will delight many who discover the basis for a worldwide enterprise, the chemical industry.

From the beginning we realized the difficulty of this undertaking. Every industrial procedure draws on knowledge from almost every aspect of chemical science and chemical engineering and requires input from many technical and financial disciplines. Even if we focus on only those considerations that are chemical in their nature it is impossible to discuss any industrial process without drawing on information from a wide range of sources. If this were *not* done, the subject would become lifeless, drained of all the complexity that makes industrial processes fascinating even though they are subject to uncertainties and empiricism. Information about the chemical industry cannot be bounded in the way a chapter in a textbook of organic chemistry can be. In a textbook the author creates a logical framework and chooses material for each chapter so that it can be explained in full detail at the level of the book. Such an approach has a powerful educational benefit.

But because the chemical industry depends on the use of chemistry in a way that transcends academic boundaries, complexities arise which force us to include aspects of attractive subjects that cannot be fully explored. We are acting like a hiker who chooses a path where there are many hills, valleys, gullies and streams and stops to look more fully at only some of these that may themselves be so complex that they cannot be fully examined in all aspects.

Accordingly, the book is not intended to provide a comprehensive view of the chemical industry or even, necessarily, a comprehensive view of the aspects we focus on. Rather, the purpose of the book is to emphasize selected principles of organic chemistry found in the chemical industry. The criterion for each choice has been that the chemical principles relate to interesting and educational aspects of industrial chemistry. We hope to generate an enhanced appreciation of these principles, to stimulate a return to their study and to spark a desire to understand even more about the chemical industry than we can describe here. Our approach is to repeat many of the same principles from chapter to chapter, thereby demonstrating how certain key ideas find their way into many useful places.

To our surprise as we wrote this book we found that each of us, chemical professionals in different areas, have come to appreciate things about the subject we were entirely unaware of. The book may be interesting to others like us who know the subject well but do not fully appreciate all the subtleties of the relationships between the academic and the industrial.

Many of the Study Guide Problems are challenging and are designed to expand on the discussion in the text and to lead the student to take another look at the organic chemistry they have studied and, very importantly, to go further. The problems can therefore be an extension of the text, asking the reader to find answers to questions that arise from the text but are not fully addressed. The reader will find information in the text and in figures which, if addressed, would take the flow of the story too far a field and which may only be addressed in questions brought up in the Study Guide Problems. Certain questions may therefore require the student to look into some aspect of organic chemistry that is not discussed in this text but which can be found in standard textbooks. However, it is not essential to undertake these problems. Students who can enjoy and derive considerable educational benefit from this book may *not* find it necessary to make these beyond-the-text explorations.

Nevertheless, we have included a list of specialized books in a section entitled "Books for Further Study and Reference," following Chapter 10.

Some of the bullets at the chapters' ends are repeated from chapter to chapter, but this is a consequence of the nature of the book, which shows how many principles and reactive patterns of organic chemistry play multiple roles in the chemical industry.

Regarding some specific characteristics of the book, our nomenclature may not be in line with IUPAC standards because we have tried to bridge industrial and academic usage. In addition, the structures in the figures are generally line drawings in which specific atoms may or may not be shown and in which geometric features such as bond angles are not meant literally as would be in computer-generated structures. The structures follow the kind of "loose" format one often finds in a blackboard presentation in which a drawing is intended to draw attention to certain information but may not be accurate at all levels of inquiry. We hope that reader may find this "looseness" friendly for educational purposes.

Readers will find that discussion of polymers form a significant part of the book. This follows the fact that polymers constitute more than 50% of the output of chemical industry. But polymers also serve our educational purposes because these macromolecules offer an excellent template for discussion of the principles of organic chemistry encountered in molecules of all sizes. At the same time, polymers demonstrate new concepts that are intimately tied to these principles.

Many parts of the book are written as tales or stories of discovery so that the reader may often be presented with facts whose basis, or mechanism, may only be revealed later. In this manner we hope the student has a chance to feel the flow of discovery, of how the facts were revealed to the people who made these discoveries and how they came to realize "what was going on."

Choosing what is possible to understand from a complex situation is akin to the nature of learning. This is how a child learns at the start and continues to learn from the world outside the classroom. It is how we all learn. It is with this philosophy of learning that we write this book. Our aim is to bring something new to the study of organic chemistry and in this way to enhance the appreciation of the beauty of a subject that is so important in our lives.

First, we are grateful to Professor Ernest Eliel who helped convince the publisher of the value of our intentions. We are much indebted for the help we received from Professors James A. Moore of Rensselaer Polytechnic Institute and Herbert Morawetz of the Polytechnic University who carefully read the manuscript and informed us of inaccuracies and lapses in the presentation of the material and gave us numerous suggestions for improvements. Their efforts made the book far better than it would have been. We are also grateful to Professor James V. Crivello of Rensselaer Polytechnic Institute for leading us to information on General Electric's early work on silicon chemistry and also to Dr. Daniel J. Brunelle of General Electric for his essential help on informing us about industrial production of polycarbonate. Thank you to Professor John Ellis of the University of Chulalongkonn, Bangkok, Thailand for his critical reading and suggestions about Chapter 7. Thank you to Professor Edward Weil of the Polytechnic University who helped to bring M.M.G. and H.A.W. together by showing one of H.A.W.'s books to M.M.G. We also thank Professor Roald Hoffmann of Cornell University who read the manuscript and encouraged us to believe we were doing something worthwhile and to Professor Das Thamattoor of Colby College for his detailed comments and encouragement concerning the book's educational value. Thank you also to Professor Qiao-Sheng Hu of the College of Staten Island of the City University of New York for his helpful comments and positive view of the manuscript. Thanks to Professor M.M. Kreevoy of the University of Minnesota for his encouragement. Thanks are also extended to Professors James W. Canary of New York University and also to Rudolf Zentel of the University of Mainz, Bob Grubbs of Cal Tech, Harold Hart of Michigan State University and Maurice Brookhart of UNC Chapel Hill for their positive comments on the manuscript.

A large measure of our gratitude goes to the students whose pictures are shown on the cover of the book. From left to right, Eugene Dimarsky, Jennifer Yim, Melodie Torres, Recia Roopnarine, Senghane Dieng, and Steven Bucholz, all juniors taking CM 3214. They used the book in manuscript form for the first class in Industrial Organic Chemistry given at Polytechnic University. All had only a single prior term studying basic organic chemistry as sophomores, and little knowledge of the chemical industry, both of which helped to make them severe critics of the clarity of the presentation. And vocal critics they were, and the book is greatly improved as a consequence of their constructive complaints.

And thank you Carla Flournoy Green for critical help as in-house editor and computer consultant, and Frank Thurston Green for figure art criticism.

Mark M. Green, New York, New York
Harold A. Wittcoff, Scarborough, New York

Recommendation of the Experts

"I have never come across such an enticing mix of stories of discovery with basic chemistry! If you want to know what paper Carothers should not have published, or what's special about Kraton, if you would like see the world of economic necessity and inventive organic chemistry come together, this is the book to read. With wit and enthusiasm, with a feeling for the molecule (and what makes polymers grow and tangle), Green and Wittcoff pull us into the exciting world of the interplay of organic chemistry principle and industrial practice."
Roald Hoffmann, Cornell University, Ithaca

"This is a highly original book filling an obvious need. It is designed to change the teaching of organic chemistry from an academic discipline hard to digest by most students of chemistry and chemical engineering to a course in which the student is excited by the demonstration how industry is providing us with materials without which modern life could not exist. The book is also important in showing the student that industrial decisions have frequently to be determined not only by scientific principles but also by economic factors and environmental concerns."
Herbert Morawetz, Polytechnic University, Brooklyn

"Simply put, this book is a gem. Green and Wittcoff have done a marvelous job of showing how organic chemistry principles, many of which are typically encountered in an introductory course, are applied in the chemical industry. Beginning with petroleum, the authors present many "delectable" samples of important industrial chemicals and processes and do this in the context of historical perspectives to give the reader an excellent sense of how the industry has evolved. The chemistry described is rigorous but the warm, humorous, and conversational writing style makes the book a joy to read. Furthermore, the interesting stories and anecdotes liberally sprinkled throughout the book are sure to enliven any classroom discussion. Indeed, as a faculty in a small liberal arts college, I am most impressed by the book's pedagogical value in illustrating the practical side of organic chemistry."
Dasan M. Thamattoor, Colby College, Waterville

"This very interesting book is going to find a unique place in the repertoire of organic textbooks. Organic chemistry is taught from a practical perspective with a view to the his-

torical development of both industrial and theoretical ideas. The book follows an approach in which the reader comes to understand how industrial problems have been solved and how organic chemistry principles played a role. At minimum, the book is likely to find its way to the shelves of many organic chemistry teachers as a rich source of illustrations to aid their presentation of the subject. It would serve as an ideal supplement for students to be bundled with textbooks that provide a more traditional approach to organic structure, reactions, and synthesis. Certainly, it would make an outstanding text for a topical course in industrial organic chemistry."
James Canary, New York University, New York

This book is a joy to read (and re-read). It bridges the gap between textbook knowledge and the trials and tribulations that have to be overcome in making a commercial product. In addition, the historical context which surrounds the text brings the reader to new levels of understanding and stimulates the development of insight into the real world of environmentally responsible commercial chemistry. Teachers of chemistry may use this book as a source of "enrichment" that will make their lectures more exciting and informative. My only regret is that the book is not longer so as to include even more examples treated in this marvelous manner.

James A. Moore, Rensselaer Polytechnic Institute, Troy

"This is a unique, fascinating book that bridges organic chemistry principles with chemical industrial applications. The story-telling style makes the reading/learning experience extremely enjoyable. The book is ideal for anyone learning or working in organic, organometallic or polymer chemistry."

Qiao-Sheng Hu, College of Staten Island, City University of New York

1

How Petroleum is converted into Useful Materials:
Carbocations and Free Radicals are the Keys

1.1
The Conflicting Uses for Petroleum: The Chemical Industry
and the Internal Combustion Engine

Petroleum is largely a complex mixture of saturated hydrocarbon molecules, the origins of which are prehistoric plants and animals. As we'll see, small numbers of molecules found in petroleum derive directly from the prehistoric precursors of biochemically active molecules found in all living species today. Chemists use these biomarkers or molecular fossils to study the history of petroleum and to decide where to drill for petroleum. But our story here is of wider interest. It is the story of petroleum as the source of trillions of kilograms of materials used all around us, the petrochemicals that provide textiles, plastics, adhesives, pharmaceuticals, rubbers and coatings to mention but a few. And to a far greater extent, it is the source of much of the energy that yields electrical power, warms our houses and propels our vehicles. Yet petroleum, as pumped from the earth, is ill-suited for these grand tasks. At the close of the nineteenth century it was useful only for illumination. To obtain the proper molecules to make everything from plastics to pharmaceuticals requires breaking petroleum into small molecules and introducing functional groups so that numerous chemical reactions can take place.

As important as petroleum is as a source of chemicals to produce products of the chemical industry, a much larger use in our modern society is the source of fuel for the internal combustion engine. In the USA, hundreds of millions of tonnes of petroleum are required annually (a tonne is 1000 kg, a metric ton). This is the same order of magnitude as the food we produce. For fuel, the molecules must be broken down to a molecular weight range consistent with the required volatility, and the structures must be rearranged so that combustion takes place with the control necessary for modern engines. The objective of the energy industry is to transform petroleum into both aromatic and aliphatic highly branched molecules with desirable combustion characteristics in automobile engines. The very different objective of the chemical industry, as indicated above, is to transform petroleum molecules into functional intermediates capable of participating in numerous chemical reactions.

1.2
How do we achieve these Two Objectives? By Using two different Kinds of Cracking: One depends on Free Radicals and the Other on Carbocations

The story of how petroleum is changed into the molecules necessary for the chemical industry on the one hand and into gasoline on the other is the story of laying the foundation of chemical technology in the twentieth century. And this remarkable story rests on two of the most common intermediates in organic chemical reactions, the carbocation and the free radical. Without these highly reactive trivalent states of carbon, the use of petroleum to support our lives would be impossible. To understand the role of carbocations and free radicals we must first explore two processes to which petroleum fractions are subjected: steam cracking and catalytic cracking.

Cracking is a well-chosen word since the large molecules of petroleum are converted to smaller molecules, just as a large object may be "cracked" into pieces. Why does steam cracking require free radicals, and catalytic cracking carbocations? And why do these intermediates change the petroleum molecules in such different ways?

In *steam cracking*, the saturated hydrocarbons in petroleum are cracked to smaller molecules with double bonds. The conversion from larger molecules to smaller ones arises from the "cracking," that is, breaking of carbon–carbon bonds. The formation of the double bonds also arises from "cracking", that is, breaking carbon–hydrogen bonds to form H_2. In the smaller molecules present in petroleum and in natural gas such as ethane, propane and butane, carbon–hydrogen bonds are cracked and the corresponding olefins are formed. For the larger molecules, cracking of both carbon–carbon and carbon–hydrogen bonds is important.

Why is the formation of double bonds in petroleum-derived molecules so valuable for the chemical industry? The answer is the same for any functional group in organic chemistry – specific chemical reactivity. Virtually all chemical reactivity in organic chemistry is based on transformation of functional groups, and the double bond is the simplest of all the functional groups. The value of the double bond therefore arises for two reasons: (i) from its ease of formation by cracking of petroleum; and (ii) from its specific reactivity. If a double bond is present, then chemists can design chemical reactions to yield specific products. For example, the double bond in propylene allows formation of isopropyl benzene or polypropylene, while the double bond in ethylene allows formation of ethylene oxide or vinyl chloride. Saturated hydrocarbons, in contrast, generally are less reactive than olefins, and when a reaction is possible it does not occur in a specific manner. Chemical reactions of saturated hydrocarbons most often yield mixtures.

In *catalytic cracking*, large molecules are also broken to smaller ones. But the intermediate carbocations favor rearrangement of the often linear or cyclo-aliphatic carbon skeletons to highly branched, but still saturated, hydrocarbons. Such molecules behave very well in the modern internal combustion engine. Every gas station tells us this when it touts high-"octane" number gasoline.

1.3
What is in Petroleum?

In Fig. 1.1, we see examples of the types of molecules found in petroleum. There is an enormous structural range, from small molecule gases to high molecular-weight substances so involatile that they cannot be vaporized at the highest accessible temperatures, even under high vacuum. In fact, petroleum contains so many different types of molecules of such varied structure that the tools of analytical chemistry are overwhelmed by the task of identifying all of them. The way that chemists deal with complex mixtures is by *fractionation*. The first task, then, is to fractionate petroleum by using, on a very large scale, a technique familiar to every chemist, fractional distillation. Some of the huge towers one observes when driving past a petroleum refinery are the distillation columns. What a difference from the small glass column used in the laboratory! But the principles

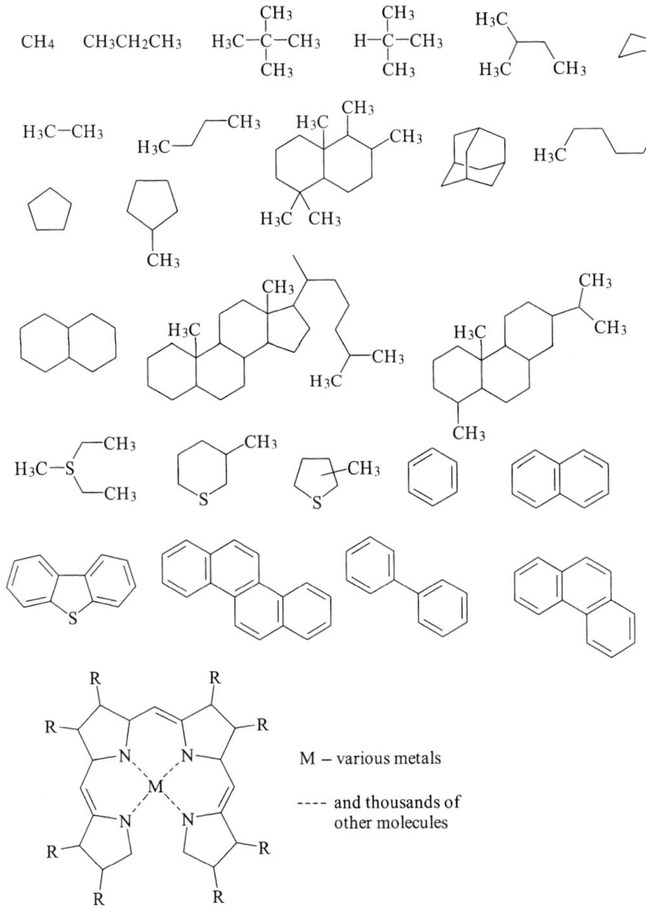

Fig. 1.1   Examples of the range of molecules found in petroleum.

Crude Oil Distillation

	Fraction	Boiling Point Range	Comments
1	Gases mostly CH_4	< 20°C	Similar to natural gas
2	Naphtha (light)	70—140°C	Largely C_5 to C_9 hydrocarbons
3	Naphtha (heavy)	140—200°C	Largely C_7—C_9 hydrocarbons
4	Atmospheric gas oil		
	Kerosene	175—275°C	Mostly C_9—C_{16} hydrocarbons
	Diesel Fuel	200—370°C	Mostly C_{15}—C_{22} hydrocarbons
5	Heavy Fractions	> 370°C	Molecules too involatile to distill under high vacuum
	Lubricating Oil Residual or heavy fuel Oil Asphalt or "resid"		

Fig. 1.2 Types of molecules found in various petroleum fractions.

applied are identical if one is fractionating 1 gram or thousands of tonnes. Fig. 1.2 relates the temperature range for distillation of the various fractions to the types of molecules present in the fraction. Certain of these fractions obtained by fractional distillation of the petroleum are then chosen for the different cracking processes.

1.4
The Historical Development of Steam Cracking

In addition to the two types of cracking already described, there is a third type, *thermal cracking*, which is less important today although it was the first type of cracking invented. The concept of thermal cracking is attributed to Professor Benjamin Silliman of Yale, who in 1855 suggested that during the distillation of petroleum (then known as rock oil, a term later replaced by petroleum, a contraction of the Latin *petra* (rock) and *oleum* (oil)) it decomposes and that some of these decomposition products might have economic value.

Economic value in the mid-nineteenth century meant fuel for lamps – a product that today we would call kerosene and which is now used to fuel airplanes. Patents issued between 1860 and 1912 show that there was considerable interest in achieving higher kerosene yields from petroleum. But it was not until 1912 that William Burton and his associates at Standard Oil of Indiana were granted a patent that defined a practical process called thermal cracking. A plant, built one year later, produced equal amounts of gasoline (which by that time was a much sought-after product), kerosene and residual or unreacted petroleum.

Coking was a serious problem, and coke – a porous, hard, involatile residue consisting mostly of carbon – was removed after every 48-hour cycle by laborious means. The need to avoid labor costs led to the Dubbs process; this was a development of a father and son who shared an intense interest in petroleum. Dubbs originally named his son simply Carbon Dubbs. Later, Carbon assumed the middle name of Petroleum. Carbon then named his own daughters Methyl and Ethyl. In engineering school, Carbon learned about partial pressures, and accordingly suggested to his father that if water were included in the thermal cracking reaction then the partial pressures of the hydrocarbons would be reduced. Because coke formation arises from the combinations of the petroleum-derived molecules and their reactive free radical fragments, reducing the partial pressure reduced coke formation by limiting the number of collisions between these fragments. In this manner, steam cracking replaced thermal cracking for the transformation of petroleum into useful chemicals.

Most important from the chemist's point of view, both thermal cracking and steam cracking produced olefins. Although not as many olefins – and certainly not in the quantities produced today from petroleum – this early production of olefins nevertheless started the ball rolling toward the connection of the chemical industry to petroleum. As we shall see when we describe the mechanism of steam cracking, free radical chain reactions were responsible for producing the olefins. The modern steam cracker is today the heart and soul of the chemical industry because it produces three of the industry's basic building blocks: ethylene; propylene; and butadiene. Also produced by steam cracking are lesser amounts of isobutene, 1-butene, 2-butene, and C_5 olefin isomers, which are all important intermediates for the chemical industry. And, under certain conditions and using fractions of petroleum containing C_7–C_{12} hydrocarbons, steam cracking also produces benzene and toluene. Worldwide, this is a major source of these aromatics. Literally, there is here a cornucopia of molecules to form the foundation of the chemical industry and therefore of our modern world.

During the 1920s, Union Carbide and Standard Oil of New Jersey were the first companies to start experiments with steam cracking and the subsequent conversion to functional products of the ethylene and propylene produced. It was in fact in 1920 that Standard Oil reacted propylene with water in the presence of an acid catalyst to provide the first petrochemical, isopropanol, a classic Markovnikov addition of water to an alkene. Its major use was not for rubbing alcohol but for oxidation to acetone, a solvent whose first use was developed in World War I as a component for the preparation of the explosive, Cordite. In Chapter 9 (see Section 9.2), we shall discover how another route to acetone for its use in Cordite played a role in world history, and in Chapter 4 (Section 4.13) how Cordite is related to dynamite, the explosive that generated the wealth behind the Nobel Prizes.

Chemists long realized that knowledge of the specific reactivity of a functional group would allow prediction of the chemical behavior of a wide range of molecules containing that functional group. And steam cracking made widely available building blocks with the most basic functional group, the alkenes. Suddenly, all kinds of new possibilities were visualized because of the availability of large amounts of these olefins. Therefore it became necessary to understand the reactivity patterns of the double bond and the mechanisms of the possible reactions. This information was gained by research motivated by the needs of the chemical industry.

1.5
What was Available before Thermal and Steam Cracking?

What kinds of molecules with what functional groups were available to the chemical industry before alkenes? The initial major building blocks were aromatics – benzene, toluene, and naphthalene – and the exceptionally dangerous explosive, acetylene. The aromatics came from coke oven distillate or coal tar and were important early in the Industrial Revolution. Coke, noted above as a hard, nonvolatile residue of petroleum cracking, was first made from coal by distilling off the volatile components. Actually, a patent for the conversion of coal to coke was first issued in England in 1590, but it was not until 1768 that a certain John Wilkinson built a practical oven for converting coal to a purer form of carbon. As development progressed it became obvious that the distillate from the operation contained a wealth of chemicals. The benzene, naphthalene, and anthracene present in the coal tar provided the basis for synthetic dyes, which had properties that rivaled and even exceeded those of natural dyes. In the manner described above, the chemical industry found the beginnings that led to the huge enterprise it is today. It was, however, not until very much later that steam cracking made ethylene and propylene available so that benzene's two most important derivatives – ethylbenzene for styrene and isopropylbenzene for phenol and acetone – could be made cheaply. These processes will be discussed in Chapter 3.

1.6
Acetylene was Widely Available before Steam Cracking and it was Exceptionally Useful but Many wanted to replace this Dangerous Industrial Intermediate.
Happily, Double Bonds replaced Triple Bonds

Acetylene, which will be discussed extensively in Chapter 10, was important in the early days of the chemical industry. Its triple bond is capable of undergoing a wide range of reactions, and in addition there is a special reactivity of the hydrogen bound to the triply bonded carbon atom. Acetylene and alkynes in general are still used today in synthetic work to make complex molecules, and a whole chapter in most organic chemistry textbooks is devoted to acetylene. But acetylene is no longer a valuable part of the bulk chemical industry in spite of the fact that large-scale organic molecules were made from it before the advent of the alkenes from petroleum cracking.

Some of the most important chemicals supporting our modern technology – vinyl chloride, vinyl acetate, acetaldehyde, 1,4-butanediol, and from it 1,3-butadiene – could be prepared from acetylene (Fig. 1.3) although all these processes are now obsolete except for 1,4-butanediol preparation. Even here however, alternate processes are in use and will eventually take over. For the chemical industry, the double bond has displaced the triple bond.

What is the problem with acetylene? Why should the industry rush to replace acetylene with the alkenes produced by steam cracking of petroleum fractions? The answer, as will be discussed in detail in Chapter 10 (Sections 10.2 and 10.3), is found in its explosive characteristics and its cost.

Fig. 1.3 Chemicals historically derived from acetylene.

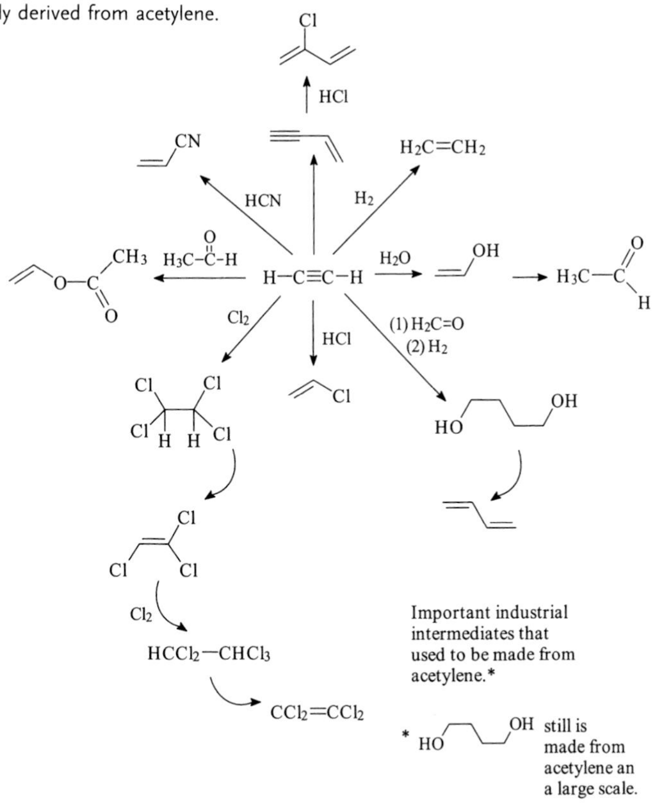

By 1920, the reaction of calcium carbide with water was the accepted way to make acetylene (Chapter 10, Section 10.4), which was used for lighting and welding torches. These applications still apply and take advantage of acetylene's high heat of combustion leading therefore to the extremely high temperatures reached when acetylene burns. In other words, the same character that makes acetylene so dangerous also, under controlled conditions, makes it so useful. At that time, The Union Carbide and Carbon Chemicals Company – a name derived from the source of acetylene – and a company name later shortened to Union Carbide, was founded to make olefins and aliphatic chemicals derived from acetylene.

George Oliver Curme, Jr., who was working at the Mellon Institute in Pittsburgh, had started working in 1913 on more economical routes to acetylene and demonstrated that acetylene could be produced via thermal cracking of hydrocarbons. Union Carbide became one of his sponsors. But ironically, while Curme attracted the attention of a company interested in acetylene production he discovered that olefins could also be produced by heating petroleum and realized that ethylene in terms of safety and cost had advantages over acetylene. Instead of focusing on acetylene, Curme started to work on developing a chemistry to replace acetylene-based processes with ethylene-based processes. Those ornery professors!

1.7
Petroleum yields Ethylene and lays the Groundwork for a New Kind of Chemical Industry

Thus, in the early part of the twentieth century it became apparent that useful industrial intermediates could be derived from alkenes and that alkenes – and especially ethylene and propylene – could be derived from petroleum. This provided Union Carbide with the impetus to start developing the petrochemical business, a commercial enterprise that would start with petroleum and through the intermediacy of olefins synthesize high value-added industrial intermediates. Standard Oil also recognized the opportunities as noted in the synthesis of isopropanol from propylene. Isopropanol was the world's first petrochemical (see Section 1.4). Suddenly, petroleum became useful not only for the combustion properties of its molecular constituents, which could be used to light lamps or power internal combustion engines, but also as a source of chemicals derived from alkenes. Witness the birth of the petrochemical industry!

The advent of polyethylene (Chapter 2, Section 2.2) immediately before World War II and the synthesis of styrene (Chapter 3, Section 3.3) as a component of synthetic rubber during World War II would not have been possible without a cheap source of ethylene. And steam cracking was a source far superior to the old method of dehydration of fermentation-based ethanol or the even more expensive hydrogenation of acetylene (Fig. 1.3).

By 1970, the steam cracker had firmly established itself as a way to produce ethylene and other alkenes, and acetylene was dropped from the list of the 50 most important chemicals in the United States. The conversion of Burton's, Dubbs' and Curme's developments into a modern cracker – the cost of which is close to one billion dollars – involved engineering work on the part of numerous companies. In 2002, the German company BASF and the Belgian company Fina, completed a cracker in the United States that produces 1.7 billion kilograms of product – 0.9 billion kilograms of ethylene and the remainder as C_3–C_4 alkenes. A world scale cracker, prior to this, produced 0.7 billion kilograms of ethylene. Even larger crackers than that of BASF are in the planning stages in 2002. In this business, bigger is better, that is, more economical.

1.8
But What about that Thirsty Internal Combustion Engine?
The Development of Catalytic Cracking

Catalytic cracking of petroleum is designed to favor rearranging the molecular structure to maximize branching, rather than cracking the molecules into smaller fragments leading to double bonds, as is the situation in steam cracking. Although fragmentation does certainly occur, the petroleum fraction (see Fig. 1.2) most desirable for catalytic cracking is in the range of volatility compatible with end use as gasoline. The idea here is to maintain the volatility of these molecular components of petroleum while rearranging their structure from linear to branched hydrocarbons.

What's wrong with the volatile small molecules produced in modern steam crackers for use in internal combustion engines? Why is gasoline not produced this way? It could be, but our automobiles would not work very well. The reason is that in steam crackers the only source of energy for breaking down the petroleum molecules is heat; and the only chemical intermediates driving the chemical processes are free radicals. As we shall discuss in detail later, these conditions can break down the petroleum molecules but cannot cause rearrangements of the structure. If the original molecule in petroleum is linear, so the fragment produced by the bond breaking will be linear. There will be, with rare exceptions, no rearrangements of the carbon skeleton in the fragmentation reactions.

Gasoline contains both aromatic and aliphatic molecules and if the aliphatic molecules are linear hydrocarbons, saturated or not, one produces poor gasoline. Only hydrocarbons with branched structures and with the volatility compatible with the internal combustion engine, molecules in the range of C_8 hydrocarbons, allow for control of the rate of combustion necessary to drive these engines at their highest performance.

If the rate of combustion in the cylinders of the internal combustion engine is not compatible with the smooth movement of the pistons, the engine "knocks." Engines running on linear hydrocarbons such as *n*-heptane (assigned octane number zero) have high knocking characteristics, while branched hydrocarbons such as 2,2,4-trimethylpentane (assigned octane number 100) cause engines to run smoothly. The octane number of a fuel is the percentage of 2,2,4-trimethyl pentane that must be mixed with *n*-heptane to give the same knocking behavior as the new fuel. Gasoline with an octane number of 86 therefore corresponds to an 86:14 mixture of the branched to the linear hydrocarbon.

How can we convert the linear hydrocarbons found in petroleum to branched hydrocarbons that will work well in the internal combustion engine? We need a chemical intermediate that can tear the molecules of petroleum apart and at the same time rearrange the carbon skeletons to branched structures (Fig. 1.4). As we shall see, the necessary intermediate is a carbocation and for its formation something more than heat is necessary. To produce a positive charge at the carbon atom, a carbocation, we need an acid catalyst – hence the name catalytic cracking.

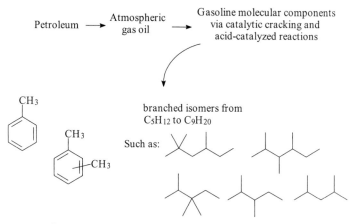

Fig. 1.4 The route to high-octane gasoline.

1.9
Discovery of the Proper Catalyst for Catalytic Cracking:
From Natural Synthetic Zeolites

As internal combustion engines increased in power, the inadequacy of thermal and then steam cracking to produce appropriate gasoline became apparent. There is a legend in the petroleum industry that the first successful catalytic cracking process was invented during the 1920s by A.J. Houdry, who was searching for a better gasoline for his race car. Chemists had tried before to use acid catalysts for cracking, but these were based on aluminum trichloride, which proved inadequate. Houdry's catalyst was a natural clay or zeolite, a complex structure based on silica and alumina. We shall encounter zeolites again in Chapter 3 (Sections 3.6 and 3.7) on electrophilic aromatic substitution where the acidic characteristics necessary for that reaction are described.

The basic building units in the zeolites are tetrahedral arrangements of silicon and aluminum. Four oxygen atoms surrounding each metal atom bridge the metals into a complex network. But the remarkable characteristic of zeolites is that this network structure is formed with a regularity giving rise to a microporosity in which channels are formed of dimensions similar in size to that of small hydrocarbon and aromatic molecules, several Angstroms to tens of Angstroms across. These critical dimensions are subject to control by the synthetic methods now used to prepare the zeolites.

What is the nature of these channels? Because silicon is neutral in its tetravalent state, there is no charge associated with silicon bonded to four oxygen atoms. However, aluminum is neutral in its trivalent state and because every aluminum atom in the zeolite is bonded to four oxygen atoms, each aluminum atom carries a negative charge requiring, therefore, a counterion to balance this charge. Fig. 1.5 shows the general structure of part of the zeolite showing the bonding between the tetrahedral aluminum and silicon oxides and the formation of the cavity. The charge character of zeolites is seen clearly in the general formula for these substances with sodium as the positive counterion: $Na_x[(AlO_2)_x(SiO_2)_y][H_2O]_y$.

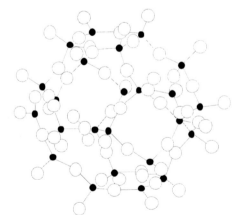

Fig. 1.5 Zeolite structure. The arrangement of AlO_4 and SiO_4 tetrahedra which gives the cubo-octahedral cavity in some zeolites and felspathoids. • represents Si or Al. (Reproduced from *Advanced Inorganic Chemistry*, 3rd edition. F.A. Cotton and G. Wilkinson, Interscience/Wiley, 1972, p. 324.)

The channels therefore are highly polar! The counterion is usually a sodium ion, but methods are available to exchange the Na^+ for NH_4^+. If strong heating follows this exchange, ammonia is released leaving behind a proton as the positive counterion for the negative charge at each aluminum atom in the zeolite network. Not only are the channels now highly polar but they are also sources of Brønstead acidity, a potent chemical mix. As in an ancient Egyptian pyramid, the molecular guest of proper size and shape is led through these channels, as in a gauntlet, to inner chambers with all the walls lined with numerous charged aluminum atoms attended by their proton counterions. This shape and size selectivity of the zeolite is one of the most fascinating aspects of industrial zeolite catalysis. The chemistry catalyzed by this environment is therefore restricted to molecules of certain sizes, in contrast to exposure to a polar and acidic environment in open reaction vessels in which any molecule with the proper functionality may be subject to catalysis.

When Houdry first used these materials, they were found in nature as clays and very little was known about them. The natural zeolites had both the proper acidity on appropriate treatment and also pores of the proper size to admit suitable molecules found in petroleum. There followed the rearrangement into the branched structures necessary for a good gasoline. The zeolites used by Houdry made catalytic cracking practicable. They were a great improvement over earlier attempts by chemists at Gulf Oil Company to use acid catalysts based on aluminum trichloride, which was a rational choice considering its use in the Friedel–Craft reaction.

But Houdry's catalyst required frequent regeneration. It was not a complete success because coke – that old nemesis of thermal cracking – had to be burned off, and the intense heat required threatened the integrity of the catalyst. Nevertheless, two companies, Socony Vacuum Oil Co. (now ExxonMobil Oil Co.) and Sun Oil Co. both built Houdry plants in 1936. These plants were tiny by today's standards, but they proved that the process worked. And in fact they were just in time to provide gasoline for the battle of Britain, that famous war in the air that doomed the air superiority that Nazi Germany counted on for their ultimate success in World War II. The Luftwaffe had the right to expect success in the air war. After all, they had superior aircraft with more powerful engines and excellent pilots. But the British had gasoline produced by the Houdry catalytic cracking process – gasoline with the branched structure associated with high octane numbers. This gave the British airplanes the greater power and maneuverability that contributed greatly to dashing the myth of Nazi superiority.

Subsequently, The Standard Oil Development Co., working closely with noted thermodynamicist, W.K. Lewis of M.I.T., conceived the idea that a powdered zeolite catalyst might be useful. And from this evolved the process known as fluidized catalytic cracking (FCC). The powdered catalyst is "fluidized" in a vessel by gas jets and thus provides not only a huge area for access of the molecules to be subject to catalysis by the zeolites but also a convenient way to remove the heat of reaction. In addition, the zeolite catalyst can be transferred easily for it actually takes on the characteristics of a fluid, which has been described as having the characteristics of quicksand. Many improvements in the zeolite catalysts have been made in the ensuing years, mainly by using synthetic zeolites in place of the acid-leached montmorillonite clays originally used by Houdry. An important modification is a combination of these synthetic clays with a zeolite designated ZSM-5.

ZSM is an acronym for Zeolite Socony Mobil, the first name for the company that became Mobil, and subsequently ExxonMobil.

1.10
Let's compare the Mechanisms of Steam and Catalytic Cracking: Free Radicals versus Carbocations

Free radicals, the important intermediate in steam cracking and carbocations, the important intermediate for catalytic cracking, are both trivalent reactive species in which one electron is missing in the former and two in the latter. Thus, they share electron deficiency. For one example of a structural characteristic shared by carbocations and free radicals, organic chemists discovered long ago that increased stability arises from higher substitution in both of these intermediates. Both tertiary carbocations and tertiary free radicals are more stable than secondary ones, which are still more stable, respectively, than carbocations and free radicals based on primary carbon. The least stable are methyl free radicals or methyl cations. The quantitative data are found in Fig. 1.6, which presents the bond dissociation energies for formation of both free radicals and carbocations for various levels of substitution. The data clearly show that the bonds that must be broken to produce either carbocations or free radicals are weaker for a higher level of substitution and moreover that substitution makes a far larger difference to carbocations than to free radicals. We'll see the great importance of this latter fact in the discussion to follow.

1.11
How are Free Radicals formed in Steam Crackers, and What do they do?

The chemical reactivities of carbocations and free radicals are very different as are the conditions for their formation. In the absence of either highly dipolar interactions on the one hand, or in polar solvents or acidic catalysts on the other hand, chemical bonds sub-

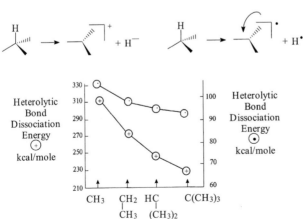

Fig. 1.6 Comparative plots of enthalpy versus degree of substitution for carbocations and free radicals.

jected to high energy tend to break homolytically. The fragments produced are free radicals. Each atom making up the covalent bond takes one of the shared electrons, and two free radicals are formed.

Chemical bonds are very strong, in the range of nearly 100 kcal per mole. Even at the high temperatures of the steam cracker, 650–900 °C, this means that enough homolytic breaking of carbon–carbon and carbon–hydrogen bonds in the petroleum molecules could not take place to account for the observed rapid breakdown of the petroleum. The petroleum molecules are exposed under the steam cracking conditions for only 30 to 100 milliseconds!

However, small concentrations of free radicals are produced. For example, reaction of oxygen with a hydrocarbon at these temperatures would lead to the abstraction of a hydrogen atom leading to both a carbon-based radical and a hydrogen peroxide radical. Homolytic breaking of carbon–carbon or carbon–hydrogen bonds does occur to a small extent, and this is another source of free radicals. The free radicals produced by these initiation steps, as seen below, can act far beyond their low concentrations to exert their effect on the petroleum. They are the initiation steps of a free radical chain process (Fig. 1.7). The propagation steps that follow these initiation steps do the major job leading to the conversion of petroleum to the low molecular weight olefins obtained.

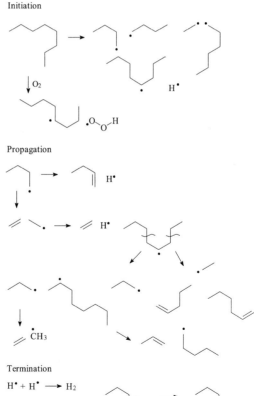

Fig. 1.7 Examples of reactions occuring in a steam cracker. Initiation, propagation and termination reactions of the chain process taking place for *n*-octane.

All the propagation steps in Fig. 1.7 are β-scissions, and each step propagates the free radical chain reaction since new free radicals are produced. Each new radical produced may then undergo another β-scission to produce another olefin and still another free radical, which may undergo another β-scission. At any step, a free radical produced may abstract a hydrogen atom from a new hydrocarbon molecule, thus further propagating the chain reaction by incorporating more molecular bystanders to enter into the chain process. Although from time to time two free radicals happen to come together and form a sigma bond and therefore reduce the overall concentration of free radicals in what is called a termination step (see Fig. 1.7), new initiation steps form new radicals to counter this loss. Remarkably, all of this happens in milliseconds as the petroleum fraction enveloped in steam passes down the hot tube to be transformed from a product pumped from the earth to low molecular-weight molecules of great value to the chemical industry.

It is clear that the initial free radicals produced by the chemical reactions shown in Fig. 1.7 are capable of wreaking havoc, so-to-speak, among many molecules, cracking them to smaller molecules. In this way, a single initiation event can produce large numbers of olefin molecules, as many as thousands to one. The free radical chain reaction plays an important role in much of industrial chemistry. This is seen in two of many examples in Chapter 2 on the discussion of the free radical polymerization of ethylene (Section 2.2) and in Chapter 3 on the production of cumene hydroperoxide (Fig. 3.11). In every instance the essential steps are the same – initiation, propagation, and termination. What has just been described is no different from the mechanism of the classic example of the free radical chain reaction that produces methyl chloride from methane.

β-Scission, shown in the propagation steps in Fig. 1.7, is an interesting reaction in which the radical at the secondary site breaks a bond once removed from it in order to form a double bond. Here, we find a classic route taken by these high-energy intermediates. The β-scission certainly removes the uncoupled high-energy radical intermediate from the site it occupies, converting the radical to a π-bond. But it does this at the cost of breaking a sigma bond and forming another free radical. And the new free radical is now not secondary but primary, a higher energy state according to the data in the graph in Fig. 1.6.

But in free radicals the energy difference between primary and secondary radicals is not very large (see Fig. 1.6), therefore easing the β-scission path. The overall enthalpic increase – the process is endothermic – is compensated for by the increase in entropy that results from the breaking of the larger molecules into smaller ones. Most important is the high temperature, which following the rule of LeChatelier increases the equilibrium constant favoring the endothermic direction, the cracking path. Because heat is absorbed in going from petroleum to the cracking products, supplying heat favors this direction.

Full naphtha, the combined petroleum fractions labeled light and heavy naphtha in Fig. 1.2, containing predominantly hydrocarbons with five to nine carbon atoms (see Fig. 1.1) is one of the feedstocks (although not the only feedstock) used for steam cracking. The steam cracking of full naphtha produces large amounts of ethylene, propylene, butadiene, isomeric butenes, toluene and xylenes. In the words of the chemical industry, the products of the cracking represent how much value is added to the petroleum (Fig. 1.8).

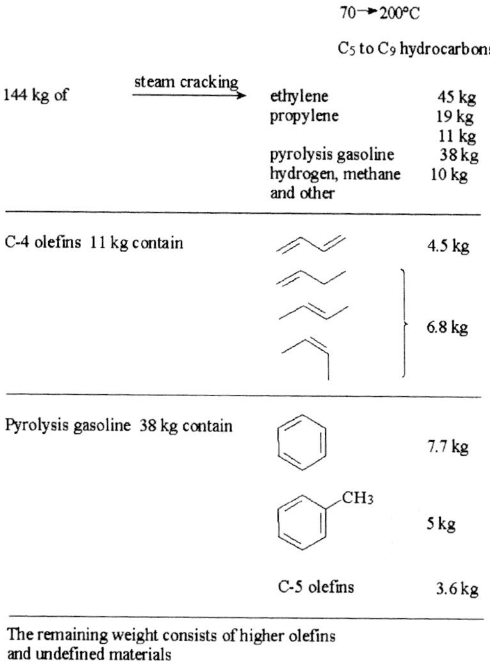

Fig. 1.8 Products created by the steam cracking of a petroleum fraction.

The remaining weight consists of higher olefins and undefined materials

1.12
Now let's look at Catalytic Cracking and the Essential Role of Carbocations and their Ability to rearrange the Structure of Organic Molecules

When the natural clays were first introduced in the 1930s to improve the quality of gasoline (see Section 1.9), there was no understanding of the mechanism. It did not, however, take long for carbocations to be identified as the responsible reactive species. Two main lines of evidence supported this supposition. First, it was discovered that acidity – specifically Brønstead acidity – was critical to the function of the zeolite catalysts. Second, it was observed that isomerizations take place leading to the production of branched hydrocarbons. Many years of research on carbocations have shown their capacity to undergo skeletal rearrangements that interconvert constitutional isomeric forms.

Much is understood about what is happening in catalytic cracking: (a) the transformations undergone by petroleum molecules in a catalytic cracker arise from the intervention of carbocations; (b) the carbocations are formed from the petroleum molecules or their thermally cracked products by reacting with the acidic sites in the zeolite catalysts; and (c) the rearrangements producing constitutional isomers that are associated with the carbocations are not observed with free radicals, and here resides the difference between catalytic cracking and steam cracking (Figures 1.8 and 1.9). Steam cracking, rather than initiating rearrangements of the carbon skeletons of the petroleum molecules, yields ethylene as the most abundant product. This is the inevitable result of a series of β-scission reactions scissoring the unrearranged linear hydrocarbon chains in the atmospheric gas oil (Figures 1.7 and 1.8).

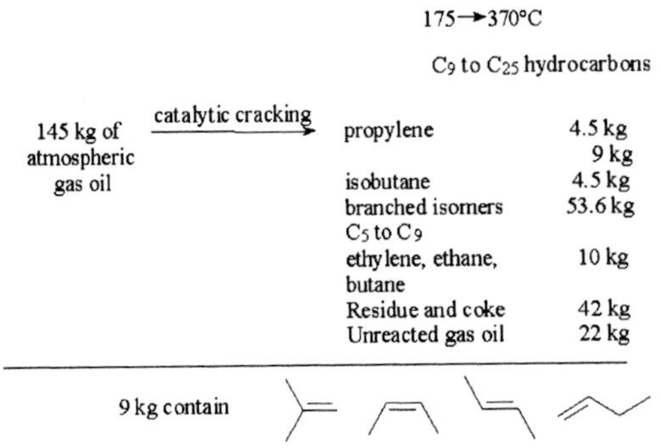

175 → 370°C

C9 to C25 hydrocarbons

145 kg of atmospheric gas oil — catalytic cracking →

propylene	4.5 kg
	9 kg
isobutane	4.5 kg
branched isomers C5 to C9	53.6 kg
ethylene, ethane, butane	10 kg
Residue and coke	42 kg
Unreacted gas oil	22 kg

9 kg contain

Fig. 1.9 Examples of the molecular components obtained via catalytic cracking of atmospheric gas oil.

The branched molecules produced via catalytic cracking, which are useful for gasoline, are of far lower average molecular weight than the petroleum molecules in the gas oil fraction, which are in the range of mostly linear hydrocarbons of 9 to 25 carbons (Fig. 1.9). Therefore, not only rearrangement is occurring but also cracking – that is, breaking the molecules in the petroleum into smaller molecules but generally larger than the molecules produced in steam cracking. The difference is the difference between the reactivity of carbocations and free radicals.

1.13
What's going on inside those Zeolite Pores?

The nature of the zeolite catalysts and their gauntlet of acidic pores of narrow molecular size were described in Section 1.9. The chemistry in the catalytic cracking process takes place hidden in these analytically inaccessible and therefore somewhat mysterious pores. Although this hinders full understanding of the details of the chemical events that convert the molecules in petroleum to the branched hydrocarbons produced in the catalytic cracking process, petroleum chemists have a good idea of the overall mechanism.

The first problem is to understand how a saturated hydrocarbon is converted to a carbocation even in the presence of strong Brønstead acidity. There is evidence that a proton can react with a saturated alkane to form a carbocation via loss of hydrogen (Fig. 1.10). In essence, this amounts to the proton abstracting H$^-$, that is, a hydride ion, from the alkane. The tricoordinate positively charged carbon produced by this reaction, which we have been calling a carbocation, is also known as a carbenium ion and one will find this nomenclature in use in the industrial literature.

Another possibility for the formation of carbocations is that some thermal cracking occurs in the zeolitic pores of the catalytic cracker producing free radicals as in steam cracking and these free radicals undergo β-scission to form alkenes (Fig. 1.7). The double bond would be protonated very rapidly in the presence of the strong Brønstead acidity in the zeolite pore to form a carbocation. This proposal not only yields a mechanism for forming the carbocation but also offers an explanation for the cracking of the large molecules in the petroleum to the smaller molecules that occur in steam cracking. In the at-

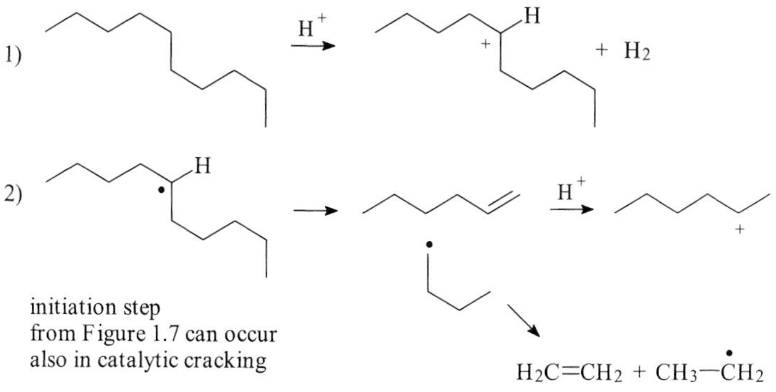

Fig. 1.10 Mechanisms for forming carbocations in catalytic cracking.

1)

$$\text{H}^+ \longrightarrow \qquad + \text{H}_2$$

2)

initiation step
from Figure 1.7 can occur
also in catalytic cracking

$$\xrightarrow{\quad} \qquad \xrightarrow{\text{H}^+} \qquad$$

$$\text{H}_2\text{C}=\text{CH}_2 + \text{CH}_3\overset{\bullet}{-}\text{CH}_2$$

mosphere of steam cracking the absence of acidity translates to stability for the olefins produced, which are then products of the steam cracking process. However, in the environment of the zeolite pores in catalytic cracking the olefins produced would be protonated by the strong Brønstead acidity. These possibilities for initiation of the carbenium ions are shown in Fig. 1.10.

The carbocations once formed – as judged from their behavior in model systems under conditions where they can be studied in detail because they are not hidden in the pores of the zeolite – are known to undergo the kinds of rearrangements that produce exactly the branched molecules isolated from the catalytic cracker (see Fig. 1.9). Fig. 1.11 shows how the rearrangement of the 2-heptyl cation produced in the model reaction demonstrated in Fig. 1.10 can yield 2,3-dimethylpentane, one of the components of gasoline (see Fig. 1.4).

The key steps in this isomerization are 1,2-shifts of both an alkyl group and hydrogen. Some of the rearrangement steps shown in Fig. 1.11 are energetically uphill, processes that are aided by the high temperature of the catalytic cracking, which takes place in the region of 500 °C. Thus, secondary carbocations form primary carbocations or tertiary carbocations form secondary carbocations as momentary intermediates on the way to forming more stable carbocations (Fig. 1.11). All the rearrangement steps are reversible and favor the more substituted carbocation (Fig. 1.11). However, the unstable less-substituted carbocation, although not favored, can rapidly rearrange to the more stable carbocations (shown in Fig. 1.11), driving the structures increasingly toward the branched isomers. In other words, although there are some intermediate steps that raise the energy of the rearranging carbocation, the overall energetic direction is toward the most substituted carbocation (Fig. 1.11). The 1,2 shifts go on repetitively in the zeolite pores for every carbocation formed, thus eventually transforming linear to branched structures.

No one knows exactly how the isomerized carbocations are transformed to the neutral molecules released from the pores but it is likely that the alkenes produced, as shown in Fig. 1.9, arise from proton loss and hybride transfer reactions (Fig. 1.12). In this manner, the saturated alkanes are thought to arise from what are called chain transfer reactions, a term which is also used in polymerization as discussed in Chapter 2 (Section 2.3). In this process the isomerized carbocation abstracts a hydride ion (H⁻) from a neutral molecule in the pore, as shown in Fig. 1.12, producing a new carbocation, which itself can then go on to

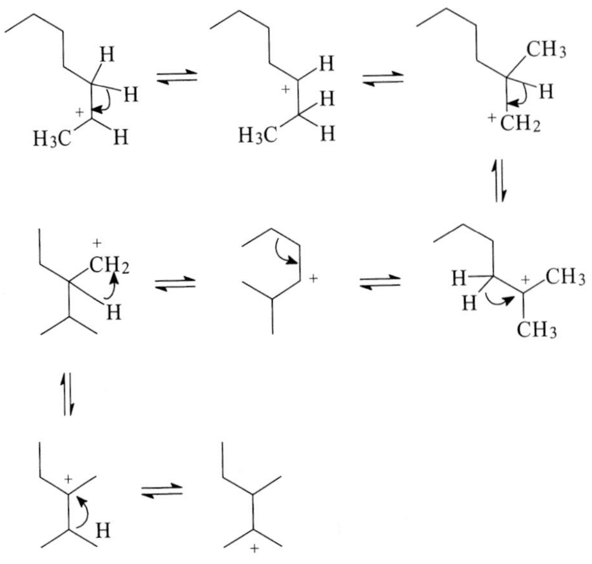
H
H
H
H3C
H
H
H
H3C
H
CH3
H
+CH2
CH2
H
H
H
CH3
CH3
H

Fig. 1.11 Mechanistic detail of isomerization of the heptyl cation.

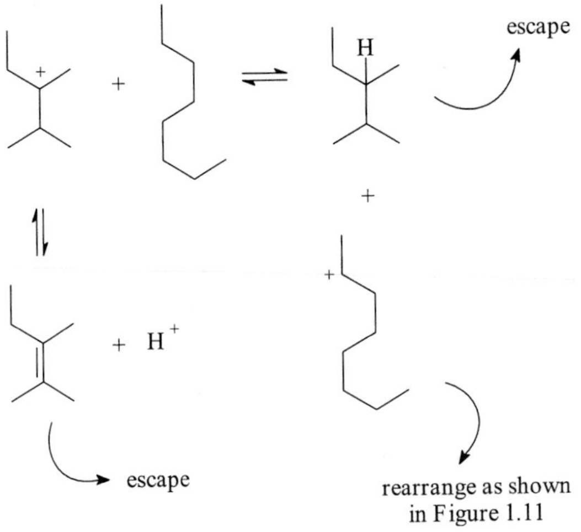
escape
H
+ H+
escape
rearrange as shown
in Figure 1.11

Fig. 1.12 Hydride transfer and proton loss can account for formation of the neutral products of catalytic cracking.

rearrange. These steps allow the release of the neutral molecules from the pore of the zeolite.

These highly energetic carbocations in the pores of the zeolite undergo other kinds of complex reactions with other molecules produced in these pores as well as with molecules directly from the petroleum. The ultimate result of much of this chemistry, which is still not clarified in its detailed steps, is the formation of residue and coke. It is not at all clear how this happens. A comprehensive picture of all the chemistry occurring in a catalytic cracker is still to be achieved, and this knowledge may well lead to a far more efficient use of petroleum. But whatever this picture shows, carbocations will certainly be the important intermediates.

1.14

Why do Steam Cracking and Catalytic Cracking produce such Different Results. Or, in Other Words, Why do Carbocations and Free Radicals behave so Differently?

What is the source of this difference between carbocations and free radicals? Although the tendency of the dependence of stability on structure is identical for carbocations and free radicals, the energy difference for structural change in carbocations is far larger than for free radicals (see Fig. 1.6). Carbocations are therefore driven to rearrange because the structural change on rearrangement gives rise to a large increase in stability. In this way, the positive charge can find a site, for example tertiary over secondary, that lowers appreciably the energy of the molecule (see Fig. 1.6). In a free radical a similar rearrangement, if it were possible (see below), would lower the energy of the molecule much less.

But most important, the β-scission path in a free radical is much more likely than this reactive path for a carbocation because the energy cost is far higher in such a scission in a carbocation (Fig. 1.6). β-Scission would produce a primary carbocation. β-Scission producing primary free radicals is far more likely (Fig. 1.7) and is the reaction responsible for cracking petroleum fractions to smaller molecules. This is the reason that steam cracking acts to break down the larger petroleum molecules to smaller molecules far more effectively than catalytic cracking (Figures 1.8 and 1.9). Here, we see how an important difference between free radicals and carbocations leads to molecules for the chemical industry, on the one hand, and fuel for the internal combustion engine, on the other hand.

In the steam cracker there is virtually no formation of branched hydrocarbons. Why not? The answer is not in the ease of β-scission, which is after all still enthalpically uphill even if the hill is not so steep (see Fig. 1.6), but rather in the fact that 1,2-shifts necessary for the rearrangements to the branched structures are restricted by the single electron occupation of the molecular orbitals in the free radicals. We cannot go into a discussion of molecular orbital theory and the source of the blocked 1,2 rearrangements of free radicals, except to try to stimulate the reader to search further by reading textbook discussions of what are called the Woodward–Hoffmann Rules and the general subject of orbital control of chemical reactions.

If nature worked otherwise, the free radical path could very well produce excellent gasoline and there would be no need for catalytic cracking. But in fact it is carbon in the positive state that is driven to rearrange its structure to satisfy the need for higher substitution, and it is allowed to carry out this rearrangement by the empty orbital associated with this positive charge. In this way catalytic cracking is the route to many of the molecules that eventually find their way into the high-octane gasoline necessary for modern engines.

1.15
Summary

The industrial drive to use petroleum to produce chemicals of greater use to society and therefore of added economic value led to the ultimate development of steam cracking and catalytic cracking. It is difficult to conceive of the development of modern technology without the transformations of the hydrocarbons in petroleum to the alkenes necessary for the chemical industry and the branched hydrocarbons necessary for the internal combustion engine. In this chapter, some key turning points in the development of petroleum chemistry are highlighted as are the essential roles played by free radicals and carbocations. The mechanistic difference between steam and catalytic cracking is essential to form the different products of these cracking processes.

The difference between free radicals and carbocations can be traced to a favored β-scission path taken by radicals while carbocation intermediates favor isomerization of the carbon skeleton. And this difference resides in the difference in the fundamental nature of these species. Producing radicals from the molecules in petroleum requires only supplying enough energy in the form of heat to overcome the homolytic bond energies and to initiate free radical chain reactions. In contrast, producing carbocations from these molecules requires a source of acidity, which was discovered when petroleum was subjected to high temperature in the presence of modified natural clays. That discovery stimulated the study of the chemistry of zeolites, which now can be synthesized industrially and are essential in the production of modern gasoline.

Some of the subjects treated in this chapter are listed below. These are key words and terms that act as reminders of the chapter's contents and should become a valuable part of your chemical vocabulary.

- Free radical chain mechanism
- Carbocation rearrangements
- Brønstead acidity
- Petroleum
- Alkenes
- Acetylene
- Branched hydrocarbons
- Isomerization reactions of carbocations
- Zeolites
- β-Scission of free radicals
- Steam cracking and catalytic cracking and their historical development
- Thermal cracking
- Gasoline

Study Guide Problems for Chapter 1

1. Predict the bonding and geometry in a free radical and in a carbocation using hybridization of atomic orbitals.

2. What evidence do you see in Fig. 1.1 for the biological source of petroleum? In what way might questions of chirality play a role in answering the question posed above and what kind of experiment would help to confirm your conclusion?

3. What are the various kinds of chemical reactions specific to carbon–carbon double bonds? Give as many examples as you can for each reaction type using propylene as the double bond-containing molecule, paying attention to the possible formation of isomers.

4. The first industrial process based on an olefin was hydration of propylene to yield isopropanol. How could you produce n-propanol from propylene? Give all reagents and show the mechanism of the reactions.

5. Propose detailed mechanistic steps for the formation of each of the industrial intermediates in Fig. 1.3.

6. Use the Kirk-Othmer *Encyclopedia of Chemical Technology* to discover what ultimate commercial use derives from each of the industrial intermediates in Fig. 1.3.

7. The reaction of calcium carbide, CaC_2, with water is highly exothermic. Offer an explanation for this high exotherm considering the pK_a of water.

8. How could acetaldehyde be synthesized from ethylene? Could propanal be synthesized from propylene in a parallel manner?

9. Propose a synthesis of styrene using benzene and ethylene as the carbon sources.

10. Why does formation of polyethylene from ethylene by any method produce large amounts of heat, a major problem for engineering polyethylene plants? Could a method be designed to produce polyethylene from ethylene that would be less exothermic?

11. How do the data in Fig. 1.6 relate to the difference between S_N1 and S_N2 reactions?

12. Chlorine reacts with methane under irradiation to yield methyl chloride among other products. Outline the mechanistic steps for this free radical chain reaction designating which are initiation, propagation, and termination.

13. When chlorine is heated with isobutane, a single monochlorinated hydrocarbon is overwhelmingly formed in spite of the fact that there are ten hydrogen atoms for which chlorine could be substituted. Use bond dissociation data from Fig. 1.6 as well as a table of bond dissociation energies in any textbook to determine the enthalpy difference for each step in the free radical chain processes that produce the product that predominates and, as well, any competitive products.

14. Propose steps that would produce 1,3-butadiene, and toluene and benzene (Fig. 1.7) in the steam cracking of full naphtha.

15. Catalytic cracking of atmospheric gas oil produces a wide variety of branched hydrocarbons varying in the range from about 5 to 9 carbons. Mechanistic proposals suggest that free radicals analogous to those formed in steam cracking give rise to olefin intermediates, which are then converted to carbocations and which then undergo rearrangements under the acid conditions of catalytic cracking. Write out detailed chemical reactions that follow these mechanistic ideas to form branched hydrocarbons with 6 carbons from linear hydrocarbons with 12 carbons.

16. Fig. 1.4 shows several 6-, 7- and 8-carbon branched hydrocarbons. Propose all steps that could take place in a zeolite pore from the respective linear hydrocarbon that could give rise to these components of gasoline.

17. Discuss how orbital symmetry ideas such as the basis of the Woodward–Hoffmann rules and the ideas of aromatic and anti-aromatic transition states relate to the fact that 1,2 rearrangements are observed in carbocations but not in carbanions. How might your discussion apply to the difference between steam and catalytic cracking of petroleum fractions?

18. Offer explanations for each of the terms in Section 1.15.

2
Polyethylene, Polypropylene and the Principles of Stereochemistry

2.1
The Thermodynamics of Addition Polymerization:
the Competition between Enthalpy and Entropy

Ethylene and propylene, among many other molecules with double bonds, form what are called "addition polymers." This means that no atoms of the polymerizable alkene, for example the ethylene, are lost on formation of the polymer. The polymer is formed simply by addition of one alkene molecule to another, and we add the prefix poly to the name of the alkene to describe the huge molecule, the macromolecule that was formed. Many thousands of ethylene molecules may bond together to form polyethylene (Fig. 2.1).

Laying aside the mechanism of the addition step for the moment, the overall process involves converting the alkene π-bonds to the σ-bonds holding the repeating units together in the polymer. Since the σ-bonds are far stronger than the disrupted π-bonds of the monomer units, a great deal of energy is released and the thermodynamic picture involves a large negative enthalpy, ΔH, favoring the polymerization, which is an exothermic reaction. On the other hand, the bonding together in the polymer chain of the many alkene molecules that had been free to move independently causes a large reduction in entropy for the process, so that ΔS is negative, disfavoring the polymerization. The polymerization is an ordering process.

These two competing thermodynamic factors control all polymerizations, and with ethylene and propylene, at the temperature of the polymerization, the enthalpic term overwhelms the entropic term. If this were not the case, a polymer could not be formed, and in fact because of these competing enthalpy and entropy factors all polymers can only be formed below a certain temperature known as the *ceiling temperature*. Above this ceiling temperature – which can be determined experimentally in many polymers – there can be no net formation of polymer. A related characteristic is the large exotherm associated with the polymerization process, which causes problems for the industrial production. Since so much heat is given off it is necessary to engineer special reaction tubes to carry the heat away to keep the temperature down. Contrast this exothermic situation with that for cracking of petroleum fractions – an endothermic reaction, where energy has to be supplied to maintain a high temperature.

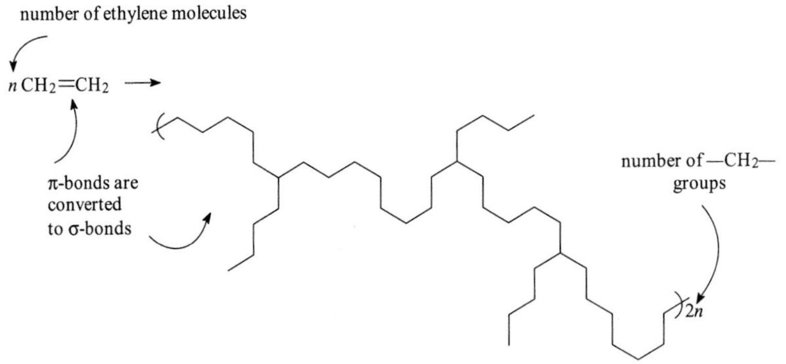

number of ethylene molecules

$n\,CH_2{=}CH_2 \longrightarrow$

π-bonds are
converted
to σ-bonds

number of —CH$_2$—
groups

$\big)_{2n}$

Fig. 2.1 Bonding changes on formation of polyethylene from ethylene.

2.2
Polyethylene is formed via a Free Radical Polymerization that involves the Classic Steps of all Chain Reactions: Initiation, Propagation, and Termination

Polypropylene and polyethylene are not only huge molecules, but they are also formed and used in our world in huge amounts. The amount of polyethylenes produced worldwide each year can be measured in the hundreds of billions of kilograms, and they find use in numerous applications. Morawetz recounts in interesting detail the discovery of polyethylene. The discovery arose from research conducted at ICI Corporation in England in the years before World War II as a consequence of a decision to explore chemical reactions under high pressure. ICI saw this as a foray into basic research and it is interesting that at this time in the late 1920s DuPont made a similar decision and hired Wallace Carothers to carry out research also without aiming at commercial products. DuPont's decision led to the invention of nylon (see Chapter 5; Sections 5.1–5.3), while that of ICI led to polyethylene. Not a bad idea to carry out basic research!

The polymerization of ethylene occurred accidentally at ICI in a misguided attempt to add ethylene to benzaldehyde under high pressure. The benzaldehyde was recovered unchanged, but a waxy solid was obtained that proved to be a relatively low molecular-weight polymer of ethylene. After several experiments to try to improve the result – and some delay caused by an explosion – it was realized that a high molecular-weight polyethylene, which is a solid material in contrast to the waxy substance originally obtained, could be made using a high pressure of ethylene with a small concentration of oxygen. The insight about oxygen arose from another mistake – a leak in the apparatus which allowed air to enter inadvertently.

We do not intend to mislead our readers to believe that experiments conducted in a sloppy manner, and aiming for impossible goals are necessary for success in chemistry. But in fact we are going to see accidental discovery again in the work that led to the Nobel Prize for Ziegler and Natta (Section 2.4). These serendipitous discoveries were only possible because these chemists, who were very much innovators and skilled in their profession, were also prepared to fully understand the surprising implications of the results of their experiments.

One is reminded of a famous quote of Louis Pasteur: "In the fields of observation, chance favors only the mind that is prepared." Pasteur, who lived from 1822 to 1895 and whose scientific work ranged widely over chemistry and biology, lives on in popular culture as the source of the name pasteurization. He was described by Partington as belonging in the history of science to a small group of great men. One of Pasteur's accomplishments has direct relevance to our studies in this chapter. In a fascinating experiment associated with his investigations in helping to solve a problem of the French wine industry, he proved the connection between molecular dissymmetry and optical activity. Later in this chapter we will discover how dissymmetry, expressed in modern terms by the word chirality, plays an essential role in the polymerization of propylene (Sections 2.6 to 2.10).

But let's return now to polyethylene and the twentieth century. A patent was issued in 1939 arising from the work at ICI with the specification that solid polymers were only obtained at pressures exceeding 1000 atmospheres with *small* amounts of oxygen. As we will come to learn below, the oxygen was there to initiate a free radical polymerization – a concept that these early investigators did not understand. Unfortunately, the oxygen also introduced some carbonyl groups into the polyethylene. During World War II, polyethylene was used by the United States and England for radar cable insulation where its low dielectric loss was of crucial importance. The dipole of the carbonyl group contributes to the property of dielectric loss requiring that the proportion of carbonyl groups be kept to a minimum. Another use for polyethylene displaced *gutta percha* (Chapter 7, Section 7.4) for the insulation of underwater telecommunication cables – a use which depended also on its low dielectric loss and, critically, for its resistance to salt water.

Even so, people who worked in the chemical industry after World War II tell stories of how no one at that time had any idea of what else polyethylene was good for. One of the first uses, to make very large balloons to hoist instruments to high altitudes for atmospheric measurements, was developed at General Mills in Minnesota. The balloons, which were held together by what we call "scotch tape" today, were commonly misidentified from the ground as flying saucers. It did not take too long, however, for many uses to be found for polyethylene. Witness its current production level – hundreds of billions of kilograms.

The characteristics found for polymerization of ethylene by the ICI researchers make perfect sense from our understanding of the nature of the polymerization process as outlined above. The polymerization is exothermic, and therefore removing the heat produced by reducing temperature favors the process. The polymerization is an ordering process in which the polymer takes less volume than the ethylene molecules that form it. Therefore, increasing the pressure favors the polymerization. And as we will see below, the oxygen is responsible for starting each polymer chain on its growth to a high polymer; less oxygen therefore means fewer chains to compete for the available ethylene molecules so that each chain can grow longer.

It was later understood that free radicals were involved, providing therefore almost as many ways to initiate the polymerization of ethylene as there are methods for producing free radicals. Although in industrial processes oxygen is used to produce the radicals that initiate the polymerization of ethylene, one could also use homolytic breaking of a peroxide bond as shown in Fig. 2.2. The addition of a radical to one end of an ethylene mole-

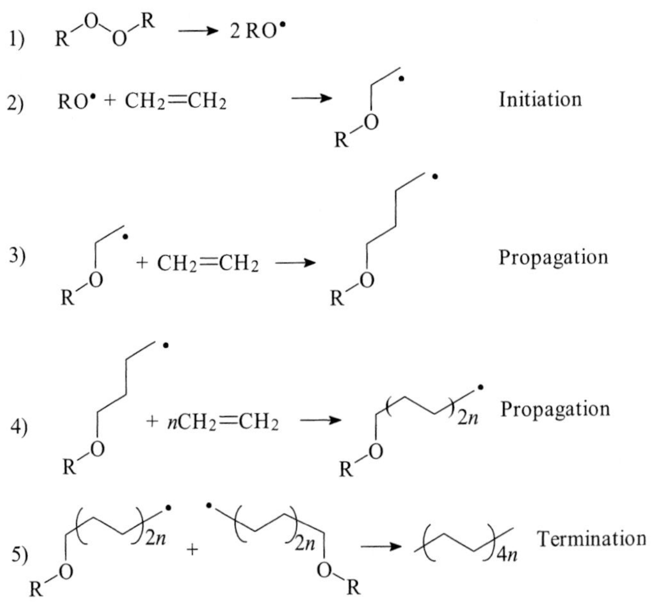

1) $R\!-\!O\!-\!O\!-\!R \longrightarrow 2\,RO^{\bullet}$

2) $RO^{\bullet} + CH_2{=}CH_2 \longrightarrow$ Initiation

3) $+ CH_2{=}CH_2 \longrightarrow$ Propagation

4) $+ nCH_2{=}CH_2 \longrightarrow$ Propagation

5) $+$ $\longrightarrow$ Termination

Fig. 2.2 Free radical chain mechanism for polymerization of ethylene.

cule produces another radical. One of the electrons of the π-bond forms half of the electron pair with the incoming radical, while the remaining electron of the π-bond remains unbonded. There is an irony in this. The addition of the radical to the π-bond was driven by the instability of the initiating radical, which could be relieved by forming a chemical bond. But the formation of this chemical bond only produced another radical. So the instability is not relieved but simply transferred, a kind of "passing the buck."

This newly formed radical will add to another ethylene molecule. This process continues, breaking in turn the π-bond of each incoming ethylene molecule and converting it to the σ-bond that adds it to the chain and acts to hold the growing chain together. In each addition step there is produced a new free radical at the growing chain end. This is a chain mechanism in which the addition of a radical to the ethylene π-bond is the initiation step while the subsequent additions of ethylene molecules are the propagation steps. Many free radical chain mechanisms are encountered in industrial chemistry. This free radical polymerization has the classic characteristics of initiation, the addition of the radical to the first ethylene; propagation, the addition of new ethylene molecules causing the degree of polymerization to increase; and finally termination.

Even if all the reactant molecules are not consumed, reactions can occur that quench or terminate the radical species that are carrying on the propagation steps. We saw this in the chain reaction occurring in the breakdown of petroleum molecules in steam cracking when two radicals combined (Chapter 1, Fig. 1.7).

In the chain mechanism leading to the formation of polyethylene, termination can occur by combinations of radicals that stop the further growth of the polymer chain. A new polymer can then begin to grow only from a new initiation. One of these processes of termination is simple combination of two chains by forming a bond between the free radical chain ends. Although this kind of termination, which doubles the molecular weight of the polymer, can be kept to a low probability by various means, inevitably the chains will terminate their growth by some process that consumes the free radical population. Nevertheless, the fact that free radical methods can produce the extremely high molecular weights necessary for the industrial use of polyethylene demonstrates that termination can be delayed until thousands of propagation steps have occurred. Fig. 2.2 shows the chain mechanisms for polymerization of ethylene.

2.3
Attempted Free Radical Polymerization of Propylene. It fails because of Resonance Stabilization of Allylic Radicals

In contrast to ethylene, polymerization is not the fate of propylene molecules exposed to free radicals. As in ethylene, a free radical can initiate a chain mechanism in propylene and therefore potential growth of the chain (Fig. 2.3, steps 1–4). But the methyl group in

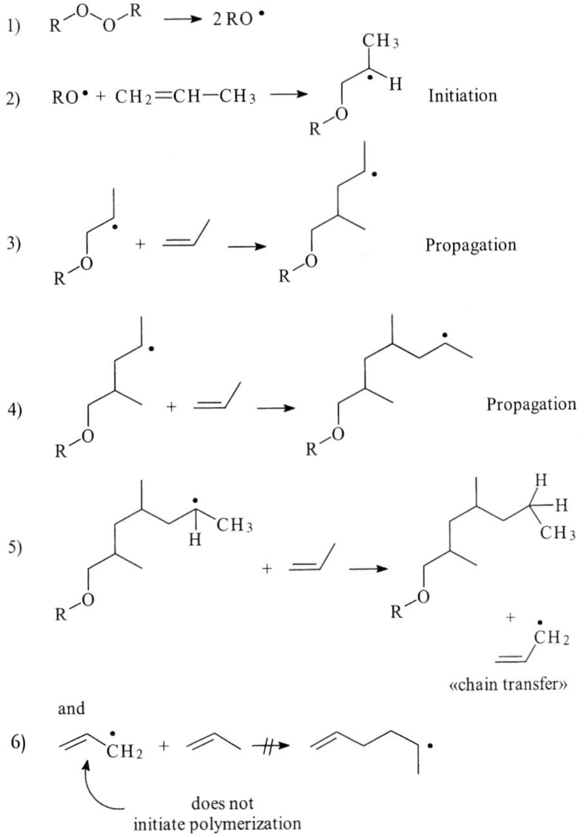

Fig. 2.3 Free radical chain growth and how it is stopped in propylene. The allylic methyl group in propylene blocks the free radical polymerization by what is called "chain transfer."

each propylene molecule is a source of easily broken carbon-hydrogen bonds and these can compete with the π-bond for reaction with a free radical. This can occur if the radical is the initiating radical or the end of a growing chain. Even if a polymerization is initiated and the chain is beginning to add propylene monomer units by addition to the π-bond, the growing chain with its free radical end can competitively abstract one of the methyl hydrogens from an incoming propylene molecule instead of adding to the π-bond of this molecule and continuing the chain growth (Fig. 2.3, step 5). This hydrogen abstraction process will terminate the growing chain by terminating the free radical via formation of a carbon-hydrogen bond. The steps in Fig. 2.3 as noted above show this characteristic competition between chain growth and termination in the attempted free radical polymerization of propylene.

This ease of breaking of the carbon-hydrogen bonds in the methyl groups of propylene is a consequence of the resonance stabilization of the resulting allyl radical. This arises from the conjugation of the resulting radical with the still existing adjacent double bond. A growing chain has been aborted. But can the allyl radical start the growth of a new chain? Unfortunately the answer is no (step 6, Fig. 2.3). You might say the allyl radical is not unhappy enough to do the job, or put in a different way its energy is too low because it is resonance stabilized. Restarting of a new free radical chain using this resonance-stabilized radical, $CH_2=CH-CH_2^\bullet$, is energetically difficult since addition of the resonance-stabilized radical to a propylene molecule would produce a less stable radical, $-CH_2-CH^\bullet-CH_3$ (step 6, Fig. 2.3). The forces of resonance stabilization work both to end an ongoing polymerization and to inhibit the initiation of a new polymerization.

While the addition of a source of free radicals to a gaseous sample of ethylene will therefore rapidly produce a white powder, polyethylene, which can be melted and molded into so many of its familiar uses, the addition of such a radical to propylene will produce a useless paste of short hydrocarbon molecules. The chain growth will be stopped before it can get going, in spite of the fact that the thermodynamic picture (Section 2.1) outlined for ethylene also fits propylene: there is no impediment from thermodynamic consequences for the formation of polypropylene.

The abortive step that stops the growth of the polymer chain by abstraction of a hydrogen atom from a methyl group is an example of what is called *chain transfer* (step 5, Fig. 2.3). The radical site that would have continued the polymer chain growth is transferred to an allyl radical. Chain transfer is not unique to propylene polymerization. Chain transfer with an entirely different mechanism from that in the attempted polymerization of propylene occurs also in the polymerization of ethylene but without, in that situation, stopping the growth to a high polymer. The structural consequences of chain transfer in the polymerization of ethylene have very important industrial consequences as will be discussed below.

2.4
So How is Polypropylene made? Organometallic Chemistry can do what Free Radical Chemistry cannot. And the Big Surprise is the Role of Stereochemistry and Specifically Chirality. This is Something No One suspected

In spite of the picture drawn above of how the free radical polymerization of propylene is blocked, polypropylene *is* widely known, to the tune of 50 billion kilograms in 2002 produced worldwide, and appears in a wide variety of materials that surround us. Someone must have figured out how to polymerize it. Let us follow the fascinating story of the discovery of how to accomplish this and the great surprise of the role of stereochemistry – a story that led to a Nobel Prize for an Italian and a German chemist.

The story actually begins with an accidental discovery of a new way to polymerize ethylene. This arose from contamination of a chemical apparatus in the laboratory of Karl Ziegler, a distinguished German professor who specialized in catalysis. In a series of experiments in Ziegler's laboratory which started just after World War II, and was intended to polymerize ethylene by methods other than those used in the ICI work (Section 2.1), it was discovered that one of the autoclaves used for the experiments was especially effective in producing butene in high yield. This was traced eventually to the presence of colloidal nickel that had been produced in the cleaning of this stainless steel autoclave and its subsequent exposure to lithium compounds used in the experiments.

The result stimulated an exploration across the spectrum of metals in the periodic table to find others that would optimize this result. In one of those ironies of scientific work the testing of many metals was not successful in achieving the objective of producing butene from ethylene but rather led to success in achieving the original objective of polymerizing ethylene. The addition of some metals, zirconium and titanium, transformed ethylene to high molecular-weight polyethylene. And it was realized immediately that the polyethylene produced by this new method differed from that produced by the ICI high-pressure process (Section 2.1).

In an early use of infrared spectroscopy, the new polymer seemed to have fewer pendant methyl groups than that produced by the ICI process, and it also softened at a higher temperature. From the work in Ziegler's laboratory, a catalyst had therefore been developed allowing a polyethylene to be produced that differed in an important way from the familiar polyethylene produced by an initiation step with free radicals. To understand this difference we have to backtrack in our story.

2.5
There are More Kinds of Polyethylene than the One produced by the Free Radical Chain Mechanism

We did not discuss one aspect of the structure of the polyethylene produced in the free radical initiation process, although perhaps you were curious about the structure of polyethylene shown in Fig. 2.1. The polyethylene macromolecules have occasional "branching" off the main chain. A great deal of research work over many years showed that these branches, which are most often four carbon units long, arise from a common reaction of carbon-based radicals in which the end of the growing chain reaches back and abstracts a hydrogen atom from a $-CH_2-$ group in the chain. In the colorful and highly descriptive language often found in industrial chemistry, this is called appropriately "backbiting." In this manner, the primary radical at the end of the growing chain is converted to a secondary radical, and this site becomes the new place for ethylene molecules to grow. The "old" end of the chain becomes the last carbon, the fourth carbon, of the branch.

The reason that four-carbon branches are formed is that the abstraction of the hydrogen from the internal position in the chain takes place through a six-membered ring transition state. Fig. 2.4 shows this process, which is another example of chain transfer (see step 5, Fig. 2.3). This process differs from the chain transfer in the attempted free radical polymerization of propylene, not only in being intramolecular instead of intermolecular but also in the stability of the new radical. In propylene, the radical site produced by the chain transfer – the allyl radical – is incapable of initiating a new polymerization because of its resonance stabilization. Conversely, the new radical site produced by this backbiting is not especially stabilized and so it acts to initiate a new site for addition of ethylene molecules (Fig. 2.4).

The kind of transfer of carbon-bound hydrogen to the free radical site in the polymerization of ethylene is a common reaction path for free radicals seen in a wide variety of reactions that play important roles in organic chemistry. Famous examples are:

- the Barton reaction, which has been used to modify steroid structures;
- the Hofmann–Loeffler–Freitag reaction, used to synthesize pyrrolidines; and
- the Norrish Type II photochemical reaction used to fragment ketones.

In every one of these reactions hydrogen is transferred in the identical manner as that leading to the four-carbon branch found in free radical-polymerized polyethylenes (Fig. 2.4), that is, through a six-membered transition state to a free radical site (Fig. 2.5).

The branching produced by the backbiting greatly affects the properties of the polyethylene in interfering with the packing and crystallization of portions of the chain. This is

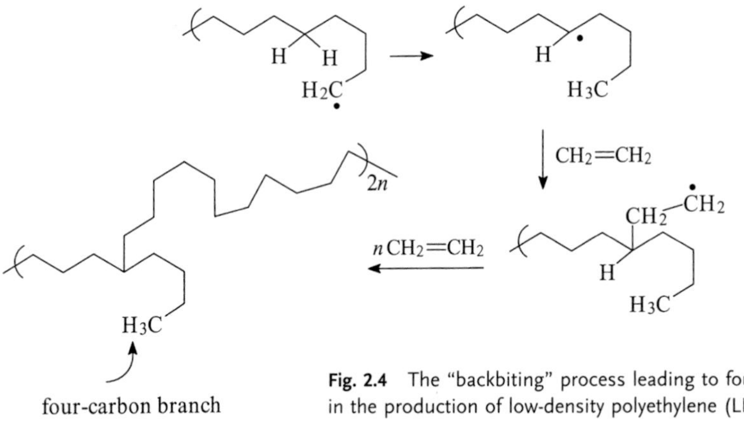

Fig. 2.4 The "backbiting" process leading to four-carbon branches in the production of low-density polyethylene (LDPE).

four-carbon branch

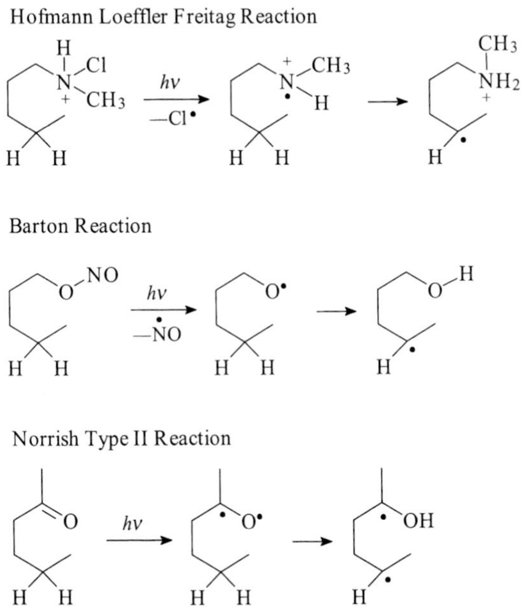

Fig. 2.5 Organic chemical reactions that involve transfer of hydrogen via a six-membered transition state to a free radical site.

Hofmann Loeffler Freitag Reaction

Barton Reaction

Norrish Type II Reaction

important to the commercial uses of this plastic. While linear portions of the polyethylene chains can pack closely together and form crystalline regions in the polymer, the branched regions – with their structural irregularities, their branches – are not able to participate in this crystallization. Since closer packing and crystallization translates to higher density, more chain branching means lower density. In fact, polyethylene synthesized via free radical polymerization is known as low-density polyethylene (LDPE). Chemists were therefore interested in finding ways to control both the population of these branches and even their length, since the density and crystallization properties of the polyethylene were closely connected to their commercial uses. But very little could be done to alter the properties of LDPE, the polymer produced in the accidental discovery in England (Section 2.2).

The polyethylene made in Germany was of great interest because, when produced from the catalyst derived under these experimental conditions, it was almost free of branches. The chains could therefore approach each other more closely, and this led to a higher level of crystallinity and therefore a higher density compared to LDPE. This caused large changes in the material properties compared to LDPE. In fact, the polyethylene produced by this method is now used to produce hundreds of billions of kilograms each year of so-called high-density polyethylene (HDPE). In addition, as we shall see below, this new method allows the production of an entirely new kind of polyethylene – one that is neither LDPE nor HDPE but something in between.

The new catalyst was found to polymerize mixtures of ethylene and 1-alkenes, to make copolymers – that is, polymers composed of two different monomers (olefins in this situation). The proportions of the two olefins can be varied so that in the case of 1-hexene mixed with ethylene the polymer will have a designed number of four-carbon branches.

Other 1-alkenes then will yield branches with a differing number of carbons, such as six-carbon branches if 1-octene were to be used. For two-carbon branches 1-butene is used, which is the most important of these copolymers although the hexane-based copolymer is not far behind. These variations of the kind and amount of the olefin to be mixed with ethylene are extremely important to adjust the material properties of the modified polyethylene. Control of how often the branches appear along the chain and the size of the branches control how well the chains can pack and therefore allow variability in the crystal properties and the density. One can tune in properties, which mix those of LDPE and HDPE. These polymers are called Linear Low Density Polyethylene (LLDPE) and are produced and sold in large amounts.

LLDPE is like LDPE and can replace it in many applications. It would be preferred because, like HDPE, it can be made at low temperature and pressure. Indeed, HDPE and LLDPE can be made in the same process vessel by simply changing the alkenes to be polymerized. LDPE survives because it is easier to use then LLDPE in extrusion (making films) and is usually mixed with LLDPE for commercial applications.

2.6
What have we learned from the Organometallic Method for Polymerizing Ethylene that leads to the Possibility of Polymerizing Propylene?

Why does the organometallic method lead to an unbranched polyethylene? Without assuming anything about the organometallic mechanisms we can conclude that there can be no free radicals involved because the branches are formed by a polymerization that is initiated by free radicals (Figures 2.2 and 2.4). If the organometallic method can form polyethylene without the intervention of free radicals, it seems reasonable that the same method might successfully polymerize propylene, since it was the intervention of free radicals that interfered with the polymerization of propylene (see Fig. 2.3).

Giulio Natta, an Italian chemist, after hearing Ziegler lecture in Germany on the polymerization of ethylene, made an agreement to take the catalyst developed for the polymerization of ethylene back to his laboratory in Milan. Here we have another irony associated with chemical research. Natta was interested in polymerizing propylene because he thought that all those methyl branches would yield a rubbery polymer. The Ziegler catalyst did in fact polymerize propylene, but there was a surprising consequence. Not only are radicals clearly *not* involved, as evidenced by forming a high polymer from propylene, but Natta was to discover that the nature of this polymerization is one of great stereochemical subtlety.

But why stereochemistry? The answer comes from that extra methyl group in propylene compared to ethylene, which causes the polypropylene chain to have a methyl branch on every third carbon – a structure we began to see developed even in the aborted free radical path to polymerize propylene (Fig. 2.3). Natta and his colleagues found the polypropylene produced by the Ziegler catalyst to be a high polymer with thousands of propylene units linked together. When the polypropylene was subjected to varying solvents of higher and higher boiling point the sample was found to be a mixture of polymer chains with varying proportions of amorphous and crystalline regions. The fractions with higher amorphous content dissolved more easily at lower temperatures than the crystalline fractions. The crystallinity amazed Natta because he thought the methyl branches would cause the polymer to be entirely amorphous by interfering with crystallization, as did the branches in LDPE (Section 2.5). The crystallinity was also, initially, a disappointment because it meant the polypropylene would not be rubbery (see Chapter 7) – a commercially valuable characteristic of amorphous polymers and the reason that originally drove Natta's interest in polymerizing propylene.

In the polypropylene sample synthesized using Ziegler's catalyst, portions of each chain passed back and forth between the crystalline and amorphous states. A fiber diagram could be obtained from an X-ray analysis, and from the repeating characteristics of such an X-ray pattern, Natta and his colleagues could determine that the methyl groups along the chain were regularly spaced in a helical arrangement, so that three propylene residues, $-(CH_2-CH(CH_3))_3-$, formed one turn of the helix. The relationship among the methyl groups can be visualized most easily by presenting the chain in a conformation it does not take naturally, that is, by stretching out the chain to make a zigzag, as shown in Fig. 2.6. In this stretched-out representation, the methyl groups all appear on the same side of the chain. But which side of the zigzag representation should the methyl groups appear on? In Fig. 2.6 both possibilities are shown.

In one zigzag chain in Fig. 2.6 the methyl pendent groups all face to the rear of the page, whereas in the other chain they face to the front of the page. Consider these two chains – they are mirror images. Given two mirror-image forms, one can ask the question if the forms can be superimposed. The answer here is not simple to come by. The mirror images shown in Fig. 2.6 could be superimposed, but only if one end of the chain could not be distinguished from the other. If the ends of the chains were different then the two chains shown in Fig. 2.6 would be enantiomers. Can we know which end of the chain started the polymerization process and which end bears the propylene group last added to the chain?

The number of units of propylene linked together in a typical polypropylene formed by this catalyst, which has become known as the Ziegler–Natta catalyst, is in the realm of thousands. Most of the units of such a chain are so far removed from the ends that they feel no influence from these end groups. This means that the properties of the overwhelming number of units in the chains are independent of the end groups, and therefore that the two chains in Fig. 2.6 are effectively identical. Only a minute proportion of

Mirror images?

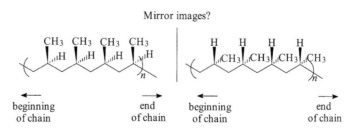

Fig. 2.6 Isotactic polypropylene showing the relationships between the cryptochiral chains and the non-role that the chain ends play.

the units – those nearest the ends of the chain – would notice, so to speak, if the beginning of one chain were laid over the end of the other chain.

However, as we shall see below, in spite of the fact that the chains are virtually identical or therefore effectively achiral, catalyst chirality is in fact essential for producing this kind of polypropylene. The fact that all of the methyl groups reside on one side of the chain (Fig. 2.6) demands that the catalyst is chiral. Let's find out why this is true and also how this conclusion guides the approach now taken in industry to improve the original Ziegler-Natta catalyst.

2.7
Do the Methyl Groups on Every Third Carbon on Each Individual Polypropylene Chain all have to be on the Same Side of the Chain?

Over the years following the creation of the Ziegler-Natta catalyst it was discovered how to create variations so as to produce polypropylene with the methyl groups not only on the same side of the chain, but also alternating from front to back or even randomly taking front and back positions along the chain. These different arrangements are shown in the planar zigzag form in Fig. 2.7. Natta coined the terms used to describe these arrangements of the methyl groups along the chain of polypropylene: isotactic (all the same); syndiotactic (alternating); and atactic (randomly arranged). Natta, working closely with Montecatini (now Polymeri Europa), a large Italian chemical company, soon discovered that these three variations of polypropylene had very different properties, with the isotactic form becoming an article of commerce.

Stereoisomers that are not mirror images of each other are called *diastereomers*, and this is certainly the case with the three kinds of polypropylene in Fig. 2.7. And as for all diastereomers, from the *cis* compared to the *trans* isomer of 2-butene or for the stereo-

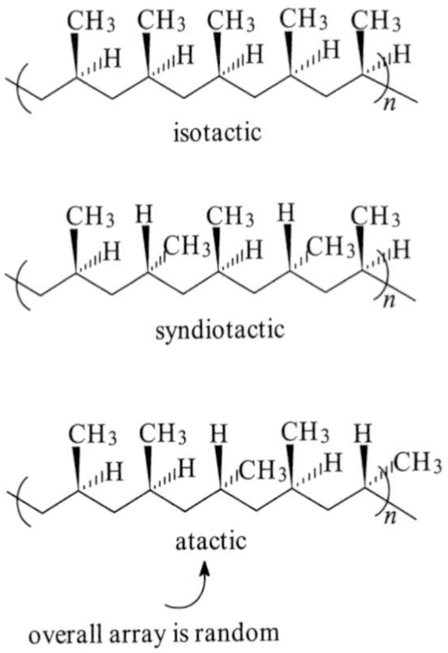

isotactic

syndiotactic

atactic

overall array is random

Fig. 2.7 Tacticities of polypropylene.

isomers of 1,4-dimethylcyclohexane, or for the *meso* form contrasted with the d- or l-isomers of tartaric acid, the properties of these stereoisomers are different. They have different spectra, heats of formation, melting points, boiling points, indices of refraction, and so on. This is in contrast to enantiomeric stereoisomers in which these properties are identical in an achiral environment.

As indicated, the isotactic form of polypropylene is an article of commercial value on the basis of its stiffness and strength. The atactic form, which is sticky for low polymer molecular weights (oligomers) is useful in adhesives, whereas the syndiotactic form – which is stiff in a different manner than the isotactic form – did not become available until certain metallocene catalysts, a variation of the organometallic method to be discussed below, were invented. Polypropylene is an example of diastereoisomerism providing a marketing consequence.

2.8
What do the Opposite "Faces" of Propylene have to do with the Formation of Isotactic Polypropylene by the Ziegler–Natta Catalyst?

Let us focus on the most important commercial diastereomer, isotactic polypropylene. How is it that the Ziegler–Natta catalyst can form polypropylenes in which every methyl group on any single chain, with a few exceptions, is on the same side of the chain, as in the isotactic polypropylene shown in Figures 2.6 and 2.7? Would this not require that every incoming propylene unit be added to the growing chain in the identical way, somehow landing on the site of polymerization only from one side of its double bond? Are the two sides – which are the two faces of the double bond of propylene – different in some way, and by what means can the catalyst tell the difference?

Draw a propylene molecule in a plane perpendicular to the plane of the paper of this page. It does not matter if the methyl group or the hydrogen faces forward toward you or back away from you. Reproduce this image so you have two identical pictures. Now place an arrow above the double bond in one image and below the double bond in the other. The two pictures are no longer identical, but are in fact mirror images of each other. Try to superimpose these images. You will not succeed. Try the same thing with ethylene. The two pictures produced in ethylene are identical. You can simply turn one of the ethylenes over and superimpose it on the other. From this simple consideration of the "faces" of the double bond in propylene versus those in ethylene we can draw the conclusion that there is some potential chiral information in propylene not present in ethylene. The two faces of the propylene are related to each other in an enantiomeric way. Approaching the double bond from one face or the other produces mirror images. Now propylene is certainly not a chiral molecule. It has a clear plane of symmetry and reflecting propylene in a mirror produces a superimposeable image. Ethylene and propylene are both achiral molecules. But if one approaches the faces of propylene, from the top and from the bottom, mirror images are produced (Fig. 2.8).

But if the two faces of propylene are related as mirror images and one is consistently chosen over the other, this must mean – based on the most fundamental principle of stereochemistry – that the catalyst must be chiral. Only a chiral entity can distinguish be-

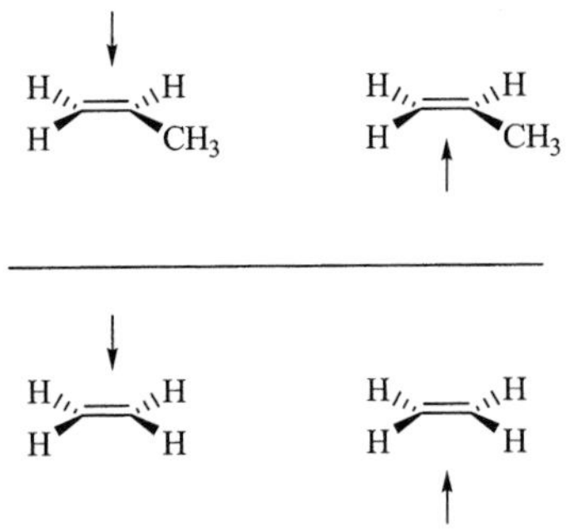

H$_{//}$,——$_\backslash$H   H$_{//}$,——$_\backslash$H

H ⎺⎺ CH$_3$ H ⎺⎺ CH$_3$

H$_{//}$,——$_\backslash$H   H$_{//}$,——$_\backslash$H

H ⎺⎺ H   H ⎺⎺ H

Fig. 2.8 Differing "faces" of propylene compared to ethylene.

tween mirror image choices. But how can the catalyst be chiral? It is made from aluminum and titanium compounds that are not chiral. The Italian chemists realized that something about the crystalline arrangements in this solid-state catalyst must be chiral. In other words, that the sites on the crystal lattice of this solid-state catalyst must allow chiral arrangements. If this were true – and they had no independent experiment to test it other than the polymerization result on propylene – they imagined that some sites were mirror images of other sites. In other words, the catalyst was a racemic mixture of sites capable of polymerizing propylene. These imagined sites would have to be racemic since the catalyst, as mentioned above, is made of achiral starting materials, which could form left- and right-handed sites equally well. What a remarkable idea this is, and as we shall see it gives rise to an explanation of how isotactic polypropylene is formed.

2.9
From the Ziegler-Natta Catalyst to Single-site Catalysts: Creating a Catalyst with a Precisely Known Structure that can Polymerize Propylene to an Isotactic Polypropylene

If one mirror form of this racemic mixture of sites on the Ziegler–Natta catalyst selects one face of the propylene, then the other mirror image site must select the other face of the propylene molecules that approach it. One site could be seen as the arrow approaching the top face, and the mirror image site could be seen as the arrow approaching the bottom face of the propylene in Fig. 2.8. This would mean that if we could tell the beginning of the chain from the end of the chain – that is, one end from the other – we would find, as expressed in the zigzag representation (Fig. 2.6), that there would be equal numbers of chains produced with the methyl groups facing in one direction as the other. Enantiomeric stereoisomers of isotactic polypropylene are in fact produced, but because the chains are so long that the ends have no effect, the difference in these chains is not perceptible. The chirality of the chains, determined by the chirality of the site of the Ziegler-Natta catalyst the chain originated from, is lost once the chain grows so long that the ends no longer matter to most units in the polymer chain.

This picture has much experimental support, and in fact X-ray structural studies of the Ziegler–Natta catalyst show how the atoms are arranged to give the necessary chiral sites, although these polymerization sites form only a small part of the overall crystal structure. It would be great to be able to study the polymerization from a catalyst that was fully understood and, to be able to isolate the site from which the polymer chain grows.

Although it has not been possible to isolate the polymerization sites in the Ziegler–Natta catalyst responsible for the formation of billions of kilograms per year of isotactic polypropylene, it has been possible to create a small molecule catalyst that produces isotactic polypropylene in exactly the same manner as the solid catalyst. This small molecule can be studied by the conventional means of structural organic chemistry: mass spectrometry, nuclear magnetic resonance spectroscopy, infrared spectroscopy and even by X-ray diffraction of single crystals, which can tell the positions of all its atoms. Indeed, this catalyst structure is known in every detail.

The story of the creation of this catalyst, this analog of the solid-state Ziegler–Natta catalyst, starts with a molecule that has nothing to do with polymerization – a molecule created many years ago by curiosity-driven basic research. Ferrocene was the first of a class of molecules called metallocenes or sandwich molecules. In this molecule, an iron atom in its ferrous or plus two oxidation state, Fe^{++}, finds itself "sandwiched" between two five-membered rings, each with one negative charge.

The charged rings arise from a neutral five-membered ring with two double bonds, cyclopentadiene, a molecule that has a surprisingly large acidity. On exposure to bases that would have no effect on most organic molecules, cyclopentadiene gives up a proton from its single CH_2 group. The ring now has a p-orbital on each of the five carbons and 6-π electrons distributed among these p-orbitals, which makes a $4n+2$ π electron aromatic stabilization with $n=1$. These are the rings that sandwich the doubly charged iron atom.

This first metallocene, ferrocene, provides an interesting story because for many years it was only a curiosity. However, when chemists replaced the iron with the elements used in the Ziegler–Natta catalyst (e.g., titanium) and activated (converted to a form where it can cause polymerization) it with organo-aluminum compounds, a powerful polymerization catalyst was created. Fig. 2.9 shows the structure of ferrocene and of the related metallocene molecule that becomes a polymerization catalyst in its activated state (Fig. 2.10). Add excess propylene under pressure and polypropylene is produced, *but not isotactic polypropylene*. This catalyst produces atactic polypropylene. Why does it produce atactic

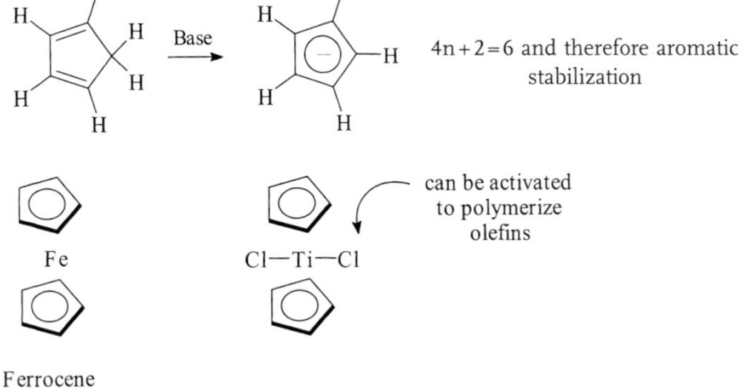

4n+2=6 and therefore aromatic stabilization

can be activated to polymerize olefins

Fe

Cl—Ti—Cl

Ferrocene

Fig. 2.9 Activated metallocene catalyst based on cyclopentadiene.

and not isotactic polypropylene? Based on the stereochemical discussion above, the catalyst shown in Fig. 2.10 cannot distinguish one face of propylene from the other. The reason is that the catalyst structure is superimposeable with its mirror image – it is achiral. But as we shall see, a minor structural variation in this catalyst structure will produce isotactic polypropylene. And as you perhaps have already figured out, that minor structural change must introduce chirality.

2.10
How does this Small Molecule Analog of the Ziegler-Natta Catalyst polymerize Propylene?

Let us try to understand the nature of the polymerization process for this organometallic molecule independent of the chiral input. This is shown in Fig. 2.10. The active form of the transition metal sandwiched between the two negatively charged five-membered rings has a bond to an alkyl group, a metal to carbon bond, and an open orbital that can coordinate with a propylene molecule. These coordinate bonds are well known in organometallic chemistry and involve donation of the π-electrons of the propylene into an empty d-orbital of the metal. This coordination of the alkene then offers a path to form a far stronger carbon–carbon σ-bond than that between the metal and the alkyl group already bonded to it, which is a weak metal-carbon σ-bond.

The alkyl group (R in Fig. 2.10) breaks the weak σ-bond to the metal and moves from its site on the metal to form a carbon–carbon bond with one of the carbons of the coordinated propylene molecule. This, however, forces the other end of the formerly coordinated double bond of the propylene to form a new σ-bond with the metal. The weak σ-bond to the metal is simply passed on, as was the free radical in the polymerization of ethylene described earlier, which was described in that case as "passing the buck" (Section 2.2). Coordination of a new propylene molecule at the now empty former site of the

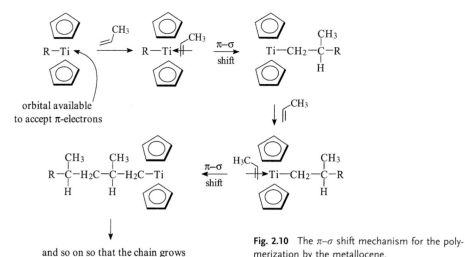

orbital available
to accept π-electrons

and so on so that the chain grows

Fig. 2.10 The π–σ shift mechanism for the polymerization by the metallocene.

Fig. 2.11 The *meso*, *d*, and *l* stereoisomers of the indene-based metallocene.

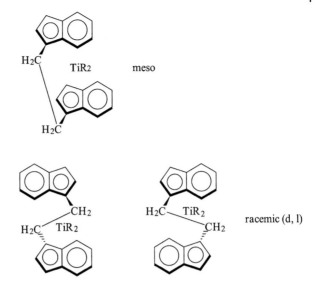

σ-bond sets up the possibility for repetition and so the chain grows, switching the chain position of σ-bond attachment to the two sites on the metal like the movement of a windshield wiper (Fig. 2.10).

In the catalyst shown in Fig. 2.11 there is now a linkage between the two coordinating rings, and this link can be introduced in two ways so that there is, or is not, a plane of symmetry between the coordinating rings. In other words, the catalyst can be designed to form an achiral or chiral arrangement respectively. Three stereoisomers, shown in Fig. 2.11, can be formed in this way, two of which are diastereomers in which one of the diastereomers can exist in mirror image forms. What happens when these different catalysts are exposed to propylene?

The achiral *meso* diastereomer (Fig. 2.11), which cannot distinguish the mirror-related faces of propylene forms atactic polypropylene on exposure to propylene gas, just as for the achiral catalyst in Fig. 2.10, while either of the chiral diastereomers forms isotactic polypropylene. In addition, it makes no difference if one uses the racemic mixture of these chiral metallocenes or either of the enantiomers, exactly as we would predict from the discussions about chain ends (see Sections 2.6 and 2.8).

One enantiomer of the catalyst in Fig. 2.11 (d, l) will yield an isotactic polypropylene with all the methyl groups facing one way, while the enantiomeric catalyst will yield an isotactic polypropylene with all the methyl groups facing the other way. Since the ends of the two chains are inconsequential in a very long polymer as we have discussed (Section 2.6), these two isotactic polymers are, for all experimental purposes, identical. That is why it makes no difference if you use the racemic mixture of the d, l catalyst in Fig. 2.11 or either of the enantiomers for the synthesis of a high molecular-weight polypropylene – that is, a polypropylene made up of thousands of propylene units.

The stereochemical hypothesis about the mechanisms of the Ziegler–Natta catalysts are precisely confirmed by the properties of these new catalysts, which now offer, for the first time, precise control of catalytic structure and therefore precise control of polymer properties. Therefore the creation of these so-called single-site catalysts (single site because the site for growth of the polymer chain is precisely defined in a single molecule) are introducing change into the industrial chemistry of olefin polymerization. This has been a great surprise to all the experts who were convinced that this area was "mature," both economically and technologically.

In the original Ziegler–Natta catalysts, changes in the polymer produced were approached by empirical changes in the catalyst. Even if the symmetry properties of the catalyst sites were suspected from the properties of the polymers produced, as we have learned above, the precise details of the chemistry of the site could not be observed directly. The single-site catalysts have now changed this picture. Precise knowledge of structure allows changes in the catalyst by the methods of organic synthesis since the position of every atom is exactly known and the chemist can hope to predict what structural changes will cause changes in the polymer produced and therefore in its commercial uses. This is already happening.

2.11
An Interesting Story concerning Industrial Conflict

At the same time that Ziegler's catalyst was invented and then used by Natta to synthesize what came to be called isotactic polypropylene, two chemists Robert Banks and J. Paul Hogan, invented a chromium catalyst for ethylene polymerization to high-density polyethylene while working for Phillips Petroleum Company. Subsequently, they tried to polymerize propylene using the same catalyst, but obtained only a very small yield of a crystalline product. The results did not merit further work, but being good chemists they wrote in a notebook that they had obtained an unexpected result – namely crystalline polypropylene – which no one before had ever described. All attempts to prepare polypropylene prior to this time had resulted in the formation of syrupy oligomers for the reasons discussed earlier (Section 2.3).

Because of work by Banks and Hogan, the discovery of crystalline polypropylene by Natta using Ziegler's catalyst (Section 2.6) led to a patent suit in the United States that attempted to define the inventor of polypropylene. No patent was allowed to issue until the suit was settled, and the litigation went on for 27 years. The litigants included among others Phillips, Amoco, DuPont, Hercules and Montedison, the company which worked with Natta after polypropylene had been invented.

Because Phillips was shown to be the first inventor of polypropylene based simply on the notebook disclosure of Banks and Hogan, Phillips was declared the victor. The US patent subsequently issued to Phillips granted them the right to collect royalties for 17 years from all the other manufacturers in the USA, but not in other countries, where Montedison had patent rights unless these countries wished to sell polypropylene in the USA. During this time, Phillips collected over one billion dollars. Have you been keeping a good record of your work in your notebook?

2.12
Summary

In tracing the path from the successful free radical polymerization of ethylene to the failure of this approach for propylene we discover the blocking role of resonance and how the possible formation of the allyl radical from propylene stands in the way of free radical polymerization. But an accidental discovery involving the chemistry of metals and organic molecules, organometallic chemistry, led to a method to polymerize ethylene in a new way and to polymerize propylene successfully with unexpected stereochemical consequences.

Polypropylene can be formed in three diastereomeric forms: isotactic; syndiotactic; and atactic. One of these, isotactic polypropylene, requires that the catalyst be chiral with equal numbers of left- and right-handed catalytic sites. We learn that the source of this conclusion is the relationship between the faces above and below the double bond of propylene. And moreover we discover that although isotactic polypropylene is, strictly speaking, chiral we come to understand the reason why the high molecular-weight polymer can never be optically active. This results because most of the units in the chain are unaffected by the fact that one end of a polypropylene chain is different from the other. All of this can be understood by considering principles of stereochemistry.

And the story then leads us to a series of new catalysts, termed metallocenes, which are derived from extensions of fundamental basic research of many years ago on so-called "sandwich compounds." The first prominent member of this family was ferrocene in which an iron atom is sandwiched between two negatively charged aromatic rings formed from cyclopentadiene. In metallocenes in which the sandwiched metals correspond to those found to polymerize propylene in the Ziegler–Natta catalysts, we find that clever structural manipulations allow confirmation of the chiral hypothesis of how the Ziegler–Natta catalyst functions. Moreover, these new single-site catalysts allow formation of new kinds of polypropylenes not possible before, which is quite a feat for a class of polymers considered a commodity produced to the extent of billions of kilograms worldwide.

And as a final point, we learn how keeping a proper notebook can be worth a billion dollars!

Some of the subjects treated in this chapter are listed below. These are key words and terms that act as reminders of the chapter's contents and should become a valuable part of your chemical vocabulary.

- Free radical chain reactions
- Thermodynamic considerations in chemical reactions
- Addition polymerization
- Free radical polymerization
- Hydrogen transfer rearrangements in free radical chemistry
- Diastereoisomerism and enantioisomerism
- Tacticity
- Enantiotopic and diastereotopic
- Resonance

- Conformational isomerism
- Aromaticity
- Organometallic chemistry
- Metallocenes
- Stereochemistry
- Chirality

Study Guide Problems for Chapter 2

1. (a) Organic chemical reactions can be broken down into classes: elimination; addition; rearrangement; and substitution. Without consideration of mechanism, give a few examples of each type of reaction and judge for each the relative sign and magnitude of ΔH and ΔS.

   (b) How do tables of bond dissociation energies help in judging the sign and magnitude of ΔH?

2. Show the steps for the free radical-initiated polymerization of ethylene that parallel steps in the chain mechanism for chlorination of methane, which was the answer to Study Guide Problem 12 for Chapter 1.

3. (a) Show all steps of initiation, propagation and termination in the free radical polymerization of styrene. Draw the structure of a portion of the polystyrene chain. Are there isomeric possibilities and of what kind? Would the polymerization process distinguish between any of the isomer possibilities to favor one?

   (b) Why might toluene be a poor solvent for the free radical polymerization of styrene? How about benzene?

   (c) Can you foresee any problems in the free radical initiated polymerization of *para-methylstyrene*?

4. (a) The diradical produced after the hydrogen transfer step in the Norrish Type II reaction in Fig. 2.5 readily breaks into two smaller molecules. What did you learn in Chapter 1 that could have predicted this behavior? What are the structures of the fragments?

   (b) The chlorine atom released in the photochemical step of the Hofmann–Loeffler–Freitag reaction shown in Fig. 2.5 forms a bond with the carbon free radical produced in the hydrogen transfer step. How could this lead to a heterocycle, that is, to a pyrrolidine?

5. What is dielectric loss and why should dielectric loss be associated with the proportion of carbonyl groups introduced by oxygen in a polyethylene chain?

6. What is the basis of the statement that *cis* double bonds in unsaturated fatty acids play the same role with regard to *trans* double bonds as branched polyethylene (LDPE) plays to unbranched polyethylene (HDPE). Can you relate your answer to the fact that biological cell membranes that work at lower temperatures have a higher proportion of *cis* unsaturated fatty acids?

7. What is linear low-density polyethylene (LLDPE), and how are the number of carbons in the branches off the main chain and the relative frequency of these branches controlled?

8. Louis Pasteur, in the early part of the nineteenth century, carried out an experiment involving salts of tartaric acid, which provided the foundation for the field of stereo

chemistry. Describe the experiment and explain what conclusions were drawn from it at the time and what implications the experiment held for work in the future.

9. A Dutch chemist interested in optical activity synthesized a saturated hydrocarbon in enantiomerically pure form with a carbon with four different groups around it, a methyl, a hydrogen and two different normal alkyl chains. He measured the optical activity as a function of the length of the two alkyl chains. What is your prediction of how the optical activity would vary as a function of the number of carbons in the normal alkyl chains? Is there some conceptual connection between this experiment and the absence of optical activity in the polypropylene produced with the Ziegler–Natta catalyst? Why was the word cryptochiral invented for these structural situations?

10. The two hydrogens on C-2 of 1-chloro-1-phenylpropane exhibit different chemical shifts in the NMR spectrum while the two hydrogens on C-2 of 1-chloropropane are identical in the NMR spectrum. The former are said to be related as diastereomers (diastereotopic) while the latter are related as enantiomers (enantiotopic). The two hydrogens in methylene dichloride also have identical chemical shifts in the NMR but they are not enantiotopic but are rather designated as identical. Can you develop a general principle to assign these designations, identical, enantiotopic and diastereotopic for relationships among groups within a molecule? For example consider the methyl groups of an isopropyl ester of acetic acid versus lactic acid?

11. Could you assign the designations in problem 10 to the space located above and below the double bonds in acetone versus acetaldehyde? How does the answer to this question relate to the stereochemical consequence of reacting ethyl magnesium bromide with acetone and with acetaldehyde?

12. Fig. 2.7 exhibits three diastereomers of polypropylene. Assign the designations, identical, enantiotopic or diastereotopic to the relationship between the two hydrogens on one of the CH_2 groups in each of the diastereomers, isotactic, syndiotactic and atactic polypropylene.

13. The catalyst structure shown in Fig. 2.11 leads to three stereoisomers, *meso, d* and *l*, as discussed in the text. There are many organic structures that lead to three such stereoisomers including tartaric acid. Show some examples of such structures. What is the common structural characteristic that leads to these stereoisomeric possibilities?

14. Offer definitions and examples for each of the terms in Section 2.12.

3
The Central Role of Electrophilic Aromatic Substitution

3.1
Materials derived from Ethylene, Propylene and Benzene are All Around Us

Petroleum fractions subject to steam cracking yield three chemicals, ethylene, propylene, and benzene, which are the basis for many of the industrially produced objects that ease and enrich our lives. Petroleum fractions, however, are overwhelmingly composed of carbon in the sp^3 hybridization state, lacking therefore the unsaturation found in these three industrial intermediates, a characteristic that is necessary to gain the reactivity that the chemical industry requires (Chapter 1, Sections 1.1, 1.2, 1.7). Chemists learned long ago how to remedy that situation using procedures that tore the molecules apart. The weapons are free radicals, and they are produced by supplying a great deal of energy in the form of heat in superheated steam, that is, steam cracking (Chapter 1, Section 1.4).

We are able therefore to reverse in an instant the reducing chemistry that formed the petroleum and reintroduce the π-bonds present in the original sources of the petroleum, the molecules derived from prehistoric life. In Chapter 1, we discovered that steam cracking produces enormous amounts of ethylene and propylene by breaking down larger saturated molecules. Benzene arises directly from petroleum by steam cracking, but is also produced in far larger amounts by a process we have not studied called catalytic reforming in which alkanes in the petroleum can be cyclized into six-membered rings. These rings then lose molecules of hydrogen to form aromatic molecules including benzene. This chapter and the one to follow focus on how much of the industrially produced materials we see and feel around us arises from reactions of ethylene or propylene with benzene. The reaction involved is one of the most important of organic chemistry, electrophilic aromatic substitution.

We gain insight into the commercial importance of electrophilic aromatic substitution in Fig. 3.1, which shows the value added as we carry out the chemical reactions necessary to convert ethylene, propylene and benzene to commercially important polymers – polystyrene, polycarbonate, and epoxy resin.

Industrial	Cents per Kg
Ethane	11
Benzene	24
Polystyrene	90
Cumene	13
Polycarbonate	363
Ethylene	42
Styrene	46
Propylene	29
Bisphenol A	231
Epoxy Resin	308

Fig. 3.1 The values in cents per kilogram for some industrial products and the common polymers derived from these intermediates.

3.2
The Carbon Atoms in Ethylene, Propylene and Benzene find their Way into Polystyrene, Polycarbonate, and Epoxy Resin

All the carbon atoms in polystyrene are derived from ethylene and benzene, while the carbon atoms in polycarbonate arise from propylene and benzene, with the two additional oxygen atoms and an extra carbon derived from a derivative of carbonic acid. Thus arises the name polycarbonate. In the epoxy resin there is again a combination of propylene and benzene but now the propylene plays more than one role. The epoxy resin structure in Fig. 3.2 shows that the two benzene-derived rings are connected in one way by a three-carbon unit derived ultimately from propylene, and in a second way through an ether linkage by another three-carbon unit derived again from propylene.

What is also very important is that the number of units in the polymer chain of epoxy resin is far less than in polystyrene or polycarbonate. Such a low molecular-weight polymer is called an *oligomer*. Put in another way, the length of the macromolecule and correspondingly the molecular weight is far less in epoxy resin than in polystyrene. In polystyrene the end groups are an insignificant part of the total molecule; thus, they contribute virtually nothing to the polymer's properties. This consideration applies also to polyethylene and polypropylene as discussed in Chapter 2 (Sections 2.7 and 2.8). The epoxy resin, which is an oligomer, is in a different category. The end groups play a critical role.

Epoxy resin requires "curing" or reaction with another molecule before becoming useful. Curing is a process with which many of us are familiar from the household use of epoxy adhesives. Two tubes are necessary, each containing a different chemical substance. Only after mixing the substances from each tube is there obtained the adhesive that binds the parts of the broken object. A chemical reaction takes place between the curing agent in one tube, an amine, and the epoxy resin in the other.

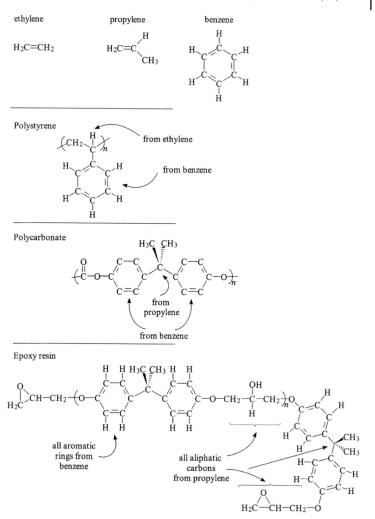

Fig. 3.2 Atoms derived from ethylene, propylene and benzene find their way into polystyrene, polycarbonate, and epoxy resin.

3.3
Industrial Synthesis of the Building Blocks of Polystyrene, Polycarbonate, and Epoxy Resin

Isopropylbenzene, cumene, is the fundamental building block for polycarbonate and epoxy resin, while ethylbenzene plays this role for polystyrene. For this reason industry produces very large amounts of both these alkyl aromatics.

Cumene, as we shall see later in this chapter, is an important intermediate to produce phenol and acetone, which then react together to form bisphenol A, which is an essential

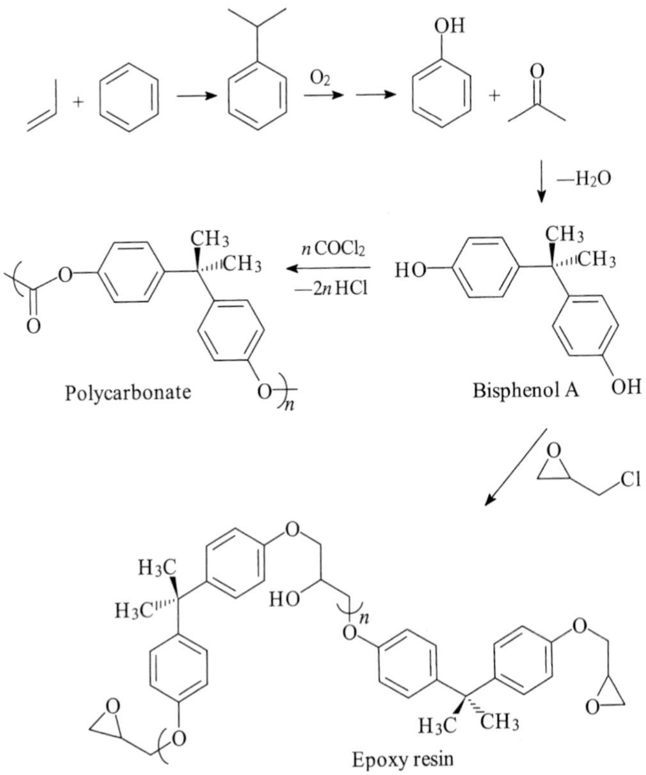

Fig. 3.3 Pathways from cumene to phenol and acetone to bisphenol A and on to polycarbonate and epoxy resin.

component of both polycarbonate and epoxy resin. The overall chemistry from cumene to bisphenol A and on to polycarbonate and epoxy resin is presented in Fig. 3.3.

Historically, the chemical industry produced styrene by a process involving elimination of water to form the double bond (Fig. 3.4). This process started with ethylbenzene, and involved functionalization of the ethylbenzene to the alcohol, 2-hydroxyethylbenzene, followed by loss of water. This is a process that was used during World War II when styrene at any cost was required to manufacture vitally needed synthetic rubber (see Chapter 7) from both styrene and butadiene.

Clearly, the direct production of styrene by elimination of H_2 from ethylbenzene (Fig. 3.4) was long ago realized to be more economical than a process requiring the intermediate 2-hydroxyethylbenzene. But, as for so many aspects of the chemical industry the more direct and economical process had to wait for the development of the proper catalyst, in this situation to eliminate H_2. Modern catalysts for production of styrene from ethylbenzene are complex, comprising iron and chromium oxides with various promoters working at high temperatures. Nevertheless, we shall learn later (Chapter 8, Section 8.4) that 2-hydroxyethylbenzene is still used, but in a situation where this alcohol is a byproduct in a route that adds value to this approach.

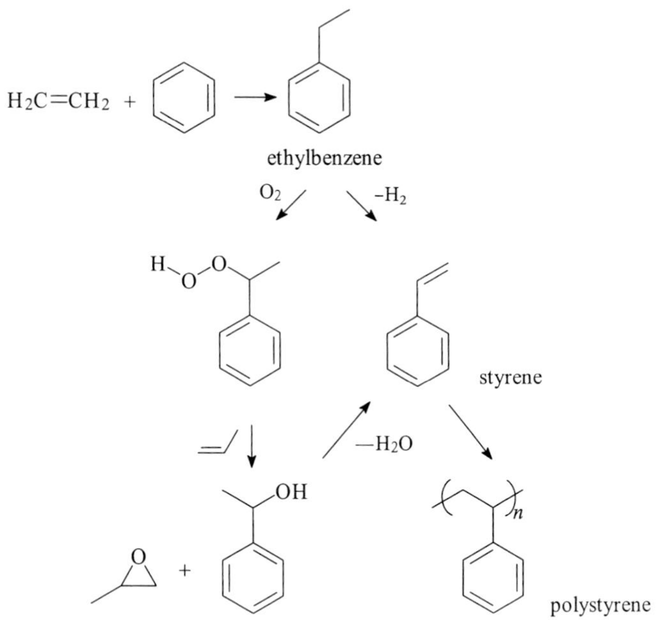

Fig. 3.4 Pathways from ethylbenzene to styrene to polystyrene.

$H_2C{=}CH_2$ +

ethylbenzene

O_2 $-H_2$

styrene

$-H_2O$

OH

+

polystyrene

Before we get into the synthesis of ethylbenzene – which is a big story on its own – let's first look into the synthesis of cumene, where we will learn of the common problems industry has had to face in producing both of these alkyl aromatics.

3.4
How Isopropylbenzene is Industrially produced and the Struggle to reduce the Di- and Triisopropylbenzene Byproducts

Industrially, the production of isopropylbenzene has used the same kinds of catalysts used in the laboratory, for Friedel–Crafts chemistry. Aluminum trichloride, with a small amount of HCl as promoter, will readily protonate propylene to form the carbocation with tetrachloroaluminate as the counterion. But there are some interesting special aspects to this industrial chemistry. In industry where cost is of the utmost importance, large excesses of benzene are used. Benzene is not free (Fig. 3.1), so why use it in very large excess? Certainly the benzene must be in a high enough concentration to capture the propylene cation rapidly to form an intermediate that loses a proton to reform the aromatic ring and yield isopropylbenzene as shown in Fig. 3.5. But this consideration is not critical. The large excess of benzene is used for a different purpose, to reduce multiple additions of propylene to the first formed isopropylbenzene.

Resonance theory and the related ideas of aromatic reactivity predict that an alkylbenzene is more reactive for further electrophilic aromatic substitution than benzene. Thus, cumene, once formed, will react faster than benzene and will add another isopropyl

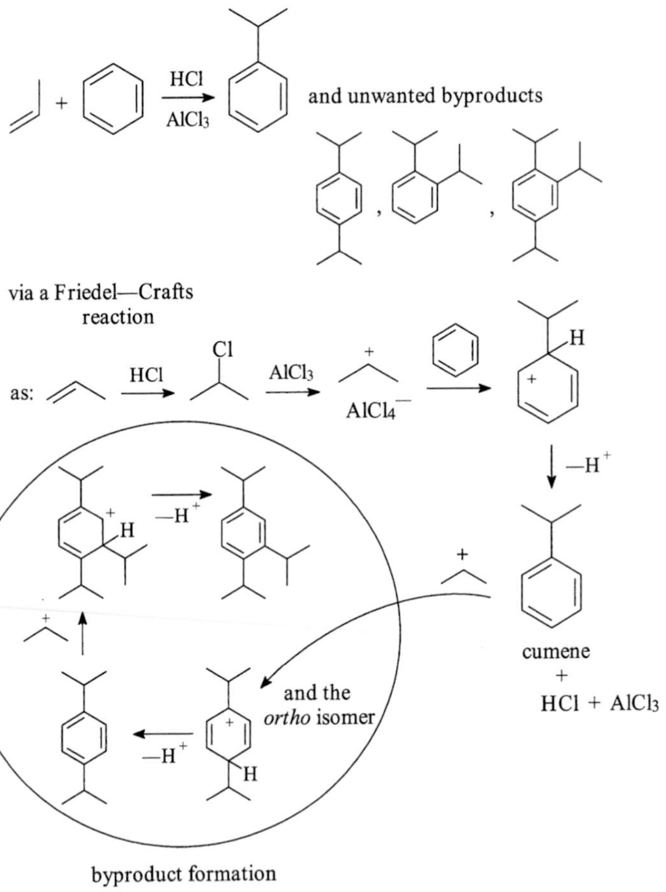

Fig. 3.5 Electrophilic aromatic substitution yields cumene from propylene and benzene, as well as, unwanted di- and triisopropylbenzene.

group predominately to the *ortho* and *para* positions. The excess benzene overcomes this inherent superior reactivity of the cumene. The subsequent addition of propylene cation to an already formed cumene is precisely what is *not* wanted (Fig. 3.5).

The problem is that the velocity of the electrophilic aromatic substitution depends on the rate constant, which is larger for addition of the propyl cation to cumene than it is for addition of this carbocation to benzene. Expressed in another way, the energy of the positively charged intermediate formed from the aromatic ring is lower for the addition of the isopropyl cation to the cumene than to benzene. Why is this?

Consideration of the resonance hybrids for the intermediate formed after addition of the isopropyl cation to the *ortho* or the *para* position of cumene demonstrates that the positive charge resides on the ring carbon atom bearing the already present isopropyl group. The alkyl group is a stabilizing influence compared to the formation of this intermediate from benzene where the positive charge must reside on carbon atoms bearing only hydrogen. The increased stability of more highly substituted carbocations that worked to favor branched hydrocarbons desirable for gasoline engines in catalytic cracking (Chapter 1, Section 1.12) provides a problem in this situation by favoring the formation of an unwanted byproduct in the synthesis of cumene. Principle can work for you or against you.

Adding more benzene can help to solve the problem discussed above in two ways. The velocity of the electrophilic aromatic substitution, the production of cumene per unit time, depends on the concentration of the reactants. The larger the number of molecules of any one reactant, the larger will be the number of collisions leading to reaction. In this way, increasing the concentration of benzene can counteract the higher intrinsic reactivity of the cumene. The probability of an isopropyl cation meeting a benzene molecule over an already formed cumene molecule will be greater by swamping the system with a large excess of benzene. However, benzene will have to be added continually, especially as the reaction proceeds, because the concentration of the cumene is continuously increasing.

A second way the excess benzene helps depends on the reversibility of the electrophilic aromatic substitution. The unwanted diisopropylbenzene reacts readily with electrophiles for the same reason that cumene is more reactive with electrophiles than benzene. The alkyl substituents stabilize the resulting positive charge. This increased reactivity leads to a high rate of reaction of the unwanted dialkylbenzenes with the high concentration of protons in the acidic reaction medium (Fig. 3.6).

As shown in Fig. 3.6, because of symmetry, the addition of a proton to p-diisopropyl-benzene must take place at either the *ortho* or *para* position of the ring with respect to

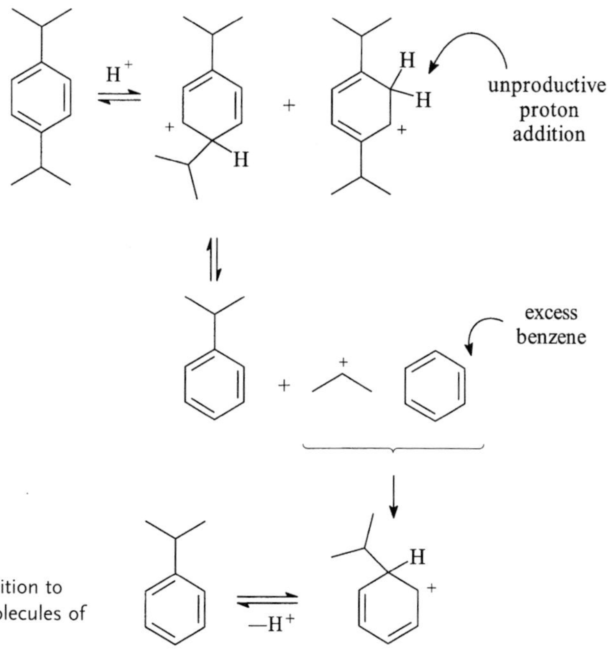

Fig. 3.6 Mechanism of proton addition to diisopropylbenzene to form two molecules of cumene.

one of the pendent isopropyl groups. No other site exists. If addition is to the *ortho* position, only exchange of one proton for another can occur and therefore no chemical change can result. But if the addition of the proton is to the *para* position two paths become possible: the newly added proton can be cast out returning the molecule to its original state, that is, to *p*-diisopropylbenzene; alternatively, one of the already-attached isopropyl groups can be eliminated as a cation yielding the desired cumene.

If there is a large concentration of benzene, the eliminated isopropyl cation can be removed by reacting with a benzene molecule producing even more of the desired cumene. In the presence of excess benzene therefore the addition of a proton takes the system from a diisopropylbenzene and a benzene molecule to two molecules of cumene (Fig. 3.6).

3.5
How Ethylbenzene is produced Industrially

The classical production of ethylbenzene from ethylene and benzene uses aluminum trichloride and HCl, precisely parallel to the production of isopropylbenzene (Figs 3.5 and 3.6). The HCl reacts with ethylene to form ethyl chloride, and the aluminum trichloride then activates the ethyl chloride for reaction with benzene. This releases the HCl and the aluminum trichloride. It all sounds like the familiar Friedel–Crafts reaction, except that the industrial scale is impressive. A world-size styrene plant can produce as much as 675 *million* kilograms of styrene per year from a similar amount of ethylbenzene produced at the same site. What does this mean in terms of the HCl and aluminum trichloride catalyst used for ethylbenzene production?

For a 675 million kilograms per year plant, 2.6 million kg (6800 kg per day) of spent catalyst is produced. The spent catalyst cannot be used again, and the cost of its disposal is high. Some of it may be used as a flocculating agent in wastewater treatment plants, and the paper industry also finds use for some of it in pulping, if it can accommodate the high shipping costs for the dilute solution. These catalyst disposal problems and their environmental costs have motivated a shift to the use of zeolite catalysts, the same catalyst type used in catalytic cracking discussed in Chapter 1 (Section 1.9). As we shall see, the use of zeolite catalysts not only solves the environmental and disposal problems but greatly reduces the formation of unwanted di- and triethylbenzene, which form for the same reason as encountered in the production of cumene (Figs 3.5 and 3.6).

3.6
Zeolites and Ethylbenzene

In Chapter 1 (Section 1.9; Fig. 1.5), we discovered that the zeolites used today contain molecular size channels formed of aluminum and silicon oxides and that the negative charge on the tetrasubstituted aluminum atoms may be compensated for by a proton so that the channels became intense sources of Brønstead acidity. It is reasonable that the acidic character of zeolites, so valuable for catalytic cracking of petroleum, could also be

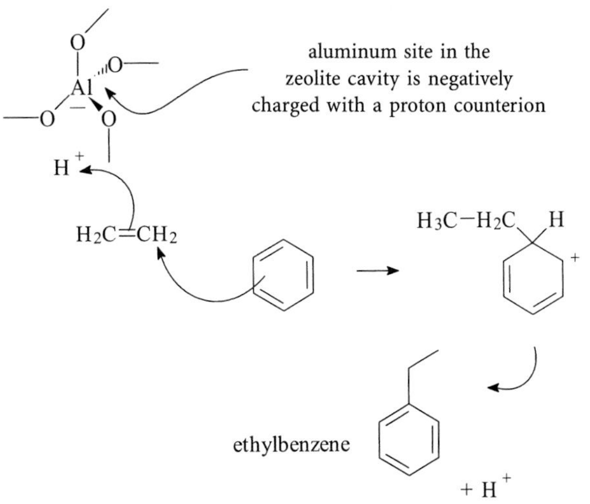

Fig. 3.7 Zeolite-catalyzed electrophilic aromatic substitution of ethylene and benzene to produce ethylbenzene.

aluminum site in the zeolite cavity is negatively charged with a proton counterion

ethylbenzene

used to catalyze electrophilic aromatic substitution of ethylene and benzene to form ethylbenzene (Fig. 3.7).

A great benefit of zeolite catalysis of the electrophilic aromatic substitution of ethylene and benzene relates to the reactor vessel. HCl and AlCl₃ are highly corrosive materials, but in the zeolite the acidity is inside the zeolite "container," within the channels of the zeolite. Therefore reactors need not be acid-resistant, and this provides a great savings in cost.

The zeolite reactor is multibed, which means that there are shelves in place to hold the catalyst. Thereafter, preheated benzene and ethylene are added via pipelines from the storage tanks that contain them. The reaction takes place at about 450 °C and a pressure of about 30 bar. The reaction time is rapid and the reaction products are separated by three stills attached in series to the reaction kettle. Most of the unreacted benzene is "flashed off" for recycling in a simple still placed between the kettle and the three stills. The residual benzene is distilled in the first column, whereas the second column distills the product ethylbenzene.

Heavy organics remain in the second still and are transferred to the third still for separation of the small amounts of di- and triethylbenzenes formed in the zeolitic method. These unwanted byproducts, *ortho* or *para* diethylbenzene, are not only produced in far smaller amounts than in the HCl/AlCl₃ process, but can also be recycled for reaction with benzene to produce more ethylbenzene parallel to the process for production of cumene from diisopropylbenzene (see Fig. 3.6). Had HCl and AlCl₃ been used, the process would be much more complex because of the disposal of the wastewater containing these corrosive materials. Bravo to the zeolites! They not only make for a better environment but also for a more economical processes.

3.7
How does the Zeolite Catalyst repress the Formation of Di- and Triethylbenzene?

In the discussion earlier in this chapter on cumene synthesis from propylene and benzene, much was made of the problem of cumene's greater reactivity compared to that of the starting benzene – a lesson in resonance theory that the industrial chemist had to face and to deal with by using large excesses of benzene. The same lesson must apply to the electrophilic aromatic substitution that produces ethylbenzene. Yet the zeolitic catalyst represses the multialkylation. In the zeolite process there is a 99.5% selectivity to the desired product ethylbenzene. Why?

The answer is in the size selectivity of the zeolitic channels. Diethylbenzene is large enough so that once formed inside a zeolite channel, its rate of diffusion through the channels to escape from the zeolite is very low. Ethylbenzene, on the other hand, can escape far more readily from the zeolite channels and be distilled away. The molecules of diethylbenzene are therefore exposed for longer periods of time to high proton concentrations in the presence of benzene molecules that can enter and leave the channel with ease because of their small size. The diethylbenzene is trapped in an "acidic sea of benzene" where the identical mechanism presented in Fig. 3.6 for conversion of diisopropylbenzene to cumene takes place. Two molecules of ethylbenzene are produced from one molecule of diethylbenzene and one of benzene. For the chemical industry, zeolites have been a revolutionary catalyst!

3.8
So Why are Zeolites not used for the Formation of Cumene?

In fact, zeolites are such an improvement that over 90% of cumene production is now via zeolite catalysis. But historically, cumene was produced not by chemical manufacturers but rather in a reaction conducted in the petroleum refinery. This historical precedent influences the situation today. The refinery is the place where petroleum is delivered and distilled, steam cracked for delivery to the chemical industry, and catalytically cracked to produce the molecules that can be further chemically modified to become components of gasoline (see Chapter 1).

The production of cumene arose from the addition of benzene to reactors at the refinery, where nonzeolitic acid catalysts oligomerized propylene and butenes for gasoline. The cumene led to a significant increase in octane number. The cumene produced right in the refinery had been favored because of the availability of cheap propylene produced in catalytic crackers right on the site. Thus, cumene production in the refinery became a means to increase the octane rating of gasoline and also a chemical intermediate to be sent to the chemical industry.

The reactors to which the benzene is added produce what is called *polygas*, which can be made solely from propylene and butenes or by addition of isobutylene to the propylene as shown by the carbocation chemistry in Fig. 3.8. Polygas molecules, branched hydrocarbons (Fig. 3.8), were widely used for blending with other components to make gasoline that is sold to the public. Subsequently, a variation of this chemistry has displaced the traditional polygas process by alkylates resulting from the cationic reactions of propylene and butene isomers with a saturated hydrocarbon, isobutane.

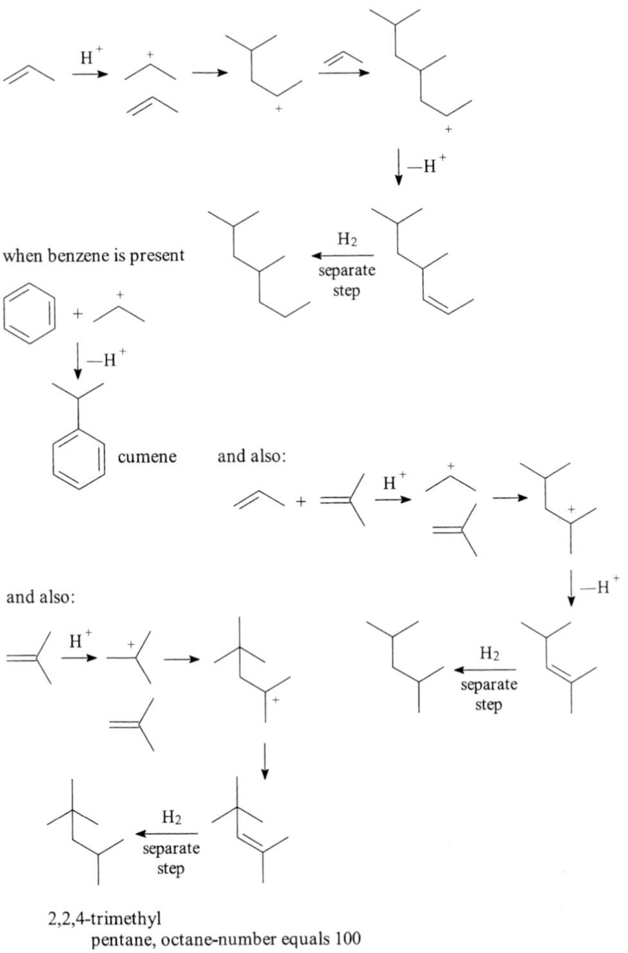

when benzene is present

−H⁺

cumene   and also:

and also:

2,2,4-trimethyl
pentane, octane-number equals 100

Fig. 3.8 Chemical reactions in the production of polygas, showing also the production of cumene and the role of carbocations.

However, as the chemical industry switched to zeolite catalysis for the formation of ethylbenzene it discovered that the same advantages pertain to the synthesis of cumene. Catalysis by zeolites in the synthesis of cumene from benzene and propylene represses multialkylation byproducts for the same reason discussed above for production of ethylbenzene. The zeolites for cumene production similarly give rise to environmental advantages that occur from the absence of Friedel–Crafts aluminum-based byproducts. And zeolite use is more economical because corrosion-free vessels become unnecessary. Moreover, there is always an advantage in simplified processes arising from the use of heterogeneous catalysts.

So the tide has shifted to production of cumene by chemical manufacturers instead of production in polygas refinery reactions. The separation of cumene and polygas production is also stimulated by the necessity to reduce aromatics such as cumene in gasoline for environmental reasons. Thus, zeolites are the future for all industrial electrophilic aromatic processes producing both ethylbenzene and cumene.

3.9
What Role does Cumene play in the Production of Epoxy Resin and Polycarbonate?

In Fig. 3.3 we saw the overall role of cumene in producing polycarbonate and epoxy resin via the intermediate bisphenol A. Now let us explore the details of this synthetic path to bisphenol A and discover the central role of electrophilic aromatic substitution. Whatever is the synthetic path, it cannot be too fancy; billions of kilograms of bisphenol A are consumed worldwide every year, and every additional penny per kilogram adds a great deal to cost.

Chemists who specialize in synthesis have learned to approach synthetic problems by working backwards. In other words, one asks the question what might be the immediate precursor for epoxy resin and polycarbonate? We know the answer, bisphenol A. Then we ask what precursor might lead to bisphenol A and thus develop a reactive path back to cumene. Let's begin with a reaction between phenol and acetone, which leads to bisphenol A. Then we can discover how to produce phenol and acetone and find out that their precursor is cumene.

3.10
How Phenol and Acetone react together to form an Isomeric Mixture of the Intermediate HOC$_6$H$_4$C(CH$_3$)$_2$OH, which then goes on to form Isomers of Bisphenol A

Acetone, like all carbonyl compounds, is subject to reactions in which electron-rich molecules add to the electropositive carbon of the carbonyl group. This is a recurring theme that we will encounter many times in the study of industrial chemistry including notably in the synthesis of methyl methacrylate from acetone and HCN (Chapter 6, Section 6.3). In general, this kind of carbonyl reactivity is catalyzed by protonation of the carbonyl oxygen. Protonation acts to emphasize the dipole, pulling electron density further away from the carbon bound to that oxygen.

In the presence of an aromatic molecule that is less primed to act as a nucleophile, such as benzene for one example, which is less reactive to electrophilic aromatic substitution than is phenol, protonation of acetone might *not* lead to reaction. But for phenol the protonated acetone is enough to initiate an electrophilic aromatic substitution. The hydroxyl group in phenol is not only highly activating for electrophilic aromatic substitution but as well is *ortho* and *para* directing. For this reason, as shown in Fig. 3.9, the reaction with protonated acetone leads to two products that are intermediates in the formation of isomers of bisphenol A.

One of the isomers shown in Fig. 3.9, **1**, the *para* isomer, is taken as an example of how the next step to bisphenol A occurs (Fig. 3.10). The aliphatic hydroxyl group connected to the tertiary benzylic carbon, on protonation with acid and subsequent loss of water forms a stabilized carbocation. The carbocation is tertiary and also derives stability

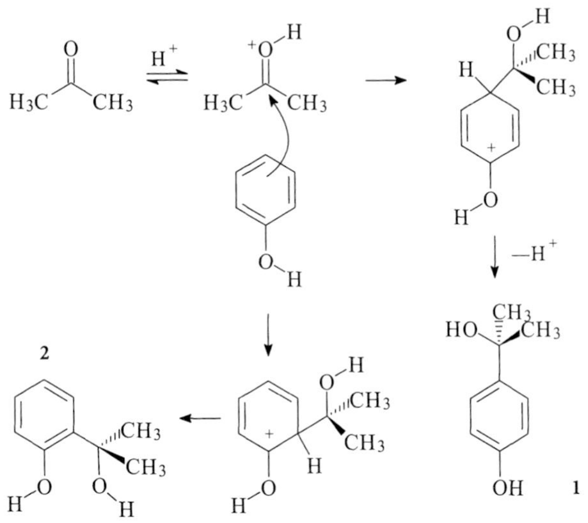

Fig. 3.9 Electrophilic aromatic substitution of phenol and acetone leading to the *ortho* and *para* isomers. The first step on the route to bisphenol A leads to isomers.

from the *para*-hydroxyl substituted aromatic ring. These are two powerful factors for stabilization of a positive charge on a carbon.

But if there were protons available in a solution of this alcohol, (**1**, Fig. 3.10) why should they be attracted only to the aliphatic hydroxyl function? Certainly under acidic conditions, the molecule could add a proton alternatively at the hydroxyl group on the ring, that is, the phenolic hydroxyl, or for that matter a proton could be added to the aromatic ring. But in both these instances there is no reasonable reaction path, and so the proton will simply be reversibly lost or exchanged with the proton already at that site. Here we see an analogy to the consequences of the variable protonation sites in the reaction of diisopropylbenzene with benzene (see Fig. 3.6). Only on proton addition to the aromatic ring carbon bearing the isopropyl group does a productive reaction occur. Similarly here, only on proton addition to the aliphatic hydroxyl group of **1** is a path opened involving the loss of water with production of the carbocation

It is certain that the carbocation formed by loss of water from **1** (Fig. 3.10) would be captured by phenol in an electrophilic aromatic substitution reaction. Moreover, the electrophilic aromatic substitution characteristics of phenol would certainly also lead to both *ortho* and *para* substitution just as for the reaction with acetone in Fig. 3.9.

We had presented the structure of bisphenol A (Fig. 3.3) as the isomer with *para* relationships between the groups on the two benzene rings and, in fact, this is the isomer absolutely necessary for the formation of polycarbonates. In polycarbonates, the shape of the units that make up the long polymer chain, as we will discuss in more detail in Chapter 4 (Section 4.19), is critical to the properties of the plastic. Therefore, if the industrial procedure for formation of bisphenol A is to follow Figs 3.9 and 3.10 it must include a separation step to obtain the pure *para-para'* isomer.

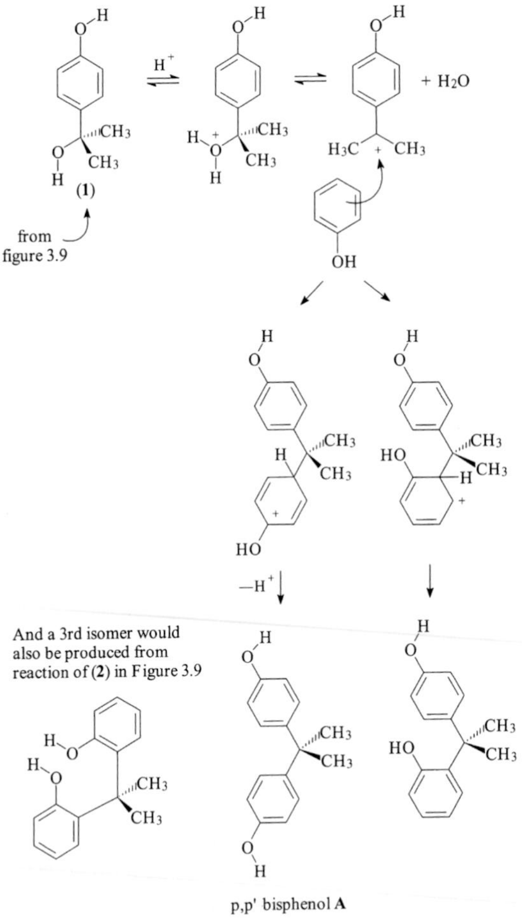

Fig. 3.10 Electrophilic aromatic substitution between $HOC_6H_4C(CH_3)_2OH$, **1**, and phenol-yielding isomers of bisphenol A.

from figure 3.9

And a 3rd isomer would also be produced from reaction of **(2)** in Figure 3.9

p,p' bisphenol A

A separation step may appear economically unacceptable when we consider dealing with billions of kilograms. However, chemical engineers have worked out conditions for large-scale crystallization in industrial processes to make such procedures quite reasonable. And, as we shall see, the structure of the desired *p,p'*-bisphenol A (Fig. 3.10) helps. And, at any rate, there is no choice. One cannot engineer away the resonance interactions that force phenol to undergo electrophilic aromatic substitution at both the *ortho* and *para* positions.

The good news for engineering a crystallization-based separation of the bisphenol A isomers (Fig. 3.10) is that symmetrical compounds often have higher melting points than less-symmetrical isomers. A higher melting substance means that more energy is necessary to destroy the crystal, that is, to transform it to the melt state. The forces holding the crystal together are larger.

There is a correspondence between melting point and solubility. Higher melting molecules tend to be less soluble because dissolving a crystal must overcome the forces holding the crystal together. If one has a mixture of molecules that can crystallize from the same solution, the more symmetrical molecule with the higher melting point will be less soluble because it forms the most stable crystal. The industrial crystallization of the bisphenol A isomers is slave to this principle. The more symmetrical *para-para'* isomer is far less soluble than any other possible isomer of bisphenol A and in a mixture of isomers in solution the *para-para'* isomer crystallizes most readily. Great!

In epoxy resins, in contrast to polycarbonate, although bisphenol A is also a starting component (Fig. 3.3), the molecular arrangements are more tangled as will be described in Chapter 4 (Section 4.2). In the epoxy resin, therefore the mixture of bisphenol A isomers can be used directly saving the crystallization step and therefore also cost.

3.11
We continue Our Backward Path. How are Phenol and Acetone formed from Cumene?

But how are we to take the very last step backwards – the step that takes us back to cumene? Our backward route now must derive phenol and acetone from this molecule.

In retrospect, the production of phenol and acetone from cumene makes a great deal of sense. But this is not the way it happened. Rather, chemists working in Germany in the midst of World War II discovered the connection entirely accidentally. It was an extension of studies on the reaction of oxygen with various hydrocarbons carried out at the Institut fuer Kohlenchemie an der Bergakademie Clausthal. And at this terrible time in 1944 the German chemists published their work on cumene. They had no way of realizing at the time that their discovery would form the foundation of one of the great industrial enterprises in the world. Others, including Distillers in England, investigated this chemistry (Fig. 3.11) after the war.

The key is the single hydrogen on the central carbon of the isopropyl group of cumene, the benzylic carbon. Both characteristics, tertiary and benzylic characteristics point to a weak bond with low bond energy. This bond between carbon and hydrogen is easily broken to form the tertiary benzylic radical and the air is full of just the reactant to do this job, oxygen, a molecule that exists in its ground state, as a triplet, a diradical.

It is only necessary to stir the cumene with very hot aerated water and some basic substances and an emulsifying agent to produce cumene hydroperoxide. Cumene is hardly soluble in water, and the emulsifying agent is there to facilitate its dispersion in the aqueous phase – a parallel to the role of soap in dispersing grease. The base is there to keep the solution from becoming acidic, thereby blocking the next step – a step we wish to delay until the right moment in the industrial procedure.

The overall result is that the oxygen dissolved in the hot water inserts itself between the benzylic carbon and its weakly bound hydrogen. This conversion occurs via a free radical chain reaction in which the key propagation step is formation of the tertiary benzylic radical via abstraction of the target hydrogen atom. As for all free radical chain reactions, from the textbook classic reaction of chlorine with methane to the free radical chain reaction underlying steam cracking discussed in Chapter 1 (Section 1.11), or the free radical polymerization of ethylene (Chapter 2, Section 2.2) the basic steps are the same, initiation, propagation and termination.

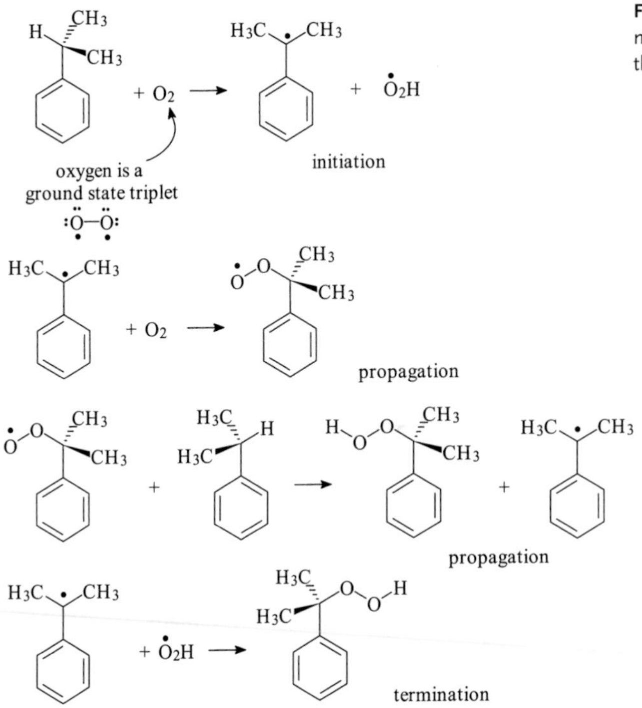

Fig. 3.11 Free radical chain mechanism for formation of the hydroperoxide of cumene.

In the industrial process, the contents of the reactor containing cumene hydroperoxide (Fig. 3.11) are acidified with sulfuric acid when the concentration of cumene hydroperoxide is at a maximum. Phenol and acetone are produced in the ratio of 1 tonne of phenol to 0.6 tonne of acetone, in exact proportion to their molecular weights and via a very surprising version of electrophilic aromatic substitution. What happened?

3.12
A Remarkable Rearrangement

Consider a molecule like the hydroperoxide of cumene shown in Fig. 3.11 in the presence of a strong acid in water. The concentration of protons in the solution is high, and there are several sites of Brønstead basicity that can interact with these protons. We have seen before that various sites for proton addition may be possible but that only one site leads to reactivity (Section 3.10). A proton can add and then be released from the benzene ring, or it can behave similarly with the oxygen closest to the aliphatic carbon. Adding the proton to the oxygen with the bound hydrogen is, however, a different story. Here the molecule has the capacity to release a molecule of water by breaking the relatively weak oxygen–oxygen bond (Fig. 3.12).

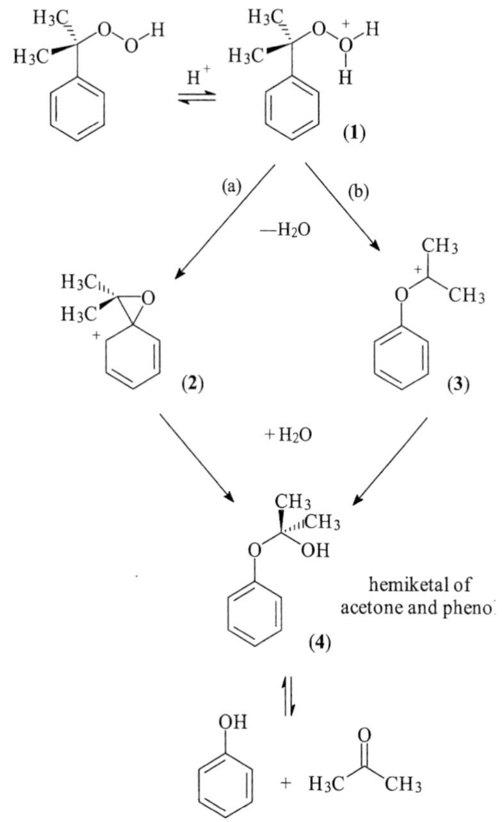

Fig. 3.12 Two possible mechanisms for the acid-catalyzed rearrangement of cumene hydroperoxide, and how the product of the rearrangement is identical to the hemiketal formed from phenol and acetone.

There are many examples where the release of small molecules drives a reaction forward for entropic reasons, a characteristic encountered throughout the book. The stability of water expressed in thermodynamic terms by its large negative heat of formation is an additional factor driving elimination. Therefore loss of water via a protonated hydroxyl group bonded to carbon is a facile reaction as we have just seen in Fig. 3.10. However, with protonated hydroperoxide (Fig. 3.12, 1) the remaining oxygen of the broken hydroperoxide bond would be left with a full positive charge, an unlikely state for oxygen. One way to avoid this state but still allow loss of H_2O would be for the aromatic ring electrons to respond in the same way as to all strong perturbations by electrophiles, by setting up the first step of an electrophilic aromatic substitution. This is shown in Fig. 3.12, pathway (a). The intermediate that would be formed is called a spiro compound. In such molecules, a single atom forms the junction between two different rings. This intermediate is not at all stable. It suffers from the loss of the aromatic character of the ring, and if this were not enough this structure also bears the strain of the three-membered ring.

Another way to avoid the formation of the positively charged oxygen formed via loss of water from the protonated hydroperoxide of cumene, (**1**, Fig. 3.12) could be via the rearrangement shown in Fig. 3.12, pathway (b), which does not disrupt the aromatic ring but does involve the breaking and making of a δ-bond. In either mechanistic pathway, Fig. 3.12 (a) or (b), leading respectively to **2** and **3** (Fig. 3.12) (truthfully we are not certain which one or some variation of the two paths is the fact) the positively charged intermediate is resonance stabilized. Whatever the molecules are really doing makes no difference to the outcome. In either mechanism, addition of water to the intermediate, and loss of a proton, as shown in Fig. 3.12, leads to the identical result, formation of an intermediate, **4** (Fig. 3.12), with an interesting structure. This molecule, **4**, is the hemiketal that would be formed from reaction of phenol and acetone catalyzed by acid.

Although the path to the hemiketal **4** in Fig. 3.12 is quite unusual, there is nothing unusual about the acid-catalyzed reaction of phenol and acetone. Parallel reactions take place under the same conditions for mixtures of all alcohols and carbonyl compounds catalyzed by acid. Moreover, hemiketals are formed reversibly in acid solution, which means that any hemiketal is in a dynamic equilibrium with the alcohol and carbonyl compounds that formed it. This means that as the hemiketal forms via the addition of water to the positively charged intermediate, **2** or **3** (Fig. 3.12) it instantly in turn forms phenol and acetone (Fig. 3.12). Once formed, the properties of **4** are identical to the hemiketal formed directly from phenol and acetone.

Because the hemiketal **4** (Fig. 3.12) is in equilibrium with phenol and acetone, the industrial process can recover the phenol and acetone by simply distilling away one or both of them from the solution. This will drive the equilibrium to produce more phenol and acetone until the hemiketal is exhausted – another example of Le Chatelier's principle. We have arrived at the beginning of our reverse synthetic path from bisphenol A all the way back to cumene and therefore still further back to petroleum, the ultimate source of cumene from benzene and propylene (Section 3.4).

The overall process adds a great deal of value. Phenol was consumed in the USA in 2002 at a level of 2.2 billion kg, worth US$2 billion. Approximately 1.3 billion kg of acetone was needed by the USA in 2000, and this is worth about US$1.1 billion. Globally, 5.4 billion kilograms of phenol worth about US$ 5.1 billion will be used, while about 3.4 billion kilograms of acetone, the value which is about US$3 billion, will be required. Why are they produced in such large amounts? Why is so much money involved? A great deal of the answer to this question is found in bisphenol A and its use in the preparation of epoxy resin and polycarbonate. Chapter 4 will deal with these plastics.

3.13
Summary

Remarkably, three molecules – ethylene, propylene, and benzene – which are one step removed from petroleum, make up, with the addition of a derivative of carbonic acid, all the carbon atoms in the three plastics, polystyrene, polycarbonate, and epoxy resin. In the formation of ethylbenzene and isopropylbenzene (cumene), the ultimate precursors of these plastics, we encounter the classic Friedel–Crafts reaction. But these electrophilic aromatic substitutions must be conducted so as to minimize the increased reactivity of the product over the starting material, which leads to undesired multialkylation products. And this requires addition of excess benzene to the reactions to take advantage of principles of reaction rate theory that work to favor the desired product.

However, the best answer to the problem of multialkylation is found in the use of zeolites whose narrow channels work to narrow the product distribution to monoalkylation. This also helps environmental considerations by avoiding the formation of aluminum-containing byproducts produced in the conventional Friedel–Crafts approach. In the formation of the polycarbonate and epoxy resins the intermediate bisphenol A is involved, and we see that this can be formed from acetone and phenol by a classic electrophilic aromatic substitution, which however forms isomers. But in a lucky break the desired isomer for polycarbonate formation has the highest melting point of all the isomers produced allowing therefore for the use of crystallization for purification.

And stepping back in the overall process, the acetone and phenol necessary to form bisphenol A may be formed from cumene by another kind of electrophilic aromatic substitution, one that was discovered accidentally.

Benzene, the central player in all electrophilic aromatic substitution chemistry, is the starting material for polystyrene, epoxy resin, polycarbonate and, another polymer of great importance, nylon. But only the first three involve electrophilic aromatic substitution, and that is our focus here. The commercial importance can be seen in just one of these monomers, styrene. In the year 2002, styrene, the immediate precursor of polystyrene, was consumed at the level of 4.5 billion kilograms in the USA. It will be almost as much in Western Europe, and the rest of the world consumed another 4.5 billion kilograms. These 14 billion kilograms are worth billions of dollars, and provide jobs for about 7000 people.

So, propylene and ethylene and benzene have come a long way from their source from petroleum to this exalted position to be used as molecular precursors of various polymers found in the sun glasses, CD cases, golf clubs and other common objects all around us. And the essential chemical reaction is electrophilic aromatic substitution, a process in which the extreme stabilization of the 6π electrons in a six-membered ring represses the usual addition to unsaturated bonds and forces the chemistry to turn to electrophilic aromatic substitution, which we now see has such enormous consequence for industry and society.

Some of the subjects treated in this chapter are listed below. These are key words and terms that act as reminders of the chapter's contents and should become a valuable part of your chemical vocabulary.

- Various aspects of electrophilic aromatic substitution
- Resonance
- Brønstead acidity
- Addition and condensation polymerization
- Thermodynamic and kinetic control of reactivity
- Hemiketals
- Free radical chain reactions
- Steric effects in zeolites

- Displacement of equilibrium
- Relationship of crystalline stability to structure
- Friedel–Crafts reaction
- Carbocation chemistry

Study Guide Problems for Chapter 3

1. Show all steps for the synthesis of bisphenol A from phenol.
2. In the conversion from cumene to phenol and acetone, the first step involves reaction of oxygen with cumene to produce a hydroperoxide, that is, an oxygen molecule is inserted between the benzylic carbon–hydrogen bond. Propose a free radical chain mechanism for the formation of this hydroperoxide.
3. Phosgene, $COCl_2$, reacts with bisphenol A to produce polycarbonate. Propose detailed mechanistic steps for this example of acyl nucleophilic substitution.
4. How does resonance theory combined with the fact that mainly *ortho* and *para* diisopropylbenzene are formed from cumene as byproducts reveal the basis of the faster reactivity of cumene compared to benzene with isopropyl cation?
5. Explain how increased concentration of benzene reduces multialkylation in the reaction of either propylene or ethylene with benzene.
6. Offer an explanation for the fact that the amount of diethylbenzene is greatly reduced in proportion to ethylbenzene via catalysis by zeolites versus HCl, $AlCl_3$ catalysis.
7. All aspects of the formation of polygas demonstrate Markovnikoff addition. Explain this statement.
8. As noted in the text, gasoline additives, alkylates, are derived from the acid-catalyzed reaction of propylene and the isomeric butenes with a saturated hydrocarbon, isobutane. Propose a series of mechanistic steps to describe these reactions and propose structures for the alkylates that arise.
9. If the branched hydrocarbons produced in the polygas reactor are stored without the hydrogenation step, great care must be taken to exclude the presence of free radicals and therefore oxygen. Why?
10. Use resonance theory to explain why the formation of isomers of bisphenol A is unavoidable in the reactions shown in Figs 3.9 and 3.10.
11. Show all the isomers of bisphenol A that could be formed by the electrophilic aromatic substitution reactions in Figs 3.9 and 3.10.
12. In the reaction of phenol with the carbocation formed via loss of water from protonated **1** in Fig. 3.10, another reactive path could be taken that would lead to an ether. What could this path be, and what would be the structure of the ether it would provide?
13. Propose other possibilities for termination of the free radical chain reaction that forms cumene hydroperoxide in Fig. 3.11. What factors might favor or disfavor the termination reaction you propose compared to termination via reaction of hydroperoxide radical with the cumene radical shown in Fig. 3.11.
14. Alcohols and ketones form ketals, while alcohols and aldehydes form acetals. Show all the acid-catalyzed equilibrium steps in the reaction of phenol with acetone and phenol with acetaldehyde. In the laboratory procedure for formation of these derivatives of alcohols and carbonyl compounds organic solvents are used and water is removed as it is formed. Why?

15. The most noted example of hemiacetal formation is involved in the mutarotation of glucose. Recount the details of mutarotation of glucose and explain how this phenomenon is related to the fact that glucose is a reducing sugar although existing in a six-membered ring form.

4
From Nucleophilic Chemistry to Crosslinking, with a Side Trip to Glycerol, in the Synthesis of Commercially Important Plastics

4.1
The Structure and Use of Epoxy Resins

Fig. 4.1 shows a complex tangle of chemical bonds and functional groups. These are some of the chemical details of a portion of a cured epoxy resin (see Chapter 3, Fig. 3.2), the substance of great strength and adhesiveness that forms by mixing the contents of the two tubes for sale at retail stores all over the world. One tube has no odor, while the other may have a vague fishy smell arising from the amine functional group necessary for the curing process. How can a structure like the one in Fig. 4.1 arise? What is curing? The answers come from substitution reactions via nucleophiles reacting with electrophiles. This classic interaction in chemistry in which there is attraction between the electron-rich seeking the positive and the electron-poor seeking the negative provides us with a road that leads to epoxy resins, a material of great value in modern life.

Epoxy resin provides one of the best adhesives known. It causes many different materials to adhere to each other by acting as a mutual bond between them. The dipoles and hydrogen bonding functions in the structure in Fig. 4.1 play key roles in the adhesive properties because attractive interactions are formed between the resin and the surface of a wide variety of materials. Which materials will adhere to epoxy resins, and which will not? What can be glued together? The answer is, those that can interact physically or chemically with the functional groups the epoxy resin has to offer. Look carefully at the label of the epoxy adhesive you've purchased and you'll find mention of the very few materials that won't work. In most cases the material that cannot be glued is incapable of strong interactions via hydrogen bonding or dipolar forces. Teflon is one example, and polyethylene is another.

Most materials, however, adhere to epoxy resin exceptionally well. Epoxy resin is, therefore, useful in modern society to provide not only adhesives but even more important, coatings. Moreover, when the epoxy resin is reinforced with glass or carbon fiber, it can form a very tough but pliant material. Golf club shafts, for example, are made in this way as are doors for space shuttles, both reinforced with carbon fiber.

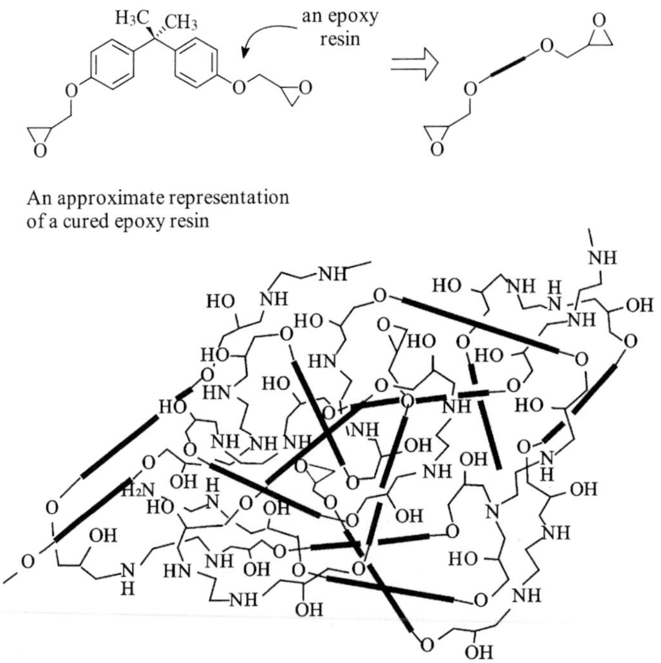

Fig. 4.1 The structure of a portion of an extensively crosslinked epoxy resin.

4.2
Epoxy Coatings and their Curing (Crosslinking) and Pot Life

Epoxy coatings take several forms, but the one that makes the greatest economic contribution involves the structure shown in Fig. 4.1, which is found in paints for metal surfaces to protect them from corrosion. An outstanding example is the application of an epoxy resin-based paint to the legs of offshore oil well rigs to protect them from corrosive salt water. Remarkably – even unbelievably – the paint can be applied under water, its physical adhesive forces being so great that the water is pushed away from the metal surface so that the paint can attach itself.

This use of epoxy coatings may add value to many of our lives but does not make life easier for the people who coat the surface to be protected. The components must be packaged separately and mixed before application. Then the mixture has a "pot life" of 3 to 18 hours depending on temperature and other factors. During this time, the coating must be applied. Once the components are mixed they begin to react immediately, following the chemistry leading to the structure in Fig. 4.1.

Inspection of Fig. 4.1 shows that many linear polymer chains can be traced with many links among these chains. The links among the chains are appropriately called crosslinks, and the process of forming the crosslinks in epoxy resin is called *curing*. Once the three-dimensional molecular network arising from the crosslinks forms, painting becomes impossible because the material is incapable of flow. Fast work is required – as many of us have experienced in our own use of epoxy resin.

The adhesive and crosslinked character of an amine-cured epoxy resin is taken advantage of in many ways. In one of these applications the epoxy composition is mixed with sand. A concrete analog results that is highly abrasion and impact-resistant and will not be destroyed by acid or base. You will find it in roller skating rinks, factory floors and in areas approaching tollbooths where the constant braking of trucks and automobiles would otherwise destroy the pavement.

Another type of coating involving epoxy resins is used for the linings of cans, but here the curing reaction does not involve the amine functional group that led to the fishy smell in one of the tubes purchased in hardware stores. The curing agent is rather a phenolic resin. The phenolic hydroxyl is far less nucleophilic than an amine group and therefore is slower to react, as we shall see shortly when we look at the chemistry of the curing in detail. There is virtually no reaction at room temperature. Higher temperatures are required for the curing process to proceed, so the components can be mixed together long before they are used. The metal, which will form the cans, can be precoated. This is a process called coil coating. Huge coils of metal are uncoiled on a moving belt and sprayed with the coating. The coating is a mixture of the two components that on curing will form the epoxy network.

The metal continues on the belt through a heated zone where enough energy is supplied so that even the reduced nucleophilic character of the phenol group can be made reactive and the coating is cured. After this, the metal is again coiled and sent to the can factory. Obviously the coating must possess remarkable resilience to withstand the stresses of coiling and fabrication between the curing process and the actual can fabrication. At the same time in its life as part of the can it must not mix with any of the "solutes" presented by the food the can is to contain. And if the can should be dented, the coating must not crack. Wonderful stuff!

An approach requiring heating, however, would hardly work for the kind of extended surfaces painted outdoors. For this application, curing agents that react at room temperature are essential. The price paid for this is the necessity to work faster after the mixing of the components.

As was mentioned above, the tangle of chemical bonds in Fig. 4.1 is a consequence of crosslinking, the term used by polymer chemists to describe formation of covalent bonds linking otherwise separate polymeric molecules. If the crosslinking is at a high level all the polymer chains in the container or mold may be linked giving rise to a supermolecule, that is, the entire object is a single molecule. This is the situation for elastomeric rubbery polymers except that in elastomers, in contrast to epoxy resin, the density of crosslinks is far less. This reduced crosslink density is necessary in an elastomer to maintain the necessary flexibility. This subject will be treated in Chapter 7 where the first commercial crosslinking reaction ever used, vulcanization, involves tying all the rubber polymer molecules together with sulfur.

In the epoxy resin, the extensive crosslinking is caused by the curing agent, the amine that reacts at room temperature or the phenol groups used in the can coating process that cures only on heating. The strength of a crosslinked structure, as demonstrated by resistance to deforming forces, is determined largely by the degree of crosslinking and this in turn is largely a function of the degree of reaction between the functional groups of the polymer and those of the curing agent.

There is an interesting limit to the degree of crosslinking because as the crosslinks form they lock the chains into place and prevent many of the unreacted groups from reacting further. Many of the potentially reactive groups that would contribute to the cure are "left at the altar" so-to-speak. Several such unreacted groups can be found in Fig. 4.1. This is a complicated subject. As we shall see below, the crosslinking process in epoxy resins involves a variety of chemical reactions. As the extent of crosslinking grows some functional groups may not have the freedom of motion to encounter a reactive partner. This kind of restriction on reactivity found in polymers is not encountered in small molecules. Completed reactions of small molecules do not hinder ongoing reactions among the remaining molecules.

Crosslinking is of fundamental interest to polymer science and the consequences of crosslinking – insolubility and inability to melt and therefore to flow – were known long before it was clearly understood what was happening in molecular detail. Now we understand that insolubility and inability to flow occur when the entire object is a single molecule. All the polymer chains are interconnected by crosslinks.

We have seen in Chapter 2 how polyethylene forms from ethylene. Here, each ethylene molecule is only capable of linking with two other ethylene molecules so that only a linear chain can form. But what if each molecule that is able to undergo polymerization can link with more than two other molecules? If enough of these extra reactive functions are present, and this is understood in quantitative detail, then not only the links necessary for a linear chain are present. In addition the chains can link with each other since reactive functions that can connect with each other are left over, not used to form the linear chains. This is the prerequisite for crosslinking, and we shall be seeing a great deal of this kind of reactivity in this chapter.

The general idea behind crosslinking was long understood, but it was not until the late 1930s that a quantitative basis evolved for the concepts associated with crosslinking. This understanding revealed the extent of reaction necessary to obtain a crosslinked structure in which all polymer chains are linked together to form a supermolecule. This was the effort of a man who will be the focus of our attention in Chapter 5, Wallace Carothers, the inventor of nylon and also by someone whom Carothers hired at DuPont in 1934, Paul Flory. These workers provided quantitative insight into crosslinking. One of Flory's earliest achievements was the development of a quantitative theory of gelation and crosslinking. It was to Carother's credit that he was responsible for stimulating Flory's lifelong interest in polymer science, an interest that eventually led to the Nobel Prize for Flory for his work on the physical chemistry of polymers.

We shall hear about Carothers and Flory again in Chapter 10 (Section 10.3) where their concepts addressed another fundamental problem in polymer science related to the fact that synthetic polymers such as those studied in Chapter 2, and all those studied in this chapter, are mixtures of chains of different length, that is, of different molecular weights. Using statistical methods these workers determined how the molecular weight distribution, that is, the relative number of chains of each molecular weight, would exist in the mixture of chains dependent on the method of synthesis.

4.3
The Molecular Source of the Toughness of Epoxy Resin

One of the marvelous things about polymers is that the nature of the molecular forms translates into the properties of the polymer. That is to say, the unimaginably small and unseen is felt and seen in the property of the macroscopic object. The bisphenol A unit in the epoxy resin discussed below and introduced in Chapter 3 (Section 3.10) helps to stiffen the epoxy to help the crosslinks accomplish the overall objective. The key characteristic of aromatic rings is that they are especially chemically stable and resistant to addition reactions. This characteristic is necessary to maintain the 6-π electron aromatic character of the six-membered ring. Another aspect of this aromatic stability is the resistance to deformation of the benzene ring. The aromatic ring resists deformation very far removed from the state in which all six carbons are in a plane since such deformations act to reduce the all-important overlap of the orbitals necessary for the aromatic character. This behavior is in contrast to that of the cyclohexanes, whose most stable conformation is not planar and a conformation that is easily able to change shape by rotation around C–C single bonds.

The stiffness of epoxy resin is greatly aided by the molecular rigidity of the aromatic rings in its bisphenol A-derived units. If the epoxy resin is synthesized with an aliphatic group in place of the bisphenol A unit the stiffness of the crosslinked structure is greatly reduced. Nevertheless, aliphatic epoxy resins are commercially useful since they resist yellowing on long exposure to light and air, which is an advantage that can make up for the lack of stiffness in certain specialty applications.

But even the considerable hardness derived from the aromatic rings in combination with the crosslinking is not enough for certain applications such as the shaft of a golf club mentioned above. Here we need a material that must be lighter than steel but must still have the strength of steel! Further strength for the golf club shaft is derived from reinforcement with carbon fibers that are combined with the epoxy resin. The strong dipoles of the resin's bonds and its positively and negatively charged groups, as well as the hydrogen bonding groups as seen in Fig. 4.1, cause the epoxy to "wet," that is, to be attracted to the carbon fibers. These kinds of forces act cooperatively to add up to a powerful overall force that binds the epoxy to the carbon fibers giving greater strength to the shaft of the golf club. On a microscopic level the carbon fibers act like the familiar steel rods imbedded in concrete to reinforce it. The molecular resistance to deformation of benzene rings in combination with the resistance to deformation of the carbon fibers, coupled with the network of crosslinks, all work together to gain the desired property – namely, a material that can be used to wack a golf ball long distances down the fairway.

4.4
With Epichlorohydrin and Bisphenol A we are only One Step, a Nucleophilic Step, from Epoxy Resin. It all depends on the Reactivity of the Epoxide Ring

Fig. 4.2 exhibits the reaction of bisphenol A with epichlorohydrin, a molecule with two reactive functions, an epoxide ring and a primary carbon–chlorine bond. Both a large excess of epichlorohydrin over bisphenol A and alternatively, a slight excess of epichlorohydrin are used. As shown in Fig. 4.2, the epoxy resins synthesized in these ways are of different molecular weight. In either case the epoxy resin has the same terminal functional group from which the name epoxy is derived and the reactivity of these terminal groups makes curing possible. Let us first look at the chemistry forming the epoxy resin and then return to how industry synthesizes epichlorohydrin.

Why are epoxides so reactive? The answer is their highly strained state. Then, why does this strain not preclude their formation? The strain arises from the three membered ring. Internuclear angles of 60° for this triangle are far too small for the two four-coordinate carbon atoms and the one two-coordinate oxygen, which prefer much larger bond angles of approximately 110° and 105°, respectively. Moreover, since three points define a plane, the epoxide ring must be flat and therefore the carbon-hydrogen bonds are forced

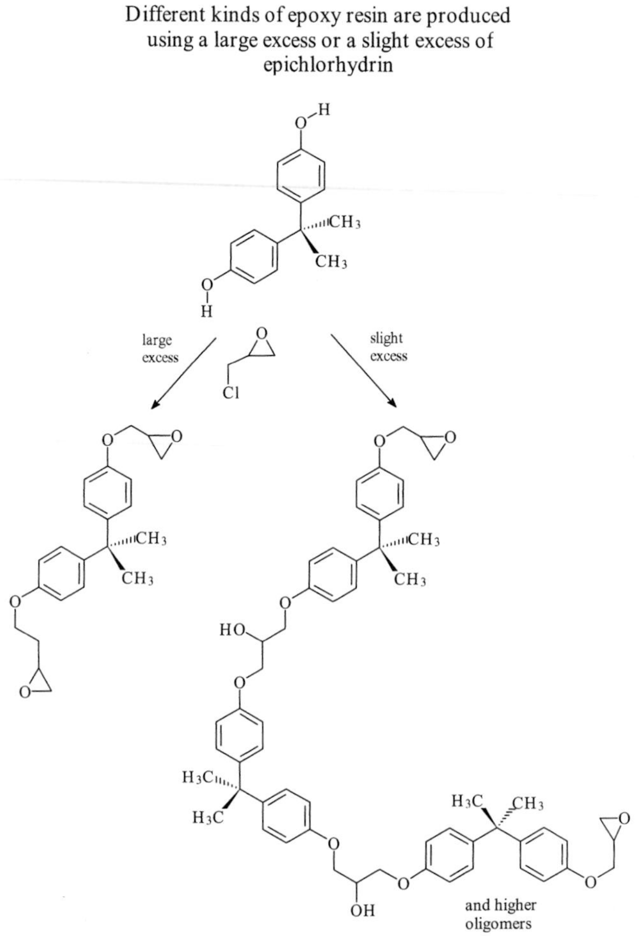

Fig. 4.2 Epichlorohydrin and bisphenol A react in different proportions to make differing oligomeric epoxy resins.

to assume eclipsed conformations. The first problem is termed angle strain and the second torsional strain.

Neither strain, however, is so severe as to preclude the formation of the three-membered ring. After all, organic chemists have figured out how to make all kinds of highly strained molecules, even ones shaped like cubes and tetrahedra. But in any strained molecule there is a tension, a "spring-loaded" situation. Certain kinds of reactants can release this strain. Nucleophiles in the case of epoxides are able to displace the carbon–oxygen bonds of the epoxide ring and this strain-derived tension is then released by opening the ring. This reaction path, as we shall see, is the chemistry that makes possible both conversion of epichlorohydrin to epoxy resin by reaction with bisphenol A (Fig. 4.2) and then later under different conditions, the curing of the epoxy resin.

The reaction of epichlorohydrin with bisphenol A, shown in Fig. 4.2, involves nucleophilic chemistry in which the phenolic hydroxyl group of the bisphenol A has two potential sites for reaction, opening the epoxide ring or displacing the chloride anion (Fig. 4.3). Nucleophilic attack of the phenolic hydroxyl group of bisphenol A at the carbon bearing the chlorine atom in epichlorohydrin is an example of the classic S_N2 reaction, which is shown in Fig. 4.3. If each of the two phenolic hydroxyl groups of a bisphenol A reacts via displacement of the chloride anion (Fig. 4.3) from two epichlorohydrins, then the epoxy resin derived from an excess of epichlorohydrin shown in Fig. 4.2 is formed. Remarkably, as also shown in Fig. 4.3, nucleophilic attack of the phenolic hydroxyl group on the epoxide ring of the epichlorohydrin, leads to the identical epoxy resin.

One of the carbons of the epoxide ring is primary, $-CH_2-$, while the other is secondary, $-CH-$. For this nucleophilic displacement of the carbon-oxygen bond of the epoxide ring, just as for an S_N2 reaction, steric hindrance is critical. Therefore attack at the methylene group, $-CH_2-$, will be the preferred site for reaction, as shown in Fig. 4.3.

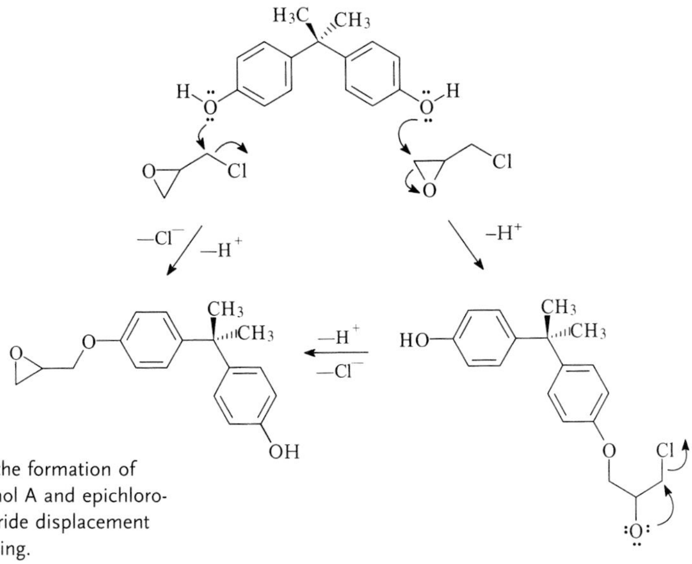

Fig. 4.3 Mechanism of the formation of epoxy resin from bisphenol A and epichlorohydrin, showing the chloride displacement route or opening of the ring.

In the intermediate formed from the step just described (Fig. 4.3), the alkoxide group that arose by breaking the carbon–oxygen bond of the epoxide ring resides on the carbon adjacent to the carbon bearing the chlorine, the vicinal carbon. This leads rapidly to an internal variation of the S_N2 reaction. The alkoxide acts as a nucleophile to displace the chloride anion, therefore forming another epoxide ring. To be sure, the new epoxide formed is not at the original position occupied by the epoxide ring in the reacting epi-chlorohydrin, but it makes no difference since the carbons involved cannot be distin-guished. The same epoxide resin is formed (Fig. 4.3).

Now we can look again at Fig. 4.2 where we see the result of a large excess of epi-chlorohydrin. The epoxy resin will be formed from one molecule of bisphenol A capped with two of the epichlorohydrin. Each of the two phenolic hydroxyl groups of a single bis-phenol A molecule will react with one epichlorohydrin via the direct chloride displace-ment, or the opening of the epoxide ring followed by displacement of chloride and refor-mation of another epoxide ring (Fig. 4.3). Both mechanistic paths yield identical results. Since there are no further bisphenol molecules, the reaction will stop.

If on the contrary, the epichlorohydrin is present in only a small excess over the bis-phenol A there will be unreacted phenolic hydroxyl groups that will not immediately be overwhelmed by excess epichlorohydrin. These phenolic hydroxyl groups can then open epoxide groups on the terminus of a bisphenol A. This allows the reaction to go further to produce a higher molecular-weight epoxy resin, as also shown in Fig. 4.2.

The higher and lower molecular-weight epoxy resins formed as shown in Fig. 4.2 have different commercial uses. When the higher molecular-weight epoxy resins are cured, they give rise to coatings with better solvent and chemical resistance, whereas the lower molecular-weight epoxy resins provide paints with far lower viscosities. There is a great advantage to the latter's flow characteristics since less solvent is necessary and even sol-ventless coatings are possible. The paint and adhesive industries are the largest users of solvents, and there is a strong drive to decrease the use of volatile organic solvents. The lower molecular-weight epoxy resin (Fig. 4.2) with its lower viscosity helps to satisfy this environmental objective.

4.5
Just as for the Formation of Epoxy Resin, the Curing of Epoxy Resin involves Nucleophilic Chemistry and the Reactivity of the Epoxide Ring

There are many curing agents, most with nucleophilic character, a chemical characteris-tic which is consistent with the opening of epoxide rings. A particular class of curing agents contains amine groups as the nucleophilic entity. Because the curing process is a method of forming a network, that is, a method of crosslinking, the curing molecule must be able to react with at least two epoxy resin molecules. Here we see the concept of the crosslink in practice in linking different polymer chains.

The nucleophilic character of the amine group is very well suited for attack at one of the carbons of the epoxide ring, therefore breaking the carbon–oxygen bond and releas-ing the ring strain. If the reaction follows the S_N2 mechanism, then nucleophilic attack at the primary center of the epoxide ring is greatly favored. This beginning step in the conversion of the epoxy resin and the steps to follow are shown in detail in Fig. 4.4.

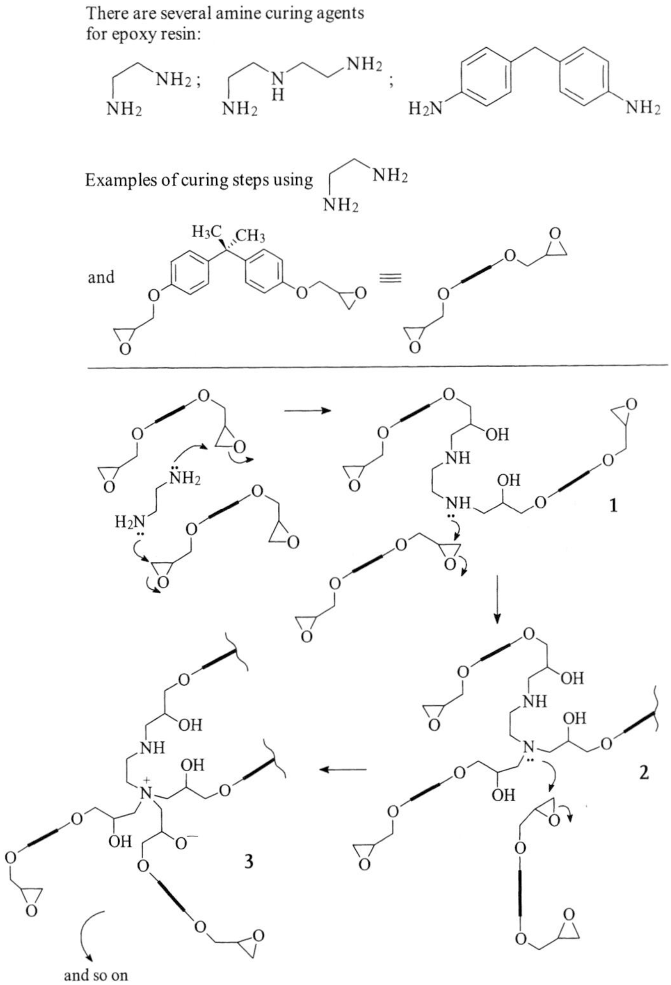

There are several amine curing agents
for epoxy resin:

Examples of curing steps using

and

Fig. 4.4 Chemical steps in detail in the curing of epoxide resin by a diamine.

The opening of the epoxide ring by the nucleophilic curing agent – an event that is in-itiated when the curing agent and the epoxy resin are mixed and starts the clock on the pot life – triggers a series of reactions leading to the final crosslinked epoxy resin. The chemical reactions that follow are complex, but we will look at these reactions in an over-simplified manner to understand, basically, what is happening. In one of several amine-based curing agents (Fig. 4.4), ethylene diamine, there are two primary amine groups to begin with; after one has reacted with an epoxide ring, the unreacted other primary amine group of the curing agent can then attack and open an epoxide ring on another epoxy resin macromolecule. This reaction forms a protonated secondary amine and an alkoxide anion. Transfer of the proton to the alkoxide anion to form a hydroxyl group forms the nucleophilic secondary amine **1** (Fig. 4.4), which is now capable of opening another epoxide ring to form **2** after proton transfer from the amine group to the alkoxide anion. The tertiary amine group in **2** may then open another epoxide ring furthering the crosslinking and producing the alkoxide anion in **3**, which may then act as a nucleophile opening still another epoxide ring (Fig. 4.4). This is a view of some of the reactions going on, all fundamentally driven by nucleophilic opening of the strained epoxide rings.

What did all this chemistry (Fig. 4.4) accomplish? We have amine groups opening epoxide rings forming secondary amine groups opening more epoxide rings forming tertiary amines that open more epoxide rings forming quaternary amines and alkoxide anions opening still more epoxide rings. It is apparent from all these chemical reactions that a network is formed in which all the epoxy resin molecules are losing their terminal epoxide ring functions and becoming linked to each other. This is classic crosslinking, and although one cannot define precisely where all the bonds are, one can be certain that a very tough material will be formed. This is the substance shown in Fig. 4.1 and described in Sections 4.2 and 4.3.

4.6
How Epichlorohydrin is synthesized from Allyl Chloride by a Classic Double Bond Addition Reaction followed by Formation of an Epoxide

Now let us see how industry produces epichlorohydrin, the other necessary component for the synthesis of epoxy resins (Figures 4.2 and 4.3). Epichlorohydrin contains three carbon atoms, as does propylene. Propylene is an inexpensive product produced by steam or catalytic cracking of petroleum fractions, as discussed in detail in Chapter 1 (Section 1.11). Therefore, it makes sense to produce epichlorohydrin from propylene. The overall scheme is shown in Fig. 4.5, where we discover that allyl chloride, an intermediate directly available from propylene plays a key role.

As seen in Fig. 4.3, one of the routes for reaction of bisphenol A with epichlorohydrin places an alkoxide group in a vicinal position, adjacent to a carbon bearing a chlorine atom. An internal nucleophilic displacement follows to form an epoxide ring. The identical reactive relationship can be derived from a double bond by addition of hypochlorous acid, HOCl, which reacts with olefins to add the OH and Cl groups. Treatment then with base to form the alkoxide from the OH group (in analogy to one of the curing steps in

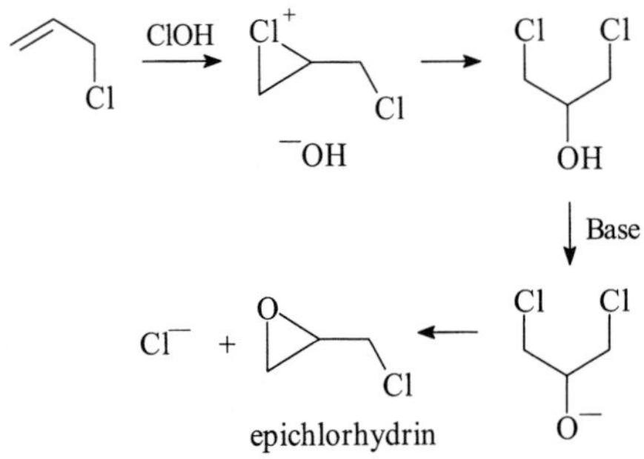

Fig. 4.5 Mechanism of the addition of hypochlorous acid to allyl chloride leading to the chlorohydrin and subsequent treatment with base leading to the nucleophilic displacement to form epichlorohydrin.

Fig. 4.4) sets up the identical relationship seen in Fig. 4.3 and yields the epoxide and epichlorohydrin is produced (Fig. 4.5). Industry has used this method to synthesize propylene oxide from propylene (Chapter 8, Sections 8.3 and 8.4).

With hypochlorous acid the difference in electronegativity between the OH and the Cl groups leads to a permanent dipole with the positive end at chlorine and the negative end at oxygen. The consequence of this is that the positive chlorine adds first to the double bond of allyl chloride to form the familiar chloronium ion intermediate encountered in additions of halogens to double bonds (Fig. 4.5).

The negatively charged hydroxide group left behind then completes the reaction by bonding to the carbon that can best bear the positive charge, placing the OH group between the two carbon-bound chlorine atoms. This fortunate reactivity pattern derives from the S_N1 character of the halonium ion ring opening, therefore leading to the hydroxide anion reacting at the carbon bearing the largest cationic character in the halonium ion, that is secondary over primary (see Chapter 1, Section 1.11).

That's great because it speeds the formation of the target epoxide ring by placing two chlorine atoms in the role of the necessary leaving group. The OH group can turn in either direction for reaction to form the epoxide ring and force chloride anion to leave (Fig. 4.5).

We are now very close to producing epoxy resin. We have both bisphenol A (Chapter 3, Section 3.10) and epichlorohydrin (Fig. 4.5) in hand. But the latter molecule has to be produced from allyl chloride as we have just seen (Fig. 4.5). How is allyl chloride produced?

4.7
How is Allyl Chloride produced industrially from Propylene?

How is methyl chloride synthesized from methane? Chlorine reacts with methane and a free radical chain reaction is initiated via the production of chlorine atoms by light or heat. As for all free radical chain mechanisms such as steam cracking of petroleum (see Chapter 1, Section 1.11) or polymerization of ethylene (see Chapter 2, Section 2.2) initiation, propagation, and termination are key reactive steps.

Applying the approach used for synthesis of methyl chloride may also be applied to ethane to produce ethyl chloride. However, this approach may not be applied to more complex hydrocarbons with different kinds of carbon–hydrogen bonds because there may be several products formed. Propylene is certainly a more complex hydrocarbon than methane, but does not have this problem. Of the six hydrogen atoms in propylene, those on the methyl group are far more reactive than those bound to the double bond. This arises from the relative instability of the sp^2 carbon radical that would be produced by abstraction of a double bond-bound hydrogen compared to the radical produced by abstraction of hydrogen from the methyl group of propylene. The radical produced by breaking one of the methyl group hydrogens is stabilized by resonance, by conjugation with the double bond. We have seen this characteristic of propylene, which here acts in our favor for synthesis of allyl chloride, acting against us to block the free radical polymerization of propylene (Chapter 2, Section 2.3).

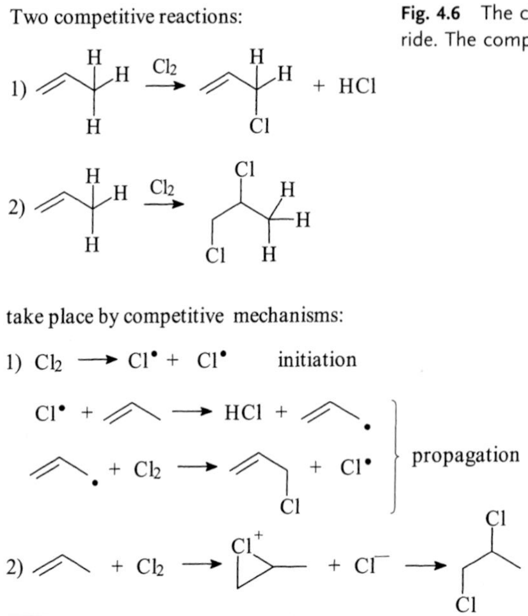

Two competitive reactions:

Fig. 4.6 The chain mechanism for formation of allyl chloride. The competition between addition and substitution.

1)

2)

take place by competitive mechanisms:

1) $Cl_2 \longrightarrow Cl^{\bullet} + Cl^{\bullet}$ initiation

$Cl^{\bullet} + \;\diagup\!\!\!\diagdown \longrightarrow HCl + \;\diagup\!\!\!\diagdown_{\bullet}$

$\diagup\!\!\!\diagdown_{\bullet} + Cl_2 \longrightarrow \;\diagup\!\!\!\diagdown_{Cl} + Cl^{\bullet}$ propagation

2) $\diagup\!\!\!\diagdown + Cl_2 \longrightarrow \overset{+}{Cl}\!\!\triangleright + Cl^{-} \longrightarrow$

There is however a problem of a different kind in synthesizing allyl chloride in this manner. Chlorine will competitively add to the double bond of propylene to produce 1,2-dichloropropane. However stable is the resonance-stabilized radical produced by the reaction path we desire, addition of chlorine to the double bond will still occur. Is there a way we can direct the chlorine to do what we wish (Fig. 4.6)?

Substituting chlorine for hydrogen bound to carbon is a process without much change in entropy since two molecules react, chlorine and propylene, and two molecules form, allyl chloride and HCl. Adding chlorine to propylene, however, involves reacting the same two molecules to get only one product, 1,2-dichloropropane. This will decrease entropy so that ΔS will be negative, which acts to make ΔG more positive thus disfavoring the chlorine addition to the double bond. Entropy change favors the desired path (1 in Fig. 4.6). However, the most important factor utilized by industry to favor the substitution path in reaction of chlorine with propylene to yield allyl chloride arises from the fact that the addition reaction leading to 1,2-dichloropropane is more exothermic than the desired path to allyl chloride. Because less heat is given off by the desired path producing allyl chloride, raising the temperature will favor this path, and the higher the temperature the larger will be ratio of allyl chloride to 1,2-dichloropropane produced.

It is no surprise therefore that the industrial reaction of chlorine with propylene is conducted at 500 °C to achieve the desired path. The high temperature also speeds this path by forming numerous chlorine radicals by the thermal splitting of the chlorine molecules, the initiation step of the free radical chain reaction shown in Fig. 4.6. Allyl chloride is formed quickly and in large amounts, with little competition.

Conducting processes in the chemical industry at 500 °C with chlorine is a challenging engineering situation. Although as we found above there is a critical reason for this high temperature, would it not be good to find another way that was safer and less of an engineering problem?

4.8
A Less Temperature-dependent Way to make Epichlorohydrin

If you count the chlorine atoms used in the processes outlined in Figures 4.5 and 4.6 you will find that one chlorine atom is lost in each of the two steps. Chlorine is too expensive to throw away and costs almost as much as propylene. A Japanese company has announced a process for epichlorohydrin, which bypasses allyl chloride by starting with allyl alcohol, which as we will see below is produced from propylene and acetic acid via a catalyzed reaction.

Although we will not go into the details of the mechanism of the formation of allyl acetate via the PdCl$_2$-catalyzed reaction of acetic acid with propylene, the catalyst is identical to that used for the formation of vinyl acetate from acetic acid and ethylene (see Chapter 10, Section 10.8). But when this catalyst is used with propylene instead of ethylene, the ease of breaking of the allylic carbon-hydrogen bonds in propylene diverts the reaction from addition of acetic acid to the double bond to substitution of the acetate for a hydrogen on the methyl group. Allyl acetate is produced, which is the critical first step of the Showa–Denko process (Fig. 4.7). Here again we see the same kind of reactive competition between the double bond and the methyl hydrogens in propylene for the reaction with chlorine. The acetic acid does not add to the double bond as it does in ethylene. Propyl acetate is not produced.

Allyl acetate is simply the ester of allyl alcohol and acetic acid, and hydrolysis reverses the esterification and yields allyl alcohol (Fig. 4.7). Addition of chlorine to allyl alcohol yields 1,2-dichloro-2-propanol, which has vicinal OH and Cl groups so that on addition of the base, Ca(OH)$_2$, epichlorohydrin is formed. This change corresponding to a difference of one chlorine atom leads to a raw material price differential of several cents a kilo-

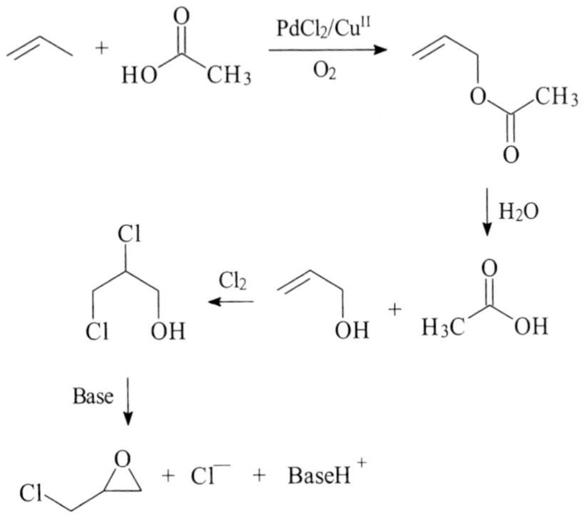

Fig. 4.7 The Showa–Denko process for production of epichlorohydrin. The consumption of chlorine and need for high temperature are both reduced.

gram, therefore favoring the process in Fig. 4.7 over that in Figures 4.5 and 4.6. This efficiency, in addition to avoiding the high temperature necessary for allyl chloride production and the great danger of chlorine under these conditions, is enough to encourage
new plants for epichlorohydrin production to use the Showa–Denko process (Fig. 4.7).

In a large-volume industrial intermediate such as epichlorohydrin, a single chlorine
atom can make a big difference in profitability. This is really what atom economy is
about.

4.9
A Final Note about Epoxy Resins

For over fifty years, epoxy resins have served well. More than one billion dollars worth of
epoxy resin is sold globally each year. The process for making epichlorohydrin, one of
the two components of an epoxy resin, as was discussed in this chapter, has been difficult, requiring the use of chlorine at 500 °C to form the prerequisite allyl chloride (see
Section 4.8). This high-temperature reaction has provided what is called a "technological
barrier-to-entry" for new companies to start producing epoxy resins, and is reflected in
the fact that there are only three major suppliers – Shell Chemical Company, the Dow
Chemical Company, and Ciba Specialties Company.

Interestingly enough, two of these major players, Shell and Ciba, sold their businesses
in 2000. The buyers were financial companies who, in due time, will probably re-sell the
companies, for a profit (they hope). Why did these companies divest themselves of a
product with such fine properties, such wide use? Because the growth of the business
based on a two-component process has been very slow. Two-component materials with
slow curing times cannot be automated. This makes for high labor costs and therefore re-
luctance to use them by industry in spite of their excellent properties. In the modern industrial world we cherish convenience and reward it handsomely. Only when nothing
else will work will epoxy resins be used. Epoxies are a highly respected product, but modern business is not interested if this is not equated to growth. Therefore, a relatively small
rate of financial return results from the considerable investment necessary for production of
these chemicals. Will someone invent a new automated crosslinking method to make epoxy
resins easier to use – that is, to give them the virtues of convenience? Will it be you?

4.10
What did the Original Shell Method for producing Epichlorohydrin have to do with Glycerol?
The Answer is Alkyd Resins and this will teach Us More about Crosslinking
and also introduce Nucleophilic Acyl Chemistry

The original method for production of epichlorohydrin via the high-temperature chlorination of propylene (Figures 4.5 and 4.6) was developed by Shell in 1936, long before epoxy
resin was invented at the Devoe–Reynolds Paint Company in the 1950s by a chemist
seeking corrosion-resistant coatings. Interestingly, Ciba in Switzerland invented them at
about the same time. Shell devised this chemistry because they were intent on develop-

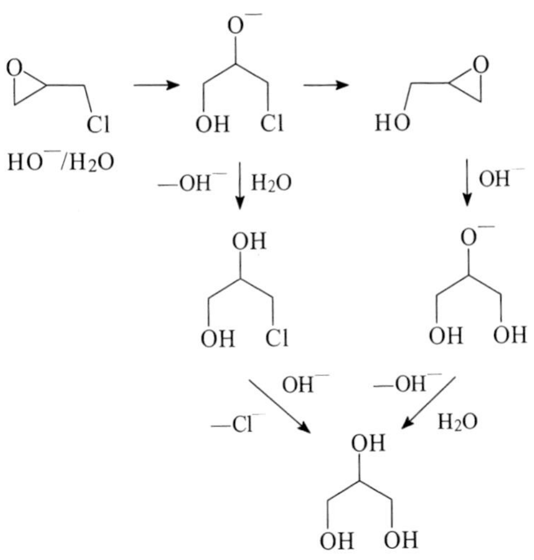

Fig. 4.8 Outline of the synthesis of glycerol from epichlorohydrin with aqueous base.

ing a synthesis of glycerol. Glycerol can be easily synthesized from epichlorohydrin by simply heating with aqueous base, as shown in Fig. 4.8. During the 1930s there were reasons to believe that the chemical industry would be faced with a shortfall of glycerol, and this was the primary stimulus for Shell's development of allyl chloride and epichlorohydrin. Let's look more closely at glycerol to understand its history, what was happening in the 1930s, and what role glycerol plays today in the chemical industry.

General Electric, in its search for insulating varnishes for electric wire, had invented a new polymer called alkyd resins. And alkyd resin synthesis requires glycerol. The alkyd resins, developed long before epoxy resins, were the first ever improvement over natural oils for coatings. Before alkyd resins were invented, unsaturated natural oils were used, known as "drying oils." The drying oils, which have been used for centuries by painters as well as by artists, are obtained from plants and are esters of fatty acids and glycerol. Since glycerol contains three hydroxyl groups (Fig. 4.8), each glycerol molecule contains three ester groups leading to the designation triglyceride. Drying oils, such as linseed, tung and soybean oils, are triglycerides in which some of the fatty acids are multiply unsaturated, containing more than one double bond per chain (Fig. 4.9).

Why are such oils as shown in Fig. 4.9 called "drying oils"? It is because, on exposure to light and heat, thin films of the oil slowly solidify. However, they did not solidify because they were drying – which normally means loss of a solvent – but rather because the double bonds in the unsaturated fatty acids slowly crosslinked, changing the oil to a solid film. The false analogy to a true drying process led to the misnomer, which stuck. An initiator, usually an oil-soluble lead or cobalt salt that forms radicals, was added to the oil. When the oil was painted onto a surface in a thin film the initiator was activated, for example, by light or heat and oxygen in the air, and the double bonds in the unsaturated oils polymerized. Just as ethylene can be polymerized by free radicals via the double bond as discussed in Chapter 2 (Section 2.2), so can the double bonds in unsaturated oils, although in the latter the reaction is much slower and only oligomers form.

α-Linolenic acid

α-Eleostearic acid

Linoleic acid

as esters of glycerol

triglyceride

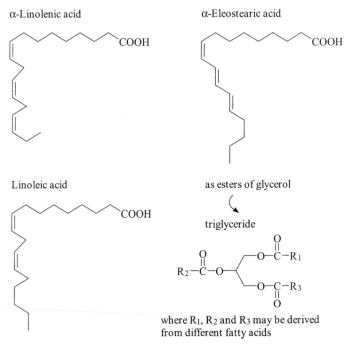

where R$_1$, R$_2$ and R$_3$ may be derived
from different fatty acids

Fig. 4.9 Three of the unsaturated fatty acids found in drying oil, and the
structure of a generalized triglyceride.

However, while each ethylene molecule contains one double bond, in the unsaturated
oils used for these coatings each fatty acid, as noted above, contains one to three double
bonds (Fig. 4.9). The free radical polymerization therefore, as shown in Fig. 4.10, leads to
a crosslinked network because there are two or three double bonds in each triglyceride.
Many or most of the molecules in the drying oil were linked together to form a giant
molecule, a supermolecule, which then covers the painted surface.

Although the network is formed with entirely different chemical reactions than the network formed from epoxy resin (Section 4.5; Figures 4.1 and 4.4), the crosslinking principle is identical. Thus, the theory of network formation by crosslinking developed by Carothers and Flory (Section 4.2) is independent of the specific nature of the chemical reactions involved and applies equally to crosslinking of drying oils and epoxy resins.

The drying oils did work as coatings, but they did not work very well. They had little
resistance to solvents, and easily broke down under the influence of oxygen or ultraviolet
light. They softened easily and although they adhered well to wood they peeled easily
from metal surfaces. Part of the problem arose because the crosslinked structure contained fatty acid chains with large numbers of methylene groups that allowed for great
flexibility. We learned earlier in this chapter how the aromatic rings in the epoxy resin
lend stiffness and accordingly strength (Section 4.3). This molecular property is completely absent in the drying oils. An attempt to solve the problem involved "cooking" oils

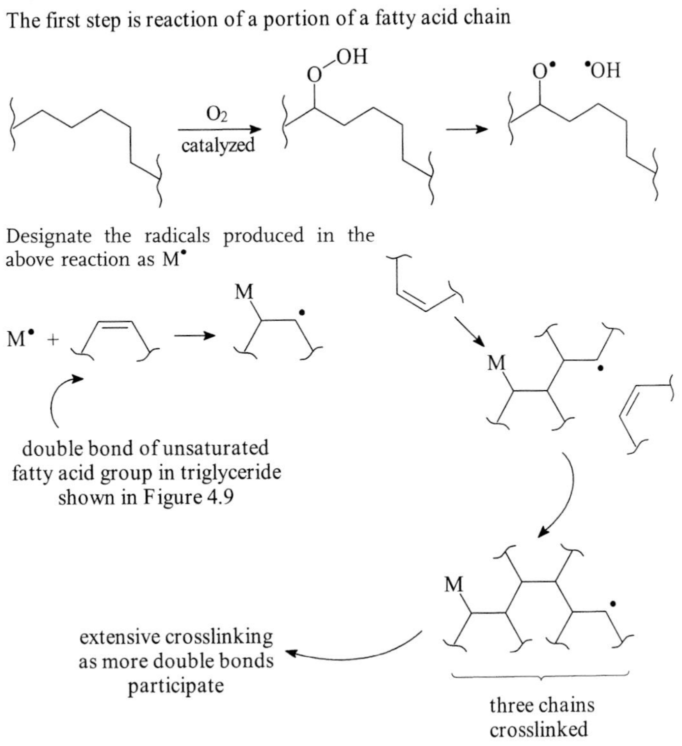

The first step is reaction of a portion of a fatty acid chain

Designate the radicals produced in the above reaction as M•

double bond of unsaturated fatty acid group in triglyceride shown in Figure 4.9

extensive crosslinking as more double bonds participate

three chains crosslinked

Fig. 4.10 The "drying" of an oil involves crosslinking.

with substances such as phenolic resins (see Chapter 5, Fig. 5.1), but this improvement had somewhat limited applicability.

We can now understand the driving force for the invention of alkyd resins by General Electric chemists. The objective was to add stiffness to the drying oils while at the same time keeping the same basic crosslinking technology. Incorporation of aromatic rings seemed the way to accomplish this objective. Thus, the same additives causing the cross-linking would yield a coating with better properties and specifically, from the point of view of General Electric, a material that could protect and insulate electric wires.

A key molecule to accomplish this objective was glycerol, the very molecule that formed the core of the natural drying oils (Figures 4.9 and 4.10). As shown in Fig. 4.11, glycerol was reacted with the drying oil in what is called a transesterification reaction to yield glycerides in which one or two of the hydroxyl groups are esterified with the unsa-turated fatty acid. Esters are subject to reactivity with nucleophiles and the General Elec-tric chemists used this characteristic in the reaction of synthetic glycerol with natural tri-glycerides to obtain the necessary mono- and diglycerides.

The reaction forming the monoglyceride, the transesterification, belongs to a class of acyl nucleophilic substitution that we will focus on in Chapter 10 (Section 10.13) as an important means of producing polycarbonates. For the transesterification of glycerol and

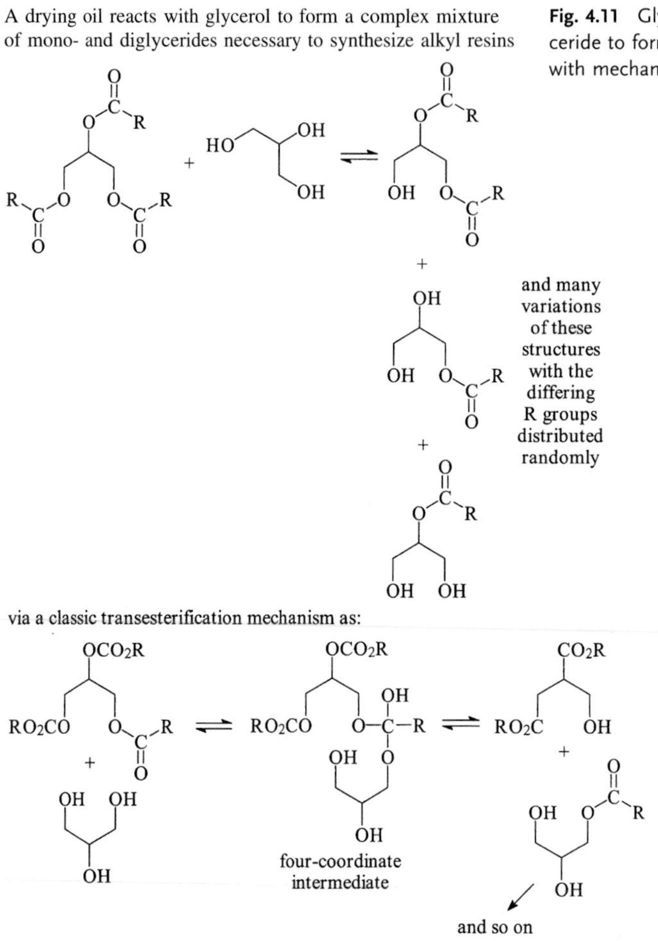

A drying oil reacts with glycerol to form a complex mixture of mono- and diglycerides necessary to synthesize alkyl resins

Fig. 4.11 Glycerol reacting with a triglyceride to form mono- and diglycerides with mechanism of the transesterification.

and many variations of these structures with the differing R groups distributed randomly

via a classic transesterification mechanism as:

four-coordinate intermediate

and so on

the triglycerides, the mechanism of the reaction exhibits the critical step of formation of a four-coordinate intermediate (see Chapter 6, Figures 6.3 and 6.8), which on reformation of the carbonyl group moves the reaction path to product as shown in Fig. 4.11.

The two remaining hydroxyl groups on the monoglyceride are then available for reaction with phthalic anhydride, as shown in Fig. 4.12, to form the alkyd resin, a polymer in which the linkages are ester groups. Why does a polymer chain form? The answer is that the two hydroxyl groups on the monoglyceride can reach out to form bonds in two directions, as do the two carbonyl functions in the phthalic anhydride. Just as links form in a chain so the ester bonds form links in the polymer, the polyester. Here we introduce a subject, condensation polymerization, to be covered in greater depth in our studies of the nylons. But that will have to wait for Chapter 5.

The linking reaction that forms the ester bonds holding the alkyd resin chain together is again an acyl substitution also involving the formation of a four-coordinate intermediate as in the transesterification reaction shown in Fig. 4.11. But for the ester linkages in

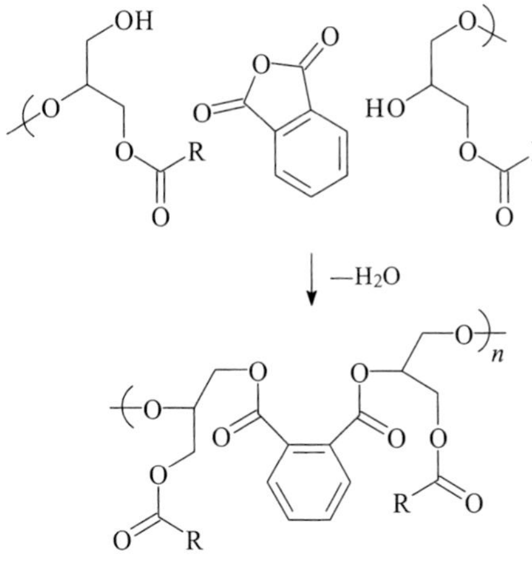

Fig. 4.12 Reaction of a monoglyceride with phthalic anhydride to form the polyester alkyd resin. The resin forms by polyester chain growth.

the polymer chain it is the reaction of the hydroxyl groups of the monoglyceride with the carbonyl groups of phthalic anhydride. One of the lessons of organic chemistry is that the reactivity of nucleophiles with the carbonyl functional group derives from the electron deficient character of the carbonyl carbon. This electron deficiency is associated with the dipole moment arising from the connection of this carbon to electronegative oxygen. This driving force is at work both in the transesterification in Fig. 4.11 and in the formation of the ester groups forming the alkyd resin in Fig. 4.12. The details are different, but the basic forces at work are identical. Anhydrides of dicarboxylic acids fit into this mold, and were therefore used by the General Electric chemists to incorporate aromatic character into drying oils.

Alkyd resins are a great idea. The unsaturated fatty acid-derived group hanging off one of the hydroxyl groups of each glycerol in the chain (the R groups in Figures 4.11 and 4.12) can crosslink in the way shown in Fig. 4.10 for the drying oils, while the aromatic groups linked together by the remaining hydroxyl groups on each glycerol as polyesters supply the stiff aromatic character necessary for the improved material properties. But at this point we have to admit that although General Electric invented alkyd resins to protect electric wires, these resins were hardly used for that purpose and instead became a major vehicle for oil-based paints. And as is the evolutionary way of the chemical industry, alkyd resins, which are a solvent-based paint, are now in disfavor for environmental reasons. Paints that do not require solvents are more environmentally friendly. Today only about 6% of glycerol use is for alkyd resins. Clever chemistry is not enough. Obsolescence is the signature characteristic of the chemical industry.

But the story of epichlorohydrin and that of alkyd resins focused our attention on glycerol. Let us use this opportunity to understand more about the use of glycerol and for this purpose to understand something of the history of this molecule. This will set the stage for understanding polyurethanes, another industrial product in which crosslinking plays an essential role.

4.11
The Earliest Production of Glycerol arose from Production of Soap

Glycerol is a time-honored byproduct from soap manufacture. All oils and fats are glycerol esters of fatty acids. Triglycerides and related molecules play an essential role in all life processes and are important components of cell membranes. One of the earliest examples of chemical production, still carried out by our fairly recent ancestors as a household task, was soap production. Waste fats and oils from the kitchen were mixed with ashes containing a high content of soda or alkali carbonates. The reaction is called saponification and, as seen in Fig. 4.13, shows still another example of nucleophilic acyl substitution involving the always critical four-coordinate intermediate as is involved in the chemistry portrayed in Figures 4.11 and 4.12.

Ash from wood with a high enough alkali content to hydrolyze fats is not commonly found in Europe and the life of certain forests of Europe was threatened. Special high soda-containing ash is restricted to certain areas in southern France and to the ash from seaweed. Therefore, obtaining proper ash caused problems including, it is said, the burning down of a great deal of forest in Europe. Many attempts were made by chemists of the 1700s to devise a synthesis of soda (sodium carbonate) until a process invented by a French chemist, Nicolas Leblanc, who held an important patent, combined the various discoveries to a practical process. As quoted from Partington, the overall equation is $Na_2SO_4 + 2C + CaCO_3 \Rightarrow Na_2CO_3 + CaS + 2CO_2$, with the process description at the time as: "charge for the black-ash furnace as 1,000 lb. saltcake, 1,000 lb. washed Meudon chalk, and 550 lb. charcoal. The black ash was exposed to the air till it effloresced, lixiviated in barrels, and the solution evaporated till the soda crystallized, the mother liquor being evaporated to dryness and the calcined residue used for glass making." Imagine the labor

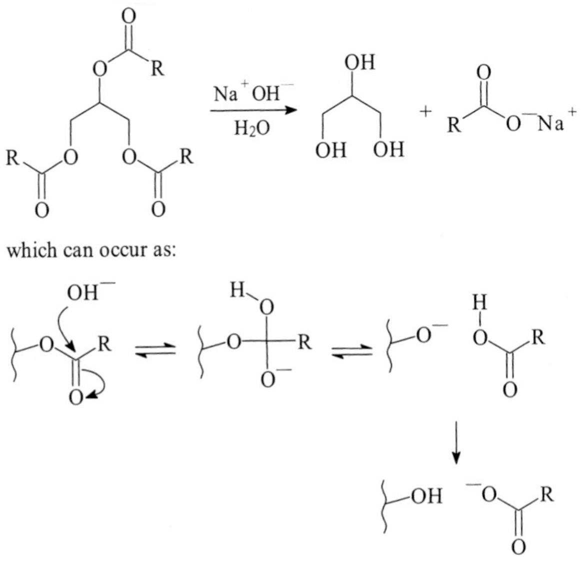

which can occur as:

Fig. 4.13 Soap is manufactured by aqueous base hydrolysis of glycerides to produce salts of fatty acids and glycerol

and shovels involved in such a process and compare that to the modern chemical indus-try! The soda, Na_2CO_3, made by this process was widely used to make soap, with glycerol as the byproduct (Fig. 4.13) and the forests of Europe were saved.

4.12
What Commercial Uses exist for Glycerol?

Glycerol is certainly an important chemical. The better part of a 450 million kilograms per year, worth hundreds of millions of dollars, is produced in the United States. Its use is not focused on a single or even overwhelming application, but is rather split among several, which in their breadth show the multiple characteristics of this molecule. Two uses – as a humectant for tobacco, and incorporation in cellophane for decorative wraps, for example for flowers – total about 10% of the total, and involve the water-absorbing properties of glycerol. Three hydroxyl groups in such a small molecule give rise to many possibilities for hydrogen bonding (Chapter 5, Section 5.4) between glycerol and water, which is responsible for water absorption

A large use for glycerol, in the range of 40%, involves providing body and lubricity for toiletries, cosmetics, and toothpaste. Although glycerol produced synthetically is pure from a chemist's view – as judged by chromatography and spectroscopy and the absence of color – certain cosmetic companies that advertise commitment to natural components do not use it. This causes great problems for chemists because natural glycerol obtained by hydrolysis of fats (Fig. 4.13) often has minor impurities, which can lend an unaccept-able "off-color" to the final cosmetic products. Thus, this desire to be "natural" can create the need for a great deal more complex separation and purification chemistry.

The uses for cosmetics, toiletries, and toothpaste also evolve from the physical proper-ties of glycerol, one of which is viscosity. Here again, hydrogen bonding (Chapter 5, Sec-tion 5.4) plays a role since the hydroxyl groups in glycerol molecules can hydrogen bond with each other so that flow has to overcome these hydrogen-bonding interactions. These weak interactions among glycerol molecules constitute a crosslinked network, which dif-fers in an essential feature from the crosslinks in epoxy (see Fig. 4.1) and alkyd resins (Figures 4.9 and 4.10). The hydrogen-bonded crosslinks in glycerol, as for all hydrogen-bonded interactions, are constantly breaking and reforming offering therefore a resis-tance to flow, but not a permanent restriction to flow as in the crosslinked epoxy or alkyd resins. In chemistry, one sometimes uses the term, "hydrogen-bonded network."

Glycerol's absence of toxicity allows incorporation in foods and beverages. Here most glycerol use involves reacting glycerol with food-grade natural triglycerides so that mono- and diglycerides are formed in analogy to one of the steps in alkyd resin synthesis (Figs. 4.11 and 4.12). However, some free hydroxyl groups in these glycerides are left pro-viding water-loving (hydrophilic) groups to contrast with the fat-loving (lipophilic) chain of methylene groups from the remaining fatty acid moiety (Fig. 4.9). This kind of mole-cule is a surfactant, which can act to help mix water- and fat-based components in foods and beverages. These surfactants work so well that about 17% of total glycerol production goes into foods and beverages. For one prominent example, surfactants derived in this manner are the basis of many low-calorie margarine substitutes.

4.13
The Role of Glycerol in Dynamite and the Nobel Prize

Nitroglycerin, which was to become the essential component of dynamite, was first synthesized in 1846 by nitration of glycerol (Fig. 4.14) by an Italian chemist from Turin, Ascanio Sobrero, who lived from 1812 to 1888. Professor Sobrero was lucky to live to old age considering the shock sensitivity of his nitroglycerin. Although Alfred Nobel, in Sweden, started to manufacture the neat, pure oil in 1862 to be used as a commercial explosive, it could not be used safely. Indeed, in 1864 a violent explosion killed Nobel's younger brother and several assistants. Nitroglycerin was so dangerous that its use was prohibited by legislation. Nobel, however, continued to work with nitroglycerin by shifting the work site to a canal boat anchored not far from the Swedish capital.

By 1867, Nobel had filed a patent for his discovery that nitroglycerin absorbed on kieselguhr, a silicaceous earth, could be handled safely. The mixture was called dynamite, and was sold as sticks. However, the loss of shock sensitivity meant that the release of its explosive power required a detonator. This was greatly welcomed since it meant the explosive could be safely handled and shipped and would not explode without warning. Much later, nitroglycerin mixed with another explosive, nitrocellulose, played a role in the formation of Cordite, an important explosive propellant that was used in World War I, as discussed in Chapter 9 (Section 9.2).

Nobel made a great deal of money from his discovery of how to handle nitroglycerin and from other investments and from the subsequent marketing of dynamite. It has been reported that between 1867 and 1872, world production of nitroglycerin soared from 11 to 1350 tonnes. An excellent review of this period is found in Aftalion's book.

At the time of his death Nobel left his estate as endowment for annual awards in chemistry, physics, medicine, literature and peace – the Nobel Prize. It seems reasonable that Nobel, in choosing peace as one of the prizes, was making a statement about the use of his invention for purposes other than what he originally had in mind, and also perhaps to have his name remembered for something other than a high explosive.

A footnote to this story is that nitroglycerin, among other organic nitrates and nitrites that have later been found to have similar effects, has been used for well over 100 years to relax involuntary muscles near the heart and therefore to relieve the pain of *angina pectoris*. This discovery was associated with the experience of workers in Nobel's nitroglycerin plant who were suffering from this affliction and found relief from the pain only while on the job.

Although the action of nitroglycerin is rapid, its effect is also rapidly lost after a dose is taken. And repeated large doses, as for example taken with a pill under the tongue, leads to side effects. This problem led to the development of a patch, based on the permeable

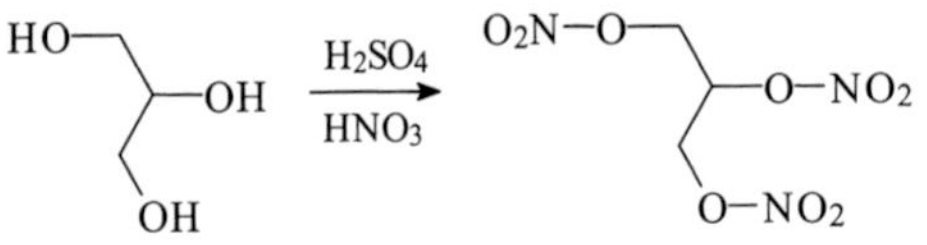

Fig. 4.14 The synthesis of nitroglycerin as it is carried out today, from glycerol via a reaction with nitric and sulfuric acids.

properties of polymers, which can be stuck to the chest allowing slow release of the nitroglycerin. How could Professor Sobrero in Turin, Italy in 1846 ever have guessed that his synthesis would lead to dynamite, the Nobel Prize, and skin patches to relieve pain from heart disease?

We have come a long way from our focus on plastics and crosslinking, but it turns out that following glycerol leads us right back to the original path of our story.

4.14
Glycerol plays a Role in the Production of Polyurethanes: Nucleophilic Chemistry and Crosslinking

In the modern world one can hardly sleep, sit down or move about without becoming involved with polyurethane. Polyurethane foams are widely used as cushioning in the furniture, transportation and the bedding industries. Foam made of polyurethane is remarkable material with the strength to rebound to stress but with a lightness that arises from a density in the range of 0.03 grams per cubic centimeter, in other words, approximately one-thirtieth the density of water. In automobiles, polyurethane foam can be made in a wide range of resiliency and density as necessary for seats, instrument panels, and headrests, not to mention its excellent characteristics for damping sound and vibration.

Polyurethanes can be found in bedding, carpet padding, packaging, sponges, scrubbers, and paint applicators. Polyurethanes can also be synthesized in a more rigid form to make boards or laminates that are rigid enough to form so-called run-flat tires, which can be used even if the air leaks from the tire. Commercial buildings are often covered with spray foams made from polyurethanes. A remarkable use, among many others, of the insulation properties of polyurethanes depends on their ability to provide temperature insulation of the main fuel tanks of the space shuttle. Or how about this property? Polyurethane is used to insulate ships transporting liquid natural gas, which must accommodate an enormous temperature range from more than 100 °F below zero to more than 100 °F above zero! And then we have surface coatings made from polyurethanes, which have excellent abrasion resistance and provide glossy coatings that reduce aerodynamic interference for airplanes. What stuff!

Spandex, a stretchy fiber widely known to the public, is a polyurethane that is not chemically crosslinked, in contrast to all the polyurethanes discussed above. But we will reserve a full discussion of spandex for Chapter 7 (Sections 7.12 and 7.13) when we focus on natural rubber and elastomers. There are whole ranges of uncrosslinked polyurethanes under the elastomers umbrella that have uses as disparate as automobile parts, skateboard wheels, and ski boots. Polyurethanes are used for solid bicycle tires, and perhaps some day automobile tires will be made from polyurethanes. Can you imagine what this would do to the rubber industry with already the run-flat tires noted above depending on polyurethane?

The ability of the urethane linkage in a polymer to provide such a large menu of uses arises simply from variability in the density of crosslinking, from no crosslinking to extensive crosslinking. Here, we will focus on the crosslinked varieties of polyurethanes to see how they fit into the general concept of crosslinking as we described it for epoxy resin, drying oils, and alkyd resins. But first let's look at the details of how urethane-based polymers can be synthesized.

4.15
Polyurethanes are a Product of the Chemical Reactivity of Isocyanates

The key functional group responsible for the synthesis of a polyurethane is an isocyanate, that is, $-N=C=O$. Organic isocyanates are far too reactive to be found in nature, and in fact they are difficult to isolate in a laboratory. Isocyanates react readily with nucleophiles such as, for one prominent example, water. A famous reaction in which isocyanates are an intermediate and react with water was reported in 1881 by a renowned nineteenth century chemist, August Wilhelm Hofmann who lived from 1818 to 1892.

In his long career, Hofmann was highly beloved as a teacher. In Partington, one finds the statement: "To his students he showed an infectious enthusiasm; his whole outlook was that chemistry is something to love, a means of arriving at truth, and not a vehicle which can be used to ride down and humiliate rivals, whose mistakes can be corrected by factual means, leading gradually but unfailingly to what is better and truer." There is insight here not only into Hofmann but also into those early days in the development of chemistry as a science where brutal arguments were the order of the day. Partington goes on to point out that: "Hofmann had much of his experimental work carried out by assistants, who idolized him, since he was rather clumsy and tended to break apparatus." There is a picture of an old German professor. Would you not like to have known him, to be his student?

The Hofmann reaction is shown in Fig. 4.15 as a way to convert a carboxylic acid to an amine with one carbon less. Here, we see that although an isocyanate is involved as an intermediate, it is not isolated. The reason for this is that the isocyanate reacts immediately with water to form what is called a carbamic acid, which is intrinsically unstable and decomposes rapidly to carbon dioxide and an amine. The high stability of carbon di-

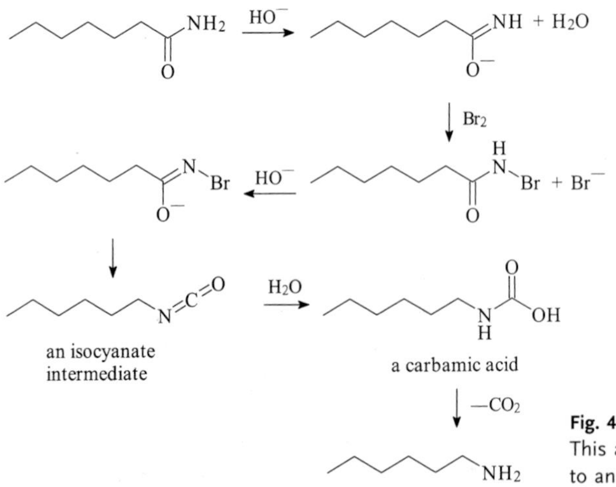

an isocyanate intermediate

a carbamic acid

Fig. 4.15 The Hofmann rearrangement. This allows transformation of an amide to an amine with one carbon less.

Fig. 4.16 R–N=C=O plus an alcohol to form a carbamate ester and R–N=C=O plus an amine to form a urea. Examples of how these derivatives of isocyanates form links in a commercial polymer, spandex.

A simplified view of the reactions of isocyanates with amines and with alcohols,

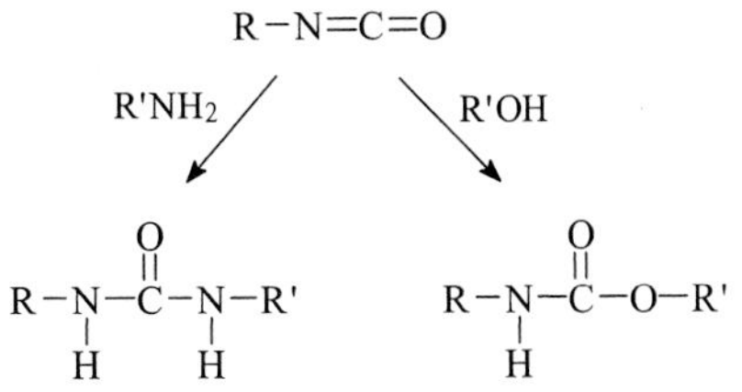

$$R-N=C=O$$

R'NH_2   R'OH

$$R-N-\overset{\displaystyle O}{\overset{\|}{C}}-N-R' \qquad R-N-\overset{\displaystyle O}{\overset{\|}{C}}-O-R'$$

H   H   H

a urea   a carbamate

allows understanding of how a polymer such as spandex is held together

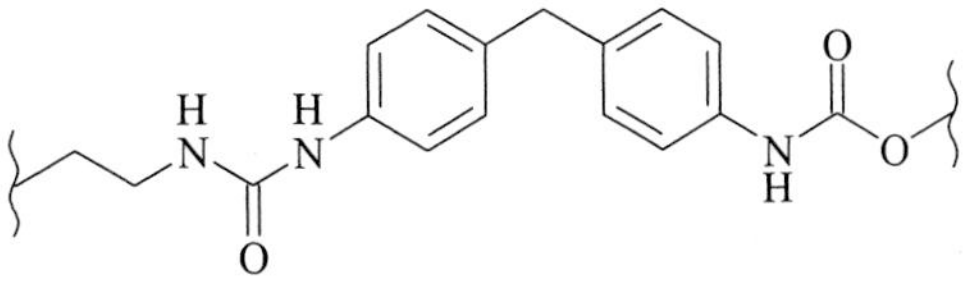

a portion of a spandex chain

oxide because of its large negative heat of formation seals the fate of the carbamic acid: carbon dioxide is expelled (Fig. 4.15). In fact, as we shall see in more detail below, it is the reaction of isocyanates with water that is responsible for the foam structure of cross-linked polyurethanes.

The isocyanate group is highly reactive with many nucleophilic groups. In Fig. 4.15 we see water playing the nucleophilic role in the step that forms the carbamic acid. The lone pair of electrons on the oxygen of water attacks the central carbon atom of the isocyanate, the site of maximum reactivity in this functional group.

Although reaction of water with an isocyanate leads to an unstable carbamic acid (Fig. 4.15), both alcohols and amines react with isocyanates to produce urethanes and substituted ureas respectively, which are stable (Fig. 4.16). As we will see, the stability of these groups forms the basis for polymers synthesized from isocyanates. This brings up one of those curious points in industrial chemistry. Even if amines are used to form polymers from diisocyanates, the polymer, which should properly be named a polyurea is often called polyurethane. Considering the role of urea in human biochemistry one might find a reason to avoid the name polyurea. Do considerations of commerce overcome proper chemical nomenclature?

Consider reacting a large number of molecules, each with two isocyanate groups with an equal number of molecules, each containing two hydroxyl groups (diols) or alternatively, two amine groups (diamines). Such reactions would yield polymers (Fig. 4.16). If you use the correct components something of great value can arise, such as the famous elastomer, spandex, whose synthesis and mode of action we discuss in detail in Chapter 7 (Sections 7.12 and 7.13 and Fig. 7.19).

Polyurethanes, like polyethylene and polypropylene discussed in Chapter 2, are addition polymers. In the case of polyurethanes, the diisocyanate and the diol and/or the diisocyanate and the diamine have identical formulas respectively to the polymers formed by these reactive monomers. In the process of forming the polymer, no atoms are lost or gained. This is not always so in the synthesis of polymers as we have seen in the formation of epoxy resins and alkyd resins discussed earlier in this chapter. Every linkage in an epoxy resin formed by reaction between bisphenol A and epichlorohydrin releases a molecule of HCl (see Fig. 4.2). The ester linkages in the polyester alkyd resins are formed with release of H_2O (Fig. 4.12). These are condensation polymers as are polycarbonates to be discussed later in this chapter (Sections 4.19 and 4.20) and extensively in the formation of the nylons in Chapter 5.

4.16
A Route to Chemically Crosslinked Polyurethanes

Consider the reaction of molecules with two isocyanate functions as discussed above (Section 4.15), but with molecules, each with *more than* two hydroxyl functions, as shown in Fig. 4.17 for toluene diisocyanate and glycerol. The identical urethane groups formed in the linear chain will now form a network linking all the chains together. Otto Bayer at Bayer AG (no relationship to the company name) worked out the basic chemistry behind the formation of the type of crosslinked polyurethanes shown in Fig. 4.17 in Germany during World War II. The work was not widely known outside of Germany until after the war, when chemists from the Allied countries inspected the German chemical industry. This led to the production of crosslinked polyurethanes in the United States, which, as we have already noted, are valuable for many applications (Section 4.14).

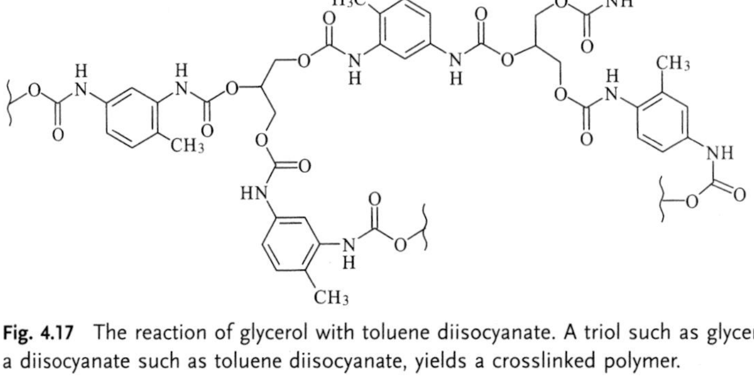

Fig. 4.17 The reaction of glycerol with toluene diisocyanate. A triol such as glycerol, when reacting with a diisocyanate such as toluene diisocyanate, yields a crosslinked polymer.

In essence, all crosslinking processes are the same. However, in the crosslinking of an epoxy resin, a curing agent that initiates the crosslinking must be added (Section 4.2). In contrast, the crosslinking of polyurethanes is inevitably based on the structures of the monomers (Fig. 4.17). Once the crosslinking begins, however, one aspect of the discussion in Section 4.2 about the curing of epoxy resin applies also to the crosslinking of polyurethanes. Just as in the epoxy resin crosslinking, some of the crosslinkable functions will be trapped in the network and will be unable to participate. This will find a parallel in the crosslinked polyurethane.

The extent of reaction necessary for locking the chains into a macroscopic crosslinked network, that is, forming the entire mass into one huge molecule, is subject to the same rules for all crosslinking. As discussed earlier in this chapter (Section 4.2), this was worked out in quantitative detail by Carothers and Flory working at DuPont over 60 years ago.

4.17
Polyether Polyols are widely used for forming Crosslinked Polyurethanes. There are many Variations on this Theme

The larger the number of hydroxyl groups per molecule that will be reacted with a monomer containing more than two isocyanate groups, the greater is the degree of crosslinking. Reacting diisocyanates in huge excess with trihydroxylated molecules, such as glycerol for one example, yields trifunctional isocyanates. Variation of the number of hydroxyl groups will therefore vary the degree of crosslinking and this is an important means for controlling the stiffness of the polyurethane. And it is stiffness that determines if the product will be used for couch cushions on the one hand or boards on the other. The larger the number of crosslinks the stiffer will be the network, and the greater will be the difficulty of deforming the overall structure.

Three examples of polyethers used for forming crosslinked polyurethanes are shown in Fig. 4.18. In general, one of the starting materials is either ethylene oxide or, more often, propylene oxide, that is, the epoxides of ethylene and propylene. These very important industrial intermediates will be the focus of our attention in Chapter 8 (Sections 8.3, 8.4) where the chemistry of ethylene and propylene will be compared. But what we have to know about the reactivity of the epoxide functional group has already been seen in this chapter. The ease of nucleophilic opening of the epoxide ring is responsible for both the formation of an epoxy resin (Section 4.4) and the curing of that resin (Section 4.5). Therefore, it is not a surprise that the triol shown in Fig. 4.18 can be formed by the reaction of glycerol or more often trimethyolpropane with propylene oxide, or for that matter that diols can be formed from ethylene or propylene oxide with water, or that an industrial intermediate with eight hydroxyl groups can be and actually is formed from sucrose with propylene oxide.

All of these hydroxyl-containing industrial intermediates (Fig. 4.18) used in the formation of various polyurethanes are formed by the same kind of reaction, the nucleophilic opening of an epoxide ring, where the nucleophile is the OH group of water or of the hydroxyl-containing molecules, such as glycerol or sucrose (Fig. 4.18). In all of the epoxide

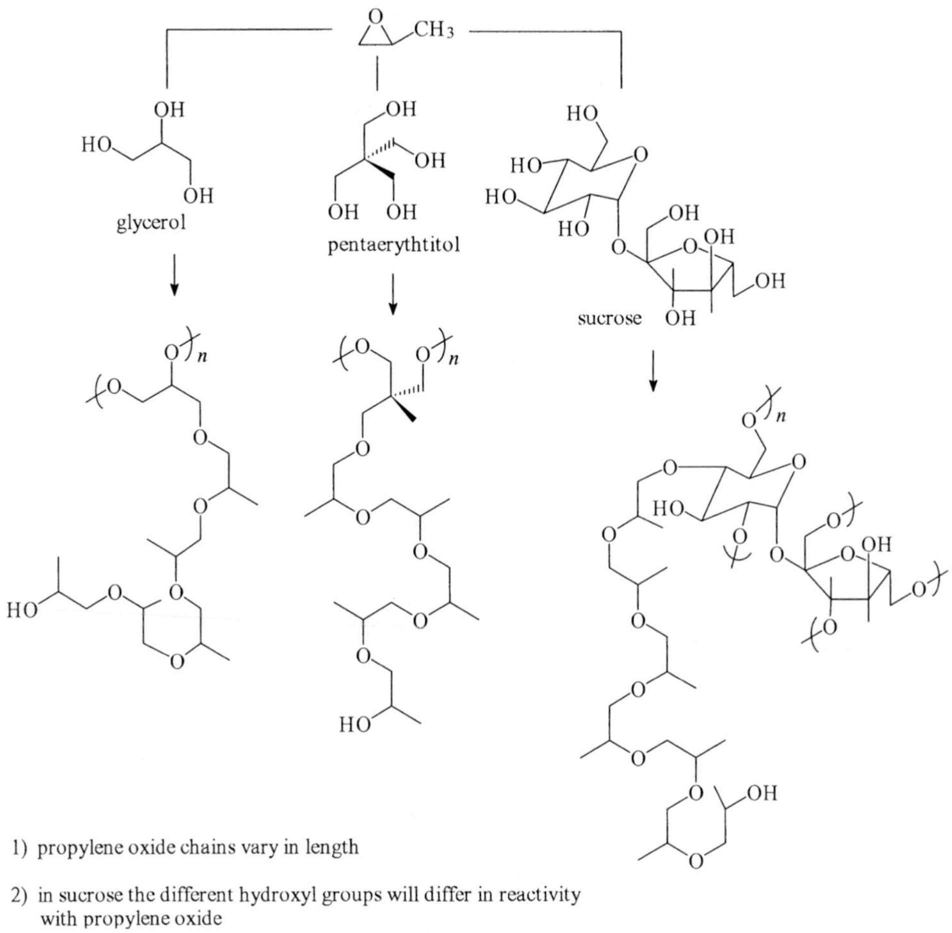
glycerol
pentaerythtitol
sucrose
1) propylene oxide chains vary in length
2) in sucrose the different hydroxyl groups will differ in reactivity with propylene oxide

Fig. 4.18 Some examples of polyols used to synthesize crosslinked polyurethanes.

ring openings seen in Fig. 4.18 the three-membered ring of the propylene oxide is attacked at the primary carbon in perfect analogy to the opening of epichlorohydrin in the formation of epoxy resin or in the curing of the epoxy resin (Figures 4.3, 4.4).

4.18 What About the Foamed Structure of the Polyurethane? Addition of a Small Amount of Water is a Common Answer

The addition of a small amount of water to the crosslinking reactions in Fig. 4.17, but certainly not enough to consume all the isocyanate groups, plays a critical role in the polyurethane produced. Each molecule of water that finds an isocyanate group will form a carbamic acid, which will rapidly expel a molecule of carbon dioxide, as we have seen in Fig. 4.15 as one of the steps in the Hofmann rearrangement (see Section 4.15).

In the viscous mass which forms as the polyurethane chains are growing longer and longer and crosslinking with each other, these carbon dioxide molecules trying to escape will act as "microscopic stirrers," causing foaming in the same way that children make foam by blowing air through a straw into a glass of milk. Eventually, the foam in the child's glass settles back to a uniform liquid. But the urethane foam is different because of the rapidly increasing viscosity arising from the continuing crosslinking. This means the foam will be trapped and locked in place forever when the crosslinking reaches a point where the entire structure sets into one macroscopic crosslinked molecule. The result is a foamed material with an exceptionally low density.

4.19
Let's return again to Bisphenol A and learn about an Entirely Different Kind of Plastic, Polycarbonate, which is Very Different from Epoxy, Alkyd Resins and Polyurethanes

Now that we have seen how bisphenol A is used to form epoxy resins (Figures 4.2 and 4.3) we are ready to learn about the other important industrial use for bisphenols, the formation of polycarbonate resins. Polycarbonates are marvelous materials. They are engineering polymers whose major function is to replace metal; but since they are clear and have good optical properties, they can also replace glass. The faceplate of a television set is polycarbonate as is the enclosure of a squash court. The side panels of a plastic car (the Saturn is an example) may be a polymer alloy of polycarbonate with other polymers, poly(butylene terephthalate) or acrylonitrile-butadiene-styrene (ABS) terpolymer. You will also find polycarbonate in an automobile in bumpers, instrument panels, lamp coverings and wheel covers. With polycarbonate, headlight lenses can be molded with sharp curves – a feat not easily accomplished with glass.

A rapidly growing market is the use of polycarbonate in compact discs, and a potential market, the virtues of which are hotly debated, is automotive glazing to replace the currently used safety glass, which is a sandwich of glass with an interlayer of a polymer called poly(vinyl butyral). Well-established is the use of polycarbonates for windows in buses, trains, aircraft, bank tellers' windows and even prisons. In areas where vandalism is a problem, shops derive some security from replacing glass with polycarbonate. Safety helmets and safety glasses use appreciable amounts of polycarbonates. Security windows and shields are polycarbonate. In the electrical area, polycarbonates are used as heavy-duty power receptacles and telephone connectors.

Polycarbonates are expensive polymers, selling for $0.80–1.00 per kilogram as compared to 35–65 cents per kilogram for the commodity polymers, such as the polyethylene and polypropylene polymers which were discussed in Chapter 2. This price difference derives from the far more complicated chemistry involved in producing the polycarbonate compared to polyethylene and polypropylene.

The price of a chemical and the volume consumed vary inversely although the relationship is not linear. Thus, about 1.5 billion kilograms of polycarbonates were sold in the world in 2002 (about 40% in the United States) as compared to almost 100 billion kilograms of polyethylenes. Bayer in Germany, and General Electric in the United States, invented polycarbonates almost simultaneously shortly after World War II. Today, the bulk of the business is enjoyed by these two companies and Dow Chemical, with General Electric having about 40% of the world's capacity. Let's see how most polycarbonate is made.

4.20
How Polycarbonates are synthesized and the Unwelcome Role of Phosgene

The structure of the polycarbonate chain is shown in Fig. 4.19. The bisphenol A units are linked together to form the polymer chain by carbonyl groups and this linkage, a carbonate, is an ester of carbonic acid, the molecule, H_2CO_3 that forms when CO_2 is dissolved in water.

Derivatives of carboxylic acids designed to speed esterification must incorporate into the structure, groups that substitute for the –OH group of the carboxylic acid. In some of the organic chemistry of carboxylic acids, the derivative used is the acid chloride so that the substituent is the familiar chlorine atom, which we saw as the leaving group chloride anion from sp^3-hybridized carbon in both the formation of epichlorohydrin (Fig. 4.5) and epoxy resins (Fig. 4.2).

The value of chlorine attached to carbon is that the bond is easily broken with the bonding electrons transferred to the chlorine. This forms the chloride anion, the stability of which is attested to by the fact that it is the conjugate base of the strong acid, hydrochloric acid. A familiar mechanism in organic chemistry, the reaction of an activated carboxylic acid with an alcohol, is shown in Fig. 4.20, and is directed to the reaction of phosgene with bisphenol A. In the formation of the polycarbonate ester linkage the chloride will be leaving from the acyl carbonyl of phosgene, which has been transformed to a four-coordinate intermediate by reaction with bisphenol A (Fig. 4.20). Phosgene is the acid chloride of carbonic acid.

The acid chloride of carbonic acid is, however, not a molecule with a happy connotation. Phosgene was vilified forever when it was chosen as a poison gas by Germany in World War I. Its activity as a poison gas is based, in fact, on the same characteristic that makes it important in the formation of polycarbonate, that is, its reactivity. But in the unfortunate use of this chemical this reactivity occurs with appropriate functional groups in the respiratory system of the intended victim. We will say more about phosgene in Chapter 10 (Sections 10.12 and 10.13) when we discover what industry is doing to rid itself of this difficult-to-work-with compound.

As we have seen in the formation of epoxy resins and polyurethanes discussed above (Sections 4.4 and 4.15), all molecules that can be polymerized must be able to react in two directions to allow chain formation. In the polycarbonate the two phenolic OH groups (one on each ring) perform this role in the bisphenol A, while the two chlorines do the job in the phosgene. In this way the reacting molecules can form a chain. The loss of the chloride anion is critical for driving the reaction forward rapidly to form more and more carbonate bonds and therefore higher and higher molecular weights. It is interesting however, that the molecular weights at which polycarbonate reaches sufficient

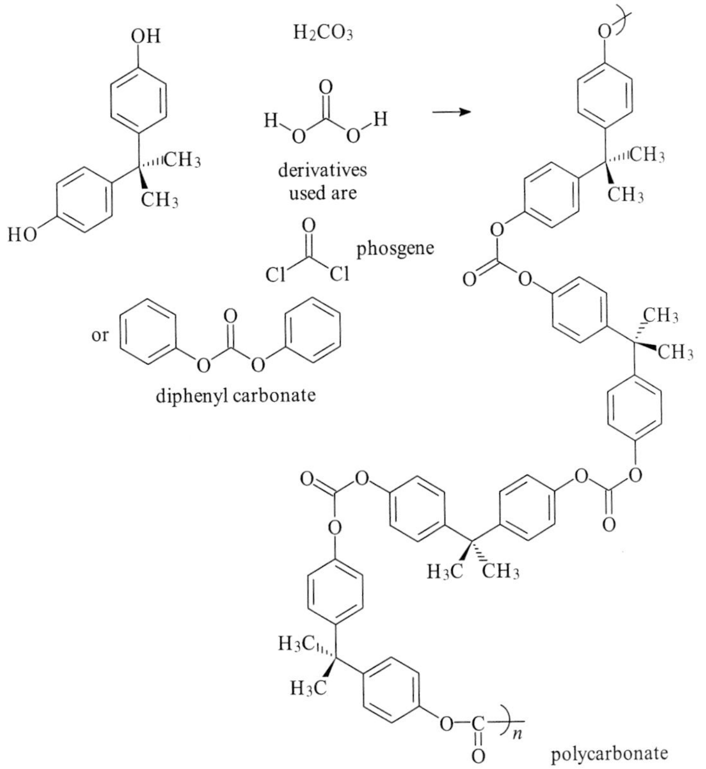

Fig. 4.19 Polycarbonate is synthesized from bisphenol A and a derivative of carbonic acid.

strength for commercial use is far lower than for polyethylene and polypropylene, many tens of thousands versus many hundreds of thousands.

Phosgene is a dangerous chemical, and therefore it is usually prepared as needed and is seldom transported. The reaction of a gas, phosgene, with a solid, bisphenol A, presents what the chemical engineer calls "mass transport problems." How do the two reactants achieve the molecular nearness needed for them to react with each other? The solution in this case seems strange, since the two reactants in the industrial process for forming polycarbonate are put into different liquid phases that are not miscible with each other, seeming therefore to make the problem of nearness even more difficult. What is going on?

Although phenols are not as acidic as carboxylic acids such as acetic acid, they are acidic enough to form a salt if one dissolves a phenol in an organic solvent and shakes the solution with water containing sodium hydroxide. A salt will be formed that will dissolve in the aqueous phase. In this way, the sodium salt of bisphenol A is dissolved in water. The phosgene, on the other hand, is dissolved in a water-immiscible organic solvent, usually methylene dichloride. The polymerization takes place at the interface between the two immiscible solvents with the process called interfacial polymerization. The two reactants – the disodium salt of the bisphenol A and the phosgene – diffuse to the interface in the heterogeneous system and react to form the carbonate bond via the steps shown in Fig. 4.20.

As the polycarbonate chains grow longer their solubility in the organic phase increases, and although the further polymerization still takes place near the interface, it occurs increasingly in the organic phase, in the methylene chloride. To aid this process, quaternary amines such as triethylbenzylammonium chloride are added, which are found to increase greatly the rate of polymerization at this stage. The reactions that occur are complex with only part of what is going on shown in Fig. 4.21.

In going from **1** to **2** in Fig. 4.21, the quaternary ammonium ion replaces the sodium ion counterion in the bisphenol A phenoxide groups and forms quaternary ammonium salts and sodium chloride. The hydrocarbon groups on the nitrogen of the quaternary amine are highly lipophilic, that is, they favor solubility in organic solvents over water

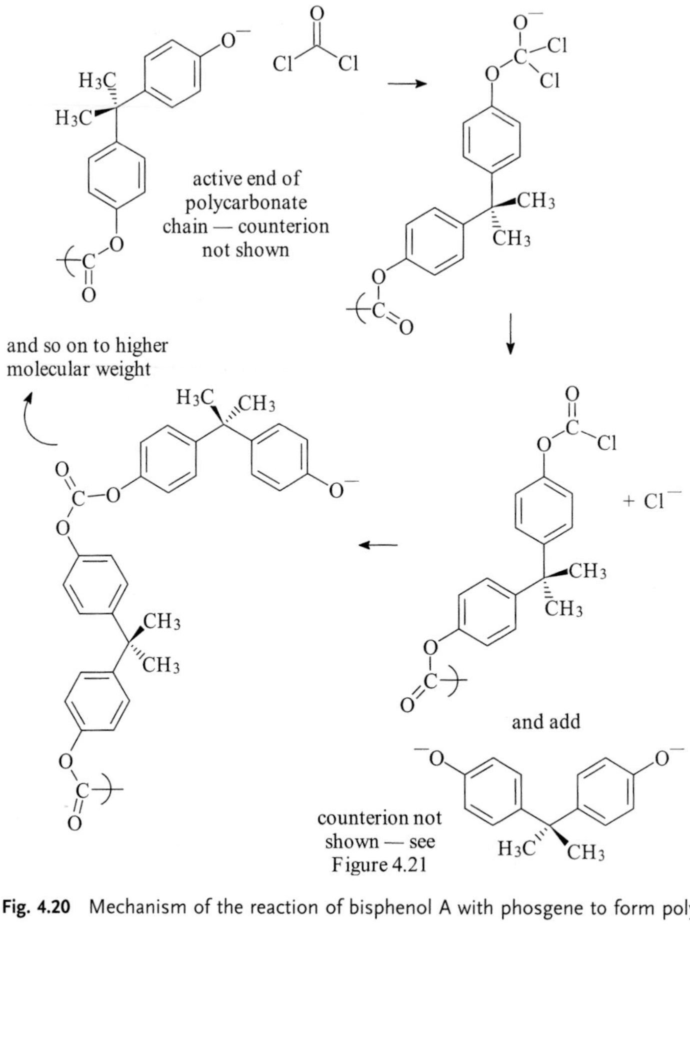

Fig. 4.20 Mechanism of the reaction of bisphenol A with phosgene to form polycarbonate.

and therefore increase the solubility of the bisphenol A salt. The solubility of the quaternary ammonium salt of bisphenol A transports it into the organic phase (3 in Fig. 4.21). Here, the growing chain with its active end group, −OCOCl, as well as the phosgene are waiting for the new arrival. After the incoming bisphenol A quaternary ammonium salt with two phenoxy groups undergoes reaction with the end of the growing chain and with the phosgene (4), the quaternary ammonium chloride salt molecule is released. The ammonium salt then returns to the aqueous phase to pick up another bisphenol A (1 Fig. 4.21) to carry it back to the organic phase and therefore to its fate to be added to a growing polycarbonate chain. As you can imagine, this transfer reaction between phases takes place countless number of times – hence the term "phase transfer catalysis."

Is there something you don't understand here? Phosgene is very reactive with water. In fact, it is this reactivity that makes it so dangerous to breathe into your preciously hydrated lungs – a fact taken advantage of in World War I as mentioned above. Why does the water, in such abundance in the aqueous phase, not react with the phosgene? The reason is that the phenoxide anion of the bisphenol A is far more reactive than the water. It reacts with the phosgene with a far larger rate constant, so much larger that the smaller concentration of the bisphenol A salt is not overwhelmed by the reactivity of the water. Some of the phosgene is indeed wasted by reactivity with the water but enough reacts with the bisphenol A salt to make a viable process.

4.21
Is there a Future in the Chemical Industry for a Chemical as Dangerous as Phosgene?

Phosgene is clearly a dangerous chemical with a history of malevolent use. Is there a way to eliminate the use of phosgene for polycarbonate formation? One positive answer lies in a process created by General Electric and also by Bayer, which not only eliminates the use of phosgene but also has proved to be more economical. It is based on transesterification chemistry in which the diacetate of bisphenol A reacts with diphenyl carbonate derived from either dimethyl or diethyl carbonate. These carbonates are produced without using phosgene. General Electric has built plants based on this chemistry in Japan and Spain, and this new approach is rapidly taking hold in the Far East. We shall look more closely at this new process in Chapter 10 (Section 10.13), where we will discover that replacing phosgene in polycarbonate synthesis is "doing well by doing good."

4.22
A few Remarks about the Double Meaning of Chloride as a Leaving Group

One of the principal themes of this chapter is nucleophilic chemistry, both aliphatic and acyl, which is expressed in the formation of epichlorohydrin, epoxy and alkyd resins and polycarbonate, as well as in the curing of epoxy resin and in the synthesis of polyurethanes. In the synthesis of epichlorohydrin and epoxy resin and the curing of epoxy resin, the nucleophilic chemistry takes place at saturated carbon. In the formation of alkyd resin, polycarbonate and polyurethane, nucleophilic chemistry takes place at acyl carbons.

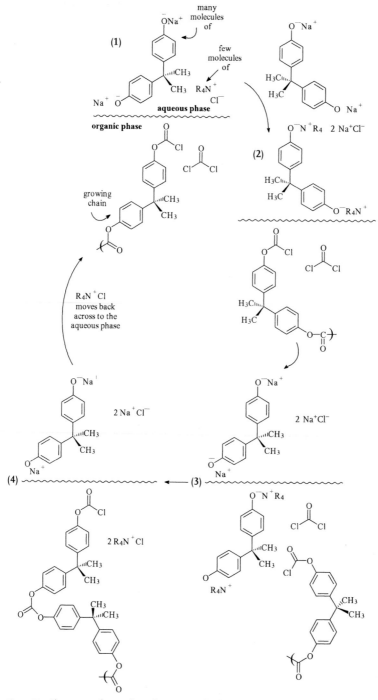

Fig. 4.21 Phase transfer catalysis formation of polycarbonate.

All nucleophilic substitution must involve a leaving group and in several instances of the nucleophilic chemistry we are studying here, chloride anion is ideally suited for this role because it is the conjugate base of a strong acid. For this reason, several intermediates are chlorinated, although we do not find this element in the final commercial products. This is a demonstration of the double meaning of the word leaving group: in chemical terms, the chlorine leaves with two electrons that had bound it to the carbon undergoing the substitution; in economic terms, chlorine atoms are necessary and paid for to advance the production. But in the end these atoms are either entirely or substantially discarded.

Compared to the price of the products containing it, the cost of chlorine is a substantial fraction of total cost. In all industrial processes any means that can be designed to eliminate use of chlorine is widely hailed. We have seen this in the Showa–Denko process for epichlorohydrin (Section 4.8), which eliminates some of the chlorine used in the older allyl chloride route. The phosgene-free processes for producing polycarbonate are taking hold in Europe, Japan and the Far East and they demonstrate the added advantage of ridding ourselves of a dangerous chemical, phosgene, in addition to reducing cost by not using chlorine, which has its own dangerous qualities (Chapter 10). We will see new approaches to the synthesis of polycarbonates in detail in Chapter 10 (Section 10.13).

4.23
Summary

This chapter is concerned with nucleophilic chemistry and crosslinking and is focused on epoxy resin, drying oils, alkyd resins, polyurethanes and polycarbonates. Bisphenol A (see Chapter 3), which is derived from benzene and propylene, is a key intermediate both in the formation of epoxy resin and polycarbonate. With epoxy resin we encounter the phenomenon of curing, or, crosslinking, in which the epoxide rings residing at the terminus of every oligomeric epoxy resin chain react with added nucleophiles to drive reactions that crosslink the system to a single supermolecule of extraordinary strength and adhesive character and with a wide variety of uses.

The synthesis of epichlorohydrin, which reacts with bisphenol A to form epoxy resin, offers a challenge to industry because of the necessity of reacting chlorine with propylene at high temperature. We learn how the high temperature is designed to solve the problem of the competition between substitution and addition chemistry. All the chemistry of epoxy resin and epichlorohydrin is tied up with the nature of the three membered epoxide rings that are the signature functional group of these molecules. We learn how such rings are susceptible to nucleophilic opening once formed and to nucleophilic chemistry to form them.

The ideas of crosslinking, so important to epoxy resin chemistry, draw us back to drying oils – natural materials based on glycerides of unsaturated fatty acids – and materials of historical interest that could be crosslinked to useful coatings. This led to the ideas of alkyd resins developed by chemists at General Electric company to solve certain material problems of the drying oils by incorporating aromatic rings and therefore strengthening these coatings. And here we came across nucleophilic acyl chemistry with phthalic anhydride and glycerol derivatives to learn about polyesters.

This led to glycerol, a molecule of historical importance to the chemical industry, a byproduct of soap manufacture, and to the role glycerol played in alkyd resin manufacture and in industrial chemistry in general. This took us from nitroglycerin to the money that finances the Nobel Prize and to many interesting historical aspects of the industry and finally back again to crosslinking. We learned that glycerol is one of the essential molecules that react to form compounds with multiple hydroxyl groups that can be used to crosslink polyurethanes. Polyurethanes are derived from the reactivity of isocyanates and have a marvelous range of uses in technology, from spandex to insulators to run-flat tires.

And finally, we arrived again at bisphenol A as the essential intermediate for synthesis of polycarbonate resin via reaction with phosgene. Again, we come to the chloride leaving group, which also plays an important role in both the synthesis of epoxy resin and its precursor epichlorohydrin. Here we discover how industry solved the mass transfer problem of bringing phosgene and bisphenol A together in a remarkable manner using a technique called phase transfer catalysis.

And throughout the chapter we learn several lessons of the economics of the chemical industry and specifically of the cost of using chlorine as an intermediate when it does not appear in the products that are sold. We learn the meaning of atom economy in the practice of industry.

Some of the subjects treated in this chapter are listed below. These are key words and terms that act as reminders of the chapter's contents and should become a valuable part of your chemical vocabulary.

- Epoxide rings
- Nucleophilic reactivity
- Leaving groups
- Phenols
- Primary, secondary, tertiary, and quaternary amines
- Crosslinking and curing
- Addition versus condensation polymers
- Oligomers
- Entropy and substitution versus addition: allyl chloride
- Epichlorohydrin
- Epoxy resin
- Polycarbonate resin
- Alkyd resin
- Polyurethanes
- Phase transfer catalysis
- Bisphenol A
- Soaps and saponification
- Glycerol
- Drying oils
- Unsaturated fatty acids and glycerides

- Nitroglycerol: heart disease, dynamite and the Nobel Prize
- Isocyanates and the Hofmann rearrangement
- Spandex
- Phosgene

Study Guide Problems for Chapter 4

1. Epoxides are subject to ring opening via nucleophilic attack, and with features of the reaction subject to the same rules as for the S_N2 reaction. Demonstrate the truth of this statement by using, as an example, the opening of propylene oxide with diethylamine.

2. Given the structure of epoxy resin in Figures 4.1 and 4.2, and the fact that "curing" of epoxy resin to a hard adhesive material could involve mixing the epoxy resin with ethylene diamine, propose what chemical reactions could be involved in the curing step.

3. Draw a portion of the polymer structure that would arise via copolymerization of styrene (Chapter 3, Fig. 3.4) with p-divinylbenzene. In what way is this structure related to a cured epoxy resin?

4. Suggest an analog of bisphenol A that could be used to synthesize an epoxy resin that would be far more deformable than epoxy resin synthesized from bisphenol A but at the same time could be cured in the identical manner.

5. Suggest a reason why epoxy resins synthesized with aliphatic units are more resistant to light and oxygen in the air than those synthesized from aromatic units.

6. Describe how both angle strain and torsional strain raise the energy of cyclohexane if it existed as a hexagon with all carbon atoms in a plane.

7. Epoxy resin can be cured with ammonia although ammonia has only a single functional group. This is in contrast to diaminoethane, with two functional groups, used for curing in Figures 4.1 and 4.4. Propose a reaction scheme for the curing of epoxy resin by ammonia.

8. In Figures 4.3 and 4.4, nucleophiles were added to three-membered epoxy rings favoring attack at the primary carbon, while in Fig. 4.5 a three-membered ring intermediate was nucleophilically opened by addition to the more substituted carbon. What does this tell you about the structures of these three-membered rings?

9. Resonance and the nature of free radicals cause problems in the attempted polymerization of propylene but helps in the formation of allyl chloride. What is the basis for this statement?

10. Predict the predominant product for reaction of chlorine with isobutane, with propane, with cumene, and with ethylbenzene.

11. Hydrogen bound to sp-hybridized carbon such as in acetylene show surprisingly high acidities. Connect this fact to the instability of carbocations and free radical sites at olefinic carbon compared to saturated carbon.

12. Count and compare the number of chlorine atoms used that do not appear in epichlorohydrin as synthesized via the allyl chloride and the allyl alcohol routes.

13. Use a bond energy table in any organic chemistry textbook to estimate the enthalpic changes in addition of chlorine to propylene to form 1,2-dichloropropane compared to reaction of chlorine with propylene to form allyl chloride and HCl.

14. What equation describes the relationship between enthalpic change and equilibrium constant as a function of temperature? Use this equation to determine the relative change of equilibrium constants between 25 and 500 °C for formation of 1,2-dichloropropane versus allyl chloride using the enthalpic changes determined in Question 13 above.

15. Offer a reason why the free radical polymerization leading to crosslinking in drying oils occurs far more slowly than the free radical polymerization of ethylene.

16. Offer a reason why the free radical crosslinking of fatty acids with more than one double bond occurs more rapidly if the double bonds are conjugated.

17. Would a triglyceride composed of glycerol esters with two saturated fatty acids such as stearic acid (n-$C_{17}H_{35}COOH$) or palmitic acid (n-$C_{15}H_{31}COOH$) and one monounsaturated fatty acid such as oleic acid, n-$C_{17}H_{33}COOH$, with a *cis* double bond at carbon-9, undergo crosslinking, that is, could be used as a drying oil? Answer the same question if linoleic acid (Fig. 4.9) replaced oleic acid in the triglyceride.

18. Fig. 4.11 shows that the reaction of synthetic glycerol with the triglyceride drying oil forms monoglyceride and diglycerides. While the monoglycerides are shown in Fig. 4.12 to form the polyester chain the diglycerides play another role in the formation of alkyd resins. What might this role be and how might it be related to viscosity control in using alkyd-based paints?

19. Carboxylic acid anhydrides as well as carboxylic acid chlorides and also *p*-nitrophenyl esters of carboxylic acids are all highly reactive for acyl nucleophilic substitution with alcohols to form esters. In what way is this reactivity related to the pK_a of carboxylic acids and HCl and *p*-nitrophenol?

20. The soap characteristics of the salts of fatty acids that are produced by hydrolysis of triglycerides (saponification) arise from the amphiphilic character of these molecules. How does soap sequester hydrophobic molecules, "grease," from aqueous media?

21. As shown in Fig. 4.14, nitroglycerin is synthesized from glycerol via reaction with sulfuric and nitric acids. Write a mechanism for this transformation. Is there any possibility that an epoxide ring is involved as an intermediate?

22. Draw structures to show how a network would form from the reactants in Fig. 4.18 allowing for this purpose that n=1.

23. Adding a small amount of water to the isocyanate and hydroxyl compounds shown in Fig. 4.17 is critical to the formation of a foamed structure. Also, addition of a small amount of water can play a role in the crosslinking. Can you offer explanations for both these statements?

24. The multiple hydroxyl functions that react with the isocyanate groups in toluene disocyanate or other diisocyanates to form the crosslinked structure are not shown in Fig. 4.18. Where are these hydroxyl groups?

25. Draw the structure of a portion of the chain of the polymer that would arise by adding water to ethylene oxide.

26. Ethylenediamine reacts with propylene oxide to form an initiator for the crosslinking of toluene diisocyanate. Draw the structure of this initiator and explain how it acts to form the crosslinked polyurethane.

27. Write out a detailed mechanism showing how one of the hydroxyl groups in sucrose reacts with propylene oxide to add three propylene oxide-derived polyether units.

28. Why do all of the ether linkages in the formation of the polyethers in Fig. 4.18 connect from a secondary carbon to a primary carbon?

29. Can you find a reason, considering the differing structures of polyethylene and polycarbonate, why the former requires far higher molecular weight to reach strength necessary for commercial use as a plastic material? From what does the polycarbonate derive its strength?

5
The Nylon Story

5.1
What was the World of Polymers like When Carothers entered the Picture?

Wallace Carothers was a young man with a doctoral degree from the University of Illinois, who wanted to use well-known reactions to create high molecular-weight polymers. He was given this opportunity when the DuPont Company hired him away from Harvard University in 1928. Carothers was promised the financial resources, the freedom and the facilities to carry out the experiments he was dreaming about. Who could resist?

Polymers were certainly known long before 1928. Polymers derived from nature such as silk, cotton, starch, and cellulose had played an important role in commerce for centuries. Purely synthetic polymers such as Bakelite, which were made from phenol and formaldehyde and, once formed, could never change their shape, or polymers from styrene, which could be melted and remolded into a variety of shapes were known commercially before 1928. In addition to polystyrene, other vinyl-based synthetic polymers such as poly(acrylic acid) and its esters were in the wings to appear as objects of commerce at about the same time (Fig. 5.1).

We have seen vinyl polymers before, in Chapter 2, where we studied ethylene and propylene. These kinds of monomers form addition polymers, polyethylene and polypropylene respectively, in which all the atoms of the monomer are found in the polymer. No atom is lost or gained in the conversion of the monomer to the polymer. The polymer is formed simply by adding together the monomer units by converting a π-bond in the monomer to a δ-bond in the polymer (Chapter 2, Section 2.1).

At the time Carothers moved to DuPont, however, polyethylene and polypropylene did not exist. The known vinyl-based polymers at that time such as polystyrene, poly(vinyl chloride), poly(vinyl acetate) and poly(acrylic acid) and its esters were formed by incompletely understood methods and their importance as we know it today (Chapter 3, Sections 3.2 and 3.3; Chapter 8, Section 8.9; Chapter 10, Sections 10.8, 10.9, 10.10) was hardly dreamed of.

It took until 1929 to discover that peroxides could initiate polymerization of these vinyl monomers, but it was not clear how this happened. The nature of the free radical intermediate and its reactivity were not well understood at this time, let alone how free radicals played a role in the formation of polymers. All of this would not be clearly spelled out until much later. There was therefore somewhat of a mystery surrounding the addition polymers formed from vinyl monomers.

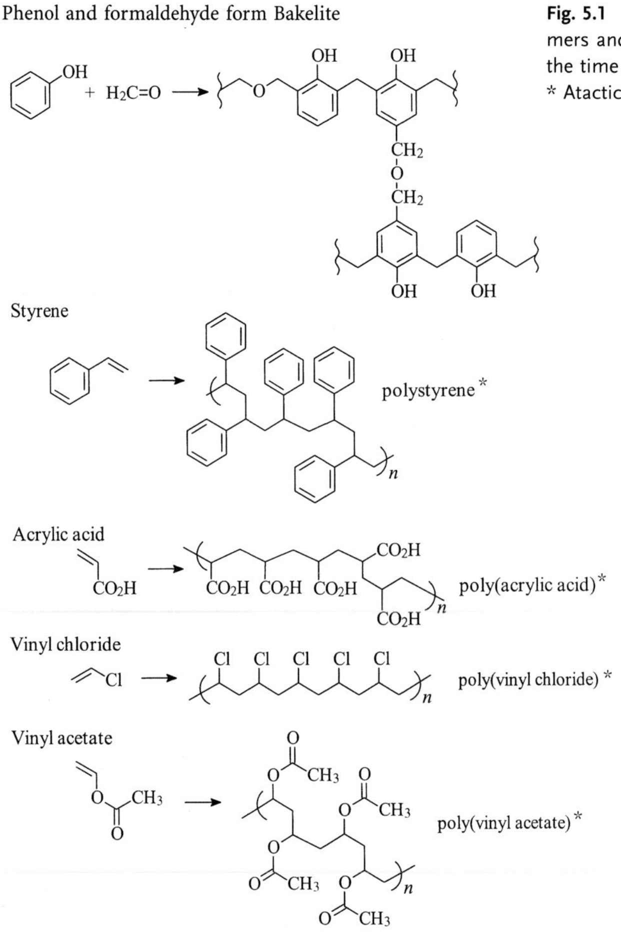
Phenol and formaldehyde form Bakelite
OH
+ H2C=O
OH
OH
CH2
O
CH2
OH
OH
Styrene
polystyrene *
n
Acrylic acid
CO2H
CO2H CO2H CO2H
CO2H
CO2H
n
poly(acrylic acid)*
Vinyl chloride
Cl
Cl Cl Cl Cl Cl
n
poly(vinyl chloride) *
Vinyl acetate
O
CH3
O
CH3
n
poly(vinyl acetate)*

Fig. 5.1 Structures of synthetic monomers and polymers known on or about the time Carothers moved to DuPont. * Atactic as in Chapter 2, Fig. 2.7.

But there was another mystery about these polymers that was not helped by the mystery surrounding their formation. In polymers of all kinds – whether derived from nature or synthesized artificially – the forces that held the polymer together were a subject of controversy which raged at the time Carothers was joining DuPont. Although the great German chemist, Herman Staudinger, who later was to win a Nobel prize for his work on the structures of polymers, insisted as early as 1920 that macromolecules consist of covalently bonded chain molecules and provided some powerful arguments for this view, many influential chemists disagreed. They suggested that such large molecules could not be bound by the normal chemical bonds known in small molecules, and that polymers formed from a combination of many small molecules held together by physical forces such as in colloids. Only later, after Carothers joined DuPont, was it generally accepted that the forces holding the chains together in polymers were no different than the forces known from better understood molecules of low molecular weight. And Carother's work was to contribute to that knowledge.

5.2
What did Carothers do at DuPont?

When Carothers joined DuPont, at the time of these controversies about the formation of polymers and their structure (Section 5.1), synthetic polymers had never been synthesized using chemical reactions that were well understood by organic chemists. It was Carothers' idea to develop a research program to synthesize these high molecular-weight materials using chemical reactions that were widely utilized and understood. This approach would have two advantages. First, there would be no question about the chemistry of how the polymer formed; and second – because the functional groups produced by these reactions would be familiar – there could be no question of what was holding the chain together in each of these very large molecules. Good idea!

Carothers considered the various possibilities and focused on esters and amides, functional groups formed by reactions between carboxylic acids with alcohols and amines. The characteristics of esters and amides were already well studied in the nineteenth century and formed part of the foundation of organic chemistry. Any chemistry focused on these functional groups could hardly be questioned.

Carothers realized that a polymer could not form unless the reacting molecules have at least two functional groups so that bonds could form in two directions, as in the links in a chain. For example, diols and dicarboxylic acids could be used to synthesize polyesters and diamines with dicarboxylic acids could form polyamides (Fig. 5.2). Carothers also realized that it was necessary for the proportions of the two monomers to be equal, to gain the highest molecular weight. To attain this equality of concentration each monomer had

Polyesters and polyamides are formed from
monomers with two functional groups such as:

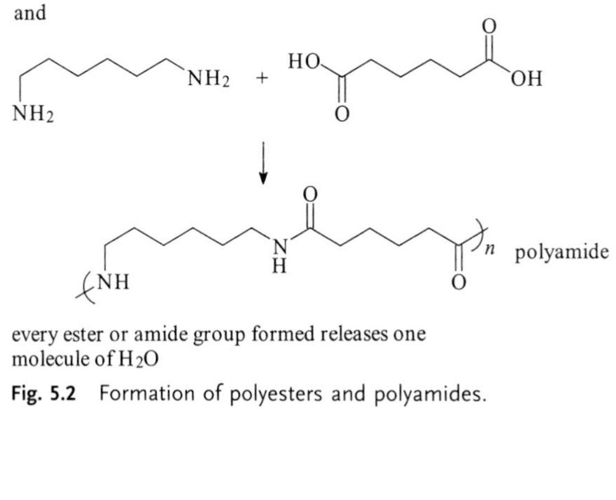

polyester

and

polyamide

every ester or amide group formed releases one
molecule of H_2O

Fig. 5.2 Formation of polyesters and polyamides.

to be of very high purity so that its precise concentration could be known and additionally no molecules would be present that would interfere with the polymerization.

He also realized that no high polymer could form unless the equilibrium constant for the bond formation holding the polymer together was very high or, alternatively, that the reaction could be driven to completion by removal of a byproduct that otherwise could reverse the reaction. Such a byproduct could be water in the case of ester or amide formation (Fig. 5.2). The chemical bonds linking the units of the chain together had to be permanent under the conditions of the polymerization.

The necessity for removing the water formed in the esterification reaction proposed to form the polymer was hardly a surprise. Bertholet, the great French chemist, about whom we will hear more in Chapter 10 (Section 10.6), in studies published in 1862, demonstrated that mixing alcohols and carboxylic acids did not lead to a quantitative conversion to esters. Moreover, if pure ester was mixed with water, the identical mixture of water, alcohol, ester and carboxylic acid resulted as if the carboxylic acid and alcohol were first mixed. These early studies on the nature of chemical equilibrium would have an important consequence on attempts to form polyesters since they demonstrated that the water formed during the esterification would have to be removed if the esterification were to proceed to completion. But removing small amounts of water turned out to be very difficult.

The water removal problem was finally solved with an invention that the DuPont workers adapted from a distillation device first created at the National Bureau of Standards. In this process, called *molecular distillation*, the path taken by an evaporating water molecule released when the ester bond was formed was long when compared to the distance to a very closely placed cold surface. This was accomplished by a very high vacuum attained by a diffusion pump powered by the condensation of mercury vapor. Thus, water molecules did not meet each other and were therefore unable to coalesce before they were trapped on the cold surface where they were kept from contact with the polyester and thus unable to reverse the esterification.

Using the molecular distillation apparatus, Julian Hill, working with Carothers, tried to form a polyester (Fig. 5.2) from an aliphatic diol and an aliphatic dicarboxylic acid and was able to gain a far higher molecular weight than previously found. And when Hill tried to remove the polyester from the apparatus, fibers formed, which could be pulled to extraordinary extensions and even form knots. As reported by Hermes in his book about Carothers, "Hill assembled his young cohorts and they ran through Purity Hall, tweezers in hand, drawing long, lustrous, continuous fibers of their first 'artificial silk' from the molten mass." It was at this moment that it became apparent at DuPont that the basic research team put together by Carothers had shown how fiber-forming polymers might be synthesized with commercial potential.

But polyesters (Fig. 5.2), although demonstrating fiber formation, were not commercially interesting. The first problem is that the molecular distillation apparatus necessary to remove the traces of water to allow for high molecular weight, while suitable for a laboratory, hardly lent itself to the technology of producing large amounts of polymer, as would be necessary if these materials became more than a curiosity.

Moreover, the polyesters made from the aliphatic dibasic acid and diol precursors were easily soluble and low melting. Such characteristics could never be useful with fibers in the marketplace. Aromatic polyesters might be better because the stiffness of the aromatic ring might increase the melting point and therefore decrease the solubility; and today we know this to be true. But these ideas were not thought of in those early days.

Carothers and his colleagues therefore turned their attention to the amide functional group. It recommended itself by a far larger equilibrium constant for its formation so that rigorous removal of the water released in the amide formation step was not necessary. Even if water were present it would not convert the amide back to its precursor amine and carboxylic acid because the equilibrium constant, $K=[-CONH-][H_2O]/[-CO_2H][-NH_2]$, was far greater than 1. Also, amides have higher melting points than esters. Working with various diamines and dicarboxylic acids they found, in analogy to the polyester formation from diols and dicarboxylic acids, that polymers – polyamides– were formed. These polymers, harboring potential for their use as commercially interesting fibers, did not have the commercially unattractive high solubility and low melting point characteristics of the polyesters. This, in essence, is how the first fully synthetic commercial fiber, nylon 6,6, came to be developed.

This is the background of the development of condensation polymerization (Fig. 5.2), an approach taken by Carothers in the United States in counterpoint to that taken in Europe, where the focus was on addition polymerization. To reiterate the contrast (see Chapter 2, Section 2.1 and Chapter 4, Sections 4.4 and 4.20), in condensation polymerization the formula of the polymer is not the same as that of the monomers from which it is made. How could it be, since in a polyester or a polyamide the water is driven off each time a link is formed? But in an addition polymerization, except for the miniscule contribution of the chain ends, the formula, the proportions of the atoms, is identical in both the monomers and the polymers. Ethylene and polyethylene have the identical ratio of carbon to hydrogen, two hydrogen atoms for each carbon atom.

5.3
Carothers' Work at DuPont had Enormous Consequences for both DuPont and the Chemical Industry

Carothers, in the very few years he had between joining DuPont and his untimely death, laid the foundation for an entirely new way to make polymers; and condensation polymerization was an enormously important accomplishment. He invented a new material, nylon 6,6, that has played a role in the fortunes of one of America's largest corporations for 60 years.

In this chapter we shall compare the various nylons that have been developed over the years. We shall see how their molecular structures determine their properties and their commercial utilities and also discover how many different principles of organic chemistry are involved in guiding the methods by which these nylons are synthesized from basic starting materials. Let's start with the first one.

Nylon 6,6 made quite a stir when it first appeared in the form of women's stockings in a store in Wilmington, Delaware on May 15, 1940. There was bedlam; the 4000 pairs of nylon stockings were sold out in a few hours. At about $1.20 a pair, nearly 5 million pairs were sold throughout 1940. But within a year, nylon went from women's legs to the needs of the war. In his book on the life of Wallace Carothers, Matthew Hermes pointed out: "The vanguard of the U.S. Army floated to earth in Normandy carried by and covered with nylon." It took almost two years after the end of the war in 1945 for the DuPont Company to produce enough material for stockings so that women could easily obtain this highly coveted item. To quote Dr. Hermes, he wrote in 1996: "DuPont sells billions of dollars worth of nylon each year now, and the profits from this single material have sustained the DuPont Company, the extended DuPont family and all the Company's stockholders for more than half a century."

5.4
The Similarities and Distinctions of the Various Polyamides that make up the Family of Nylons

There are many possibilities for forming polymers from the combination of carboxylic acid and amine functional groups. Several of these are shown in Fig. 5.3, which also contains the monomers from which the nylons are synthesized, and the melting points of the polymers. The latter is a critical parameter because it is the minimum temperature at which the nylon can be manipulated or molded, that is, processed to form a fiber or material of specific shape.

The amide functional group, which acts as the link holding the units of the nylon chain together (Fig. 5.3), is also a functional group of great importance to biochemistry. The linking group of proteins is also the amide function and so in this way nylon is a synthetic polymer that is a biological analog. Indeed, the natural polymer that nylon was competing against used earlier for women's stockings, silk, also contains the amide functional group to hold the units of the chain together. Nylon was first called artificial silk.

There is hydrogen on the nitrogen of each amide group linking together the dicarboxylic acids and diamines in nylon. It also links the amino acid derived units in silk. This hydrogen bound to nitrogen plays an essential role in the overall structures of all nylons and in proteins of all varieties. In the proteins in silk, this hydrogen is most notably involved in the formation of β-sheets, which hold the protein chains together in an arrangement that is responsible for the strength and fibrous properties of the silk. Similarly, in nylons this hydrogen – which is responsible for very strong interactions among the polymer chains – is also responsible for the fiber-forming properties of the nylons. In both silk and nylon the nitrogen-bound hydrogen participates in hydrogen bonding with the oxygen of the carbonyl group of a different amide group. Fig. 5.4 shows that hydrogen bonding plays an analogous role in both the structure of nylon 6,6 and silk.

We have seen the importance of hydrogen bonding in the properties of glycerol in Chapter 4 (Section 4.12). A variety of different kinds of covalently bound hydrogen atoms can undergo hydrogen bonding, but the essential necessity for this effect is that the electron deficiency of the hydrogen, covalently bound to an electronegative group, is relieved by forming a weak interaction between this hydrogen and an electron-rich atom. In the amide group the carbonyl oxygen plays the role of the electron-rich atom. In the nylons (Figs 5.3 and 5.4), many hydrogen bonds can form between different chains. The hydro-

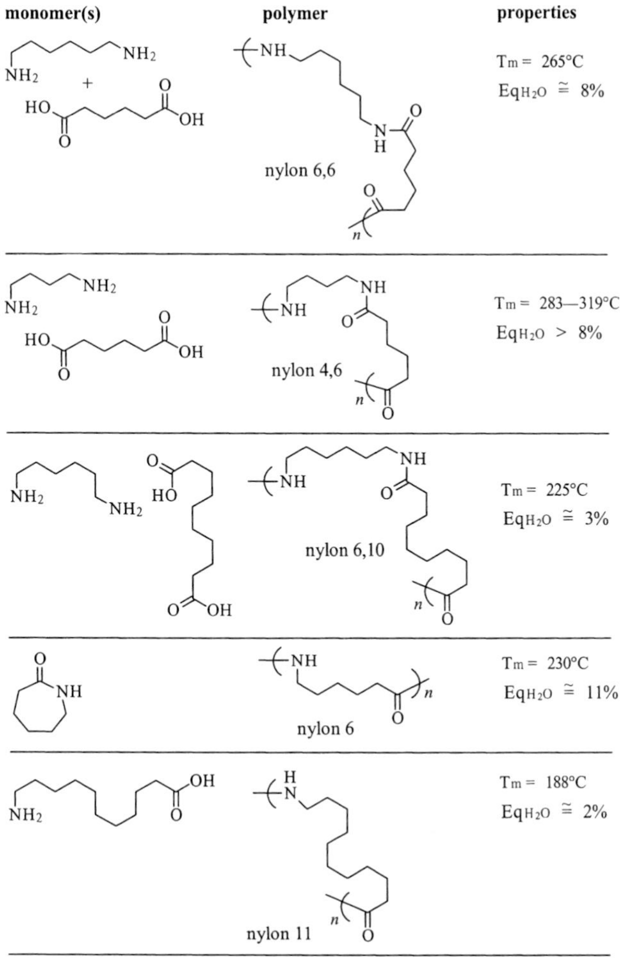

monomer(s)	polymer	properties

nylon 6,6 — $T_m = 265°C$, $Eq_{H_2O} \cong 8\%$

nylon 4,6 — $T_m = 283\text{—}319°C$, $Eq_{H_2O} > 8\%$

nylon 6,10 — $T_m = 225°C$, $Eq_{H_2O} \cong 3\%$

nylon 6 — $T_m = 230°C$, $Eq_{H_2O} \cong 11\%$

nylon 11 — $T_m = 188°C$, $Eq_{H_2O} \cong 2\%$

* Eq_{H_2O} = % absorption of H_2O at ambient temperature and 100% humidity
* T_m = melting point

Fig. 5.3 Structures of several variations of the nylons with their melting points and water-absorption properties.

gen atoms covalently bound to nitrogen are hydrogen-bonded to the carbonyl oxygen of an amide group on a different chain.

Hydrogen bonds, which are only about 1/15th as strong as normal covalent bonds, play a critical role in the properties of nylons. To understand this effect let's follow the story of nylon 6,6 to the point where commercially interesting fibers are made. Attractive as nylon 6,6 seemed, the road to a much-desired serviceable and cheaper replacement for silk stockings was a rocky one. The first problem was that the polyamide fibers lacked tensile (pulling apart) strength. Fibers must have very high tensile strength and very low elongation to perform properly. They have to give a little when pulled on, but not too much.

Fibers are usually made by extruding polymers, either as melts or as solutions, through a spinneret, a plate containing fine holes. These holes cause the fiber to form as the melt hardens or the solvent evaporates. Melted nylon has a low viscosity because in the melt the hydrogen bonds that hold the polymer molecules together (Fig. 5.4) have been broken. Thus, it goes through the spinneret easily. But the fibers made in this way from nylon 6,6 lacked the strength necessary to be useful.

Analogously to his observations with the polyester formed in the molecular distillation apparatus (see Section 5.2), Julian Hill, melted some nylon in a Petri dish, stirring it with a glass rod. When he withdrew the rod he observed a filament of the nylon attached to the stirring rod. As he walked with it, it became thinner – and, most important,

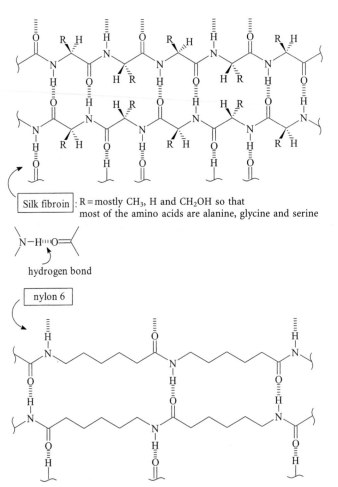

Silk fibroin : R = mostly CH_3, H and CH_2OH so that most of the amino acids are alanine, glycine and serine

N–H''''O

hydrogen bond

nylon 6

Fig. 5.4 The analogous role of hydrogen bonding in nylons and silk.

stronger. What had happened? The molten nylon molecules originally were in disarray, tangled one with the other. As the nylon was stretched, the polymer molecules lined up, came close together and hydrogen-bonded, the amide groups providing ample opportunity for such bond formation.

The regular arrangement of the chains enforced by the hydrogen bonding between amide groups along the aligned chains encouraged the regular arrangement of repetitive hydrogen bonds seen in Fig. 5.4 and therefore for crystallization to take place. This combination of crystallization among the chains guided and strengthened by the hydrogen bonding gives rise to the strength of the nylon fiber. Attempting to pull the chains apart, either perpendicular or parallel to the stretch direction, must overcome the hydrogen bonds between the chains and the crystallization forces, both arising from the stretching of the melted state. Forming fibers by stretching the melted polymer gives rise to the exceptionally strong material we know nylon to be. The process of stretching is called drawing or orientation and this became an essential step in engineering fiber formation.

Also listed in Fig. 5.3 is the amount of water that will be taken up by these various nylons at equilibrium under 100% humidity. This large absorption of water cannot act chemically to break the amide linkages holding the polymer together for the same reason that it is not necessary to remove the water rigorously when the amide bond is formed. The equilibrium favoring the amide group is overwhelming in contrast to the problem of forming esters. However, the physical absorption of water by the nylons is an unfortunate property because it changes the dimensions of the nylon and weakens it as a material. Why?

The absorption of water by nylons reveals that the amide group is a double-edged sword. The hydrogen-bonding properties are responsible for the valuable properties of the nylons, as discussed above, and make the amide group critical to the value of the nylons. But this characteristic of the amide group also attracts water, which forms strong hydrogen bonds to both the amide oxygen and the amide N–H group. The oxygen of the water is hydrogen bonded to the N–H hydrogen, while the hydrogen covalently bonded to that same oxygen is hydrogen bonded to the carbonyl group of the amide. The water in this way acts to interfere with the hydrogen-bonding network and resulting crystallization shown in Fig. 5.4 and weakens the forces necessary to enhance the fiber properties. This water absorption is unavoidable because it arises from the very characteristic of the amide group that gives nylon its valuable properties, the hydrogen bonding discussed above.

A good way to manipulate the hydrogen-bonding properties of the nylons is to reduce the number of amide groups. This can be accomplished by increasing the proportion of the structure that repels water, that is, the hydrophobic part. This effect can be seen in Fig. 5.3 in the water absorption properties of the different nylons produced industrially. The higher the proportion of the methylene groups, CH_2, to the amide groups, the less water is absorbed by the nylon. This critical relationship between the proportion of hydrophilic and hydrophobic groups is determined by the monomers used in the synthesis of the various nylons and it is this aspect of the nylon story to which we now turn our attention. At one extreme of maximizing the hydrophobic character of a nylon, there is even a nylon that is derived from castor oil, which supplies a large number of necessary methylene groups. We'll learn more about that in the sections (5.10) to follow.

5.5
The Industrial Route to Adipic Acid and Hexamethylene Diamine: the Precursors of Nylon 6,6. Benzene is the Source

Although nylon 6,6 (Fig. 5.3) was chosen for further development, in certain respects it was inferior to other polyamides examined in these experiments. The reason for choosing nylon 6,6 (the first 6 indicates the number of carbon atoms in the diamine; the second 6 the number in the dibasic acid) was that both adipic acid and hexamethylene diamine, the monomers leading to this polyamide, contained six carbons and therefore could be synthesized from a six-carbon molecule, benzene, which was (and still is) a large-volume intermediate in the chemical industry. The benzene, in supplying six carbons already connected together, would allow a simple route to the two monomers because new carbon–carbon bonds would not have to be made (Fig. 5.5).

At the time that DuPont was turning its attention to the commercial production of nylon 6,6, benzene was available in large amounts from coal via coke oven distillate or coal tar – a material recovered when coal is heated to make coke for the steel industry. Coal tar, with its many chemical components played a key role in the development of the chemical industry during the industrial revolution. Years later, the choice of benzene as the basic precursor for nylon 6,6 was reinforced further because benzene would become available from petroleum (see Chapter 3, Section 3.1), making this starting material even more available. The question is: how is benzene, a highly unsaturated molecule, converted to both adipic acid and hexamethylene diamine, which are open-chain aliphatic molecules (Fig. 5.5)?

We start with adipic acid and realize that cyclohexane can be oxidized to adipic acid via well-known oxidative carbon–carbon bond-breaking reactions. Given cyclohexane, we can produce adipic acid. Several transition metals in the presence of hydrogen under pressure, and with enough heat, are capable of overcoming the intrinsic resistance of aromatic molecules to lose their aromatic π-electrons by conversion of these electrons to carbon–hydrogen bonds. In this way benzene can be reduced to cyclohexane, and this is the first step in the industrial route to nylon 6,6.

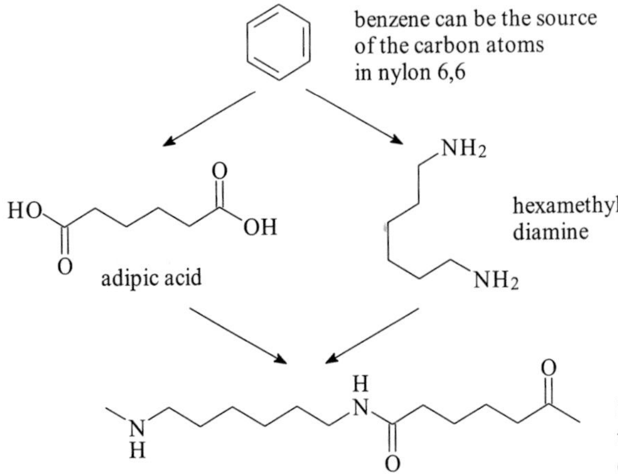

benzene can be the source of the carbon atoms in nylon 6,6

adipic acid

hexamethylene diamine

Fig. 5.5 The overall connections between benzene, adipic acid, hexamethylene diamine and nylon 6,6.

Cyclohexane is an ideal candidate for a general oxidation reaction since all the carbon–hydrogen bonds are constitutionally identical and therefore isomeric products are not formed. Contrast this with *n*-hexane, where several isomeric oxidation products could be formed. Salts of the highest oxidation state of cobalt, Co III, are appropriate oxidation agents to convert cyclohexane to what is industrially called "mixed oil." Moreover, in the presence of air the resulting reduced state, cobalt II, produced as a consequence of the oxidation of the cyclohexanes, is reoxidized to the cobaltic state, cobalt III. This cycle allows catalytic amounts of the oxidizing metal to be used, supplying the efficiency and economy necessary in a large-scale industrial procedure. A typical production unit is scaled to about 300 million kilograms per year.

Mixed oil, a mixture of cyclohexanone and cyclohexanol, is isolated in low conversion by distilling it away from the unreacted cyclohexane. If the oxidation were pushed further, the yield of mixed oil would increase but some of it would be further oxidized to adipic acid which, in the presence of this powerful oxidizing system, would be subject to further reactions producing unwanted side products. It is more economical to take a low conversion to mixed oil and then recycle the cyclohexane for reaction again with cobalt III. This eliminates separation problems that always have complex engineering requirements.

The purified mixed oil is isolated and then oxidized to adipic acid using nitric acid, which opens the cyclohexane ring at the site of the hydroxyl or ketone function. This reaction, on a laboratory scale, is often conducted in sophomore organic chemistry laboratory courses because of its specificity, although the industrial reaction is something else indeed, taking place in huge stainless-steel containers because of the corrosive nature of nitric acid. Nitric acid works so well in this role that the basic chemistry of the conversion from mixed oil to adipic acid has not changed in nearly sixty years – a record for industrial procedures, which are constantly undergoing inventive change (Fig. 5.6).

In contrast to the industrial production of adipic acid, the route to hexamethylene diamine, the other necessary precursor to nylon 6,6 (see Fig. 5.3), has been subject to constant innovation. At first, the DuPont chemists took what appeared to be the most direct path to hexamethylene diamine. Any carboxylic acid reacts with ammonia to form the ammonium salt, which on heating will eliminate water to form the amide, which on heating further will eliminate water to form the nitrile. All this simply takes place in one

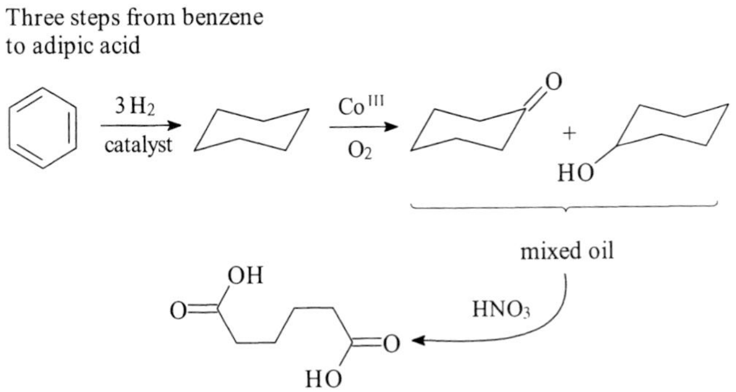

Fig. 5.6 The reactive path from benzene to adipic acid.

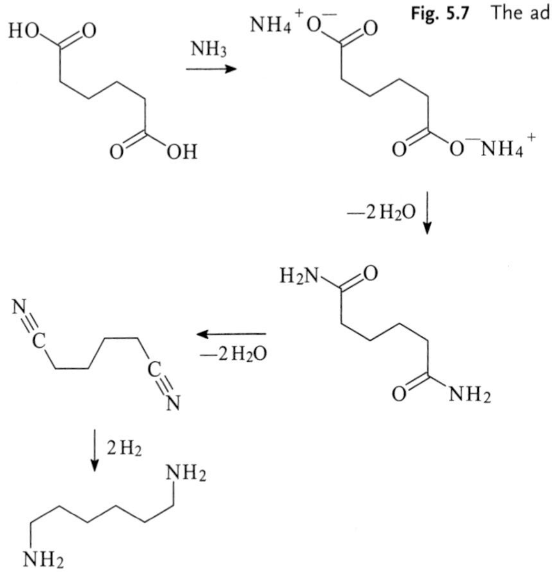

Fig. 5.7 The adipic acid route to hexamethylene diamine.

reactor. And nitriles on reduction yield amines. In the laboratory, many kinds of reducing agents can be used, which are far too expensive for a large-scale industrial process. Lithium aluminum hydride is one example. But industrially the most economical path is catalytic hydrogenation with a nickel catalyst. Apply the above steps to adipic acid and the product is the desired hexamethylene diamine (Fig. 5.7).

Certainly using the adipic acid route to hexamethylene diamine is straightforward chemistry, but it is also expensive since many steps are necessary to convert adipic acid to the hexamethylene diamine (Fig. 5.7), and as well many steps are necessary to produce the adipic acid in the first place (Fig. 5.6). In the chemical industry, the closer an industrial intermediate is to petroleum the more desirable it is. When nylon 6,6 was first put into production over sixty year ago, cracking of petroleum fractions was less sophisticated and had not replaced coal as the basic starting point for industrial production. Therefore, producing both the adipic acid and the hexamethylene diamine from benzene derived from coal was a great advantage. In fact, DuPont first advertised nylon as being made from coal, air and water. But now, given large quantities of 1,3-butadiene from steam cracking of petroleum (Chapter 1, Fig. 1.8) a more economical route to hexamethylene diamine becomes possible. Let us see how this can be done.

5.6
Hexamethylene Diamine from 1,3-Butadiene. Improving a Route to a New Kind of Rubber led to a Better Way to synthesize Hexamethylene Diamine: Industry and the Principle of Thermodynamic versus Kinetic Control of Reaction Products

There are several routes to hexamethylene diamine from butadiene, but only one is commercially important. Surprisingly, the story begins with the attempt by DuPont chemists to solve a problem entirely unrelated to nylon, that is, to find analogies to natural rubber. Carothers, in another of his contributions to DuPont's profits and to the fundamental development of polymer science, discovered that vinylacetylene, a molecule created by Father Julius Nieuwland, a professor of organic chemistry at Notre Dame, could be reacted with HCl to form chloroprene, that is, 2-chloro-1, 3-butadiene (Fig. 5.8) an analog of a molecule, isoprene, the structural residue of natural rubber.

The DuPont chemists discovered that chloroprene could be polymerized to what is now a well-known elastomeric material called neoprene (see Chapter 7, Fig. 7.12, Section 7.9) (Fig. 5.9). While neoprene rubber does not have the resilience of natural rubber and therefore cannot be used in tire manufacture, it does have one important advantage over natural rubber. Natural rubber is decomposed by air and particularly by ozone – a severe problem in cities with smog problems where ozone concentrations can be relatively high. Certain chemicals, so-called antioxidants and antiozonants, must be added to natural rubber for tire manufacture to protect against this problem. Neoprene, on the other hand, has excellent resistance to oxygen and ozone (see Chapter 7, Section 7.8). Although double bonds still decorate the backbone (Fig. 5.9), the chlorine causes a large change in their reactivity to oxidation.

Neoprene became an important industrial product with many uses for this material including roofing. Therefore synthesis of the monomer from which it is made, chloroprene, became important to DuPont. However, the industrial route to chloroprene although feasible from vinylacetylene (Fig. 5.8) could be greatly improved requiring therefore a different approach, which had the extra benefit of new synthetic opportunities for hexamethylene diamine and hence nylon manufacture.

As we have learned throughout this book, industry seeks to develop synthetic approaches that involve as few steps as possible from petroleum-derived molecules. With chloroprene, the synthetic focus was to start with 1,3-butadiene, one of the important products of the steam cracking of petroleum (see Chapter 1, Fig. 1.8). Chloroprene can be produced via loss of HCl from 3,4-dichloro-1-butene, which is one of the addition products of chlorine to 1,3-butadiene, the 1,2 addition product in which the chlorine molecule adds to one of the double bonds of the butadiene. There is, however, another addi-

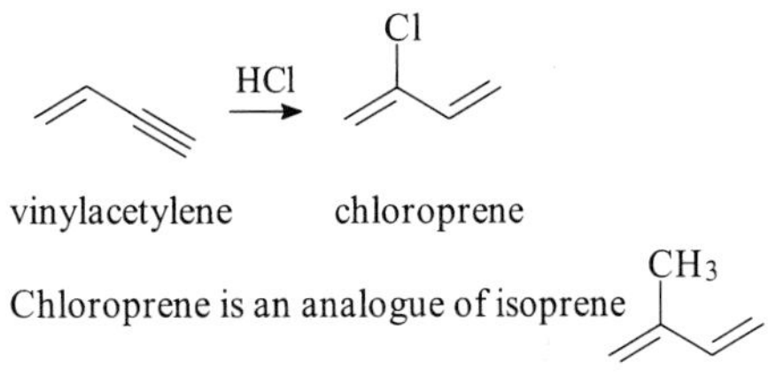

vinylacetylene   chloroprene

Chloroprene is an analogue of isoprene

Fig. 5.8 From vinylacetylene to chloroprene, and comparison to the structure of isoprene.

tion product produced in the reaction, the 1,4 addition that produces the *cis* and *trans* iso-mers of 1,4-dichloro-2-butene. Both 1,2 and 1,4 addition products are produced in the industrial procedure in the ratio of about 36 to 60%, respectively (Fig. 5.10).

This competition (Fig. 5.10) between 1,2 versus 1,4 addition to 1,3-butadiene demonstrates an important aspect of chemical reactivity. In the situation where more than one product may be formed in a chemical reaction, there is the possibility that the product formed over the lowest barrier, via the lowest energy of activation, and therefore the fastest-forming product may not be the most stable. If this fastest-forming product is predominately formed, we call this kinetic control. But alternatively, the product ultimately formed in highest amount may be the slower-forming product, which could be the most stable product. This situation is called *thermodynamic control*. How can we influence a reactive system along one or the other of these choices?

1,4-Addition of chlorine to 1,3-butadiene (Fig. 5.10) produces the thermodynamically more stable product, but via a higher energy of activation path. The 1,2-addition of chlorine to 1,3-butadiene that produces the less stable isomer is the faster reaction path with a lower energy of activation. The greater stability of the 1,4-addition product over the 1,2-addition product arises from a long-known principle of organic chemistry – more highly substituted double bonds are more stable than constitutionally isomeric less substituted double bonds (Fig. 5.10).

In the industrial situation the 1,2 versus 1,4 competition for addition of chlorine to 1,3-butadiene takes us down two entirely different paths. The former leads to chloroprene (Fig. 5.10) and therefore via polymerization to neoprene (Fig. 5.9), while, as we shall see below, the latter leads to hexamethylene diamine and therefore to nylon 6,6. Now we see that the question of how to influence this competition is more than academic. But the answer is straightforward and always the same.

If the process allows equilibration of the isomeric addition products, then the thermodynamically favored product, 1,4 addition, will be formed in largest amount. If equilibration is not possible, then the energy of activation controls the product ratio, and the 1,2 path will be favored (Fig. 5.10). Let's see how the mixture obtained on chlorination of 1,3-

Neoprene and natural rubber
have analogous structures
to some extent:

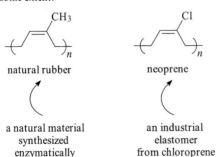

natural rubber

a natural material
synthesized
enzymatically

neoprene

an industrial
elastomer
from chloroprene

Fig. 5.9 Comparison of the structure of neoprene and natural rubber.

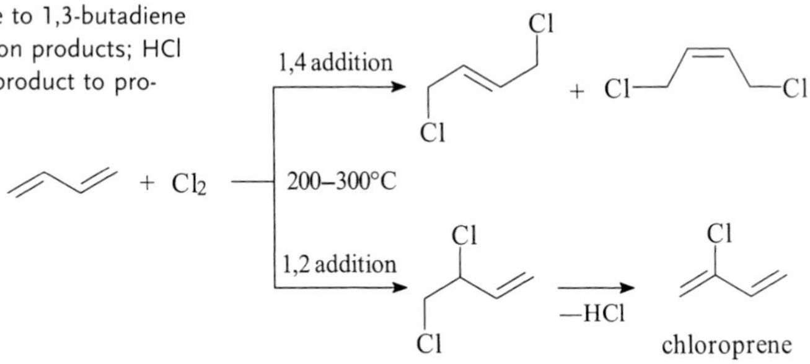

Fig. 5.10 Addition of chlorine to 1,3-butadiene
yields both 1,2 and 1,4 addition products; HCl
is lost from the 1,2 addition product to produce the neoprene monomer.

butadiene is manipulated for both production of chloroprene and hexamethylene diamine.

First let's turn our attention to using the isomer mixture in Fig. 5.10 to produce hexamethylene diamine. Cyanide anion is a nucleophile capable of displacing chloride anion
in both 3,4-dichloro-1-butene and 1,4-dichloro-2-butene so that both 3,4-dicyano-1-butene
and the *cis* and *trans* 1,4-dicyano-2-butene stereoisomers are produced. In the industrial
process however, copper-based catalysts are employed that allow equilibration of these isomeric dicyanides. Now, the higher stability of the more substituted double bond works to
our favor producing a large excess of 1,4-dicyano-2-butene, the isomer that can take us to
hexamethylene diamine (Fig. 5.11).

If we compare 1,4-dicyano-2-butene with hexamethylene diamine, we see that addition
of two hydrogen atoms to the carbon–carbon double bond to produce adiponitrile and
four hydrogen atoms to each of the cyanide groups, produces the desired nylon 6,6 precursor, hexamethylene diamine. This addition is easily accomplished on an industrial
scale by simply adding hydrogen to the unsaturated molecule in the presence of the
right catalyst. These interconversions are shown in Fig. 5.11.

In this situation for production of chloroprene from the mixture in Fig. 5.10, equilibration produces an excess of the wrong isomer. Industrial chemists and chemical engineers
found a solution to this problem based on the fact that the 1,2 addition product, 3,4-di-
chloro-1-butene, boils at 123 °C while the 1,4-dichloro-2-butene *cis* and *trans* isomers boil
about 155 °C. This wide boiling range for these constitutional isomers means that the 1,2
product can be distilled away from the 1,4 product. In the presence of copper or iron
salts that interconvert the chlorinated isomers, the higher-boiling 1,4 product remaining
in the still is constantly isomerizing regenerating the 1,2 product to compensate for that
which has distilled away – a classic LeChatelier displacement of equilibrium. And so in
this manner a large proportion of the product of chlorine addition to 1,3-butadiene can
be obtained as 3,4-dichloro-1-butene, which on loss of HCl (Fig. 5.10) yields chloroprene.
Beautiful! – a mixture produced on chlorination of 1,3-butadiene can be manipulated so
that the mixture is converted predominately to produce either the intermediate for neoprene rubber of that for nylon 6,6.

The above chemistry is an organic chemist's delight, and we include it because of its
pedagogical value, but it serves to underscore a key point in industrial chemistry –
namely, that the beauty of the chemistry may not be related directly to commercial success. This process for hexamethylene diamine was not used for long before it was discov-

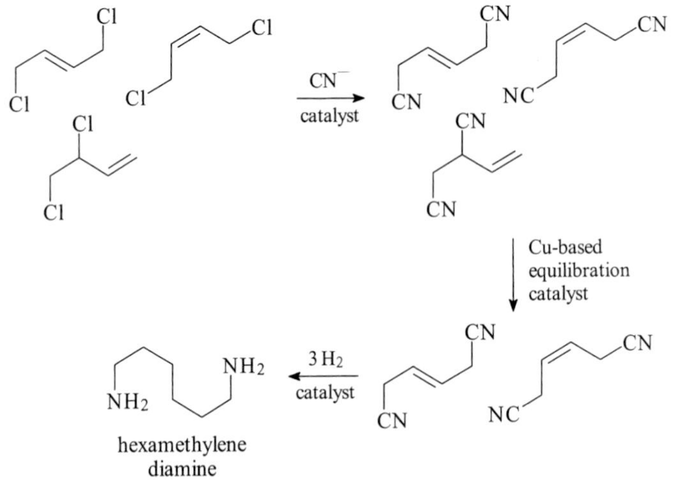

Fig. 5.11 Chemistry of the cyanide reaction with the equilibrating dichlorobutenes and the final hydrogenation to hexamethylene diamine.

ered that 1,4-dichlorobutene was carcinogenic. DuPont, accordingly, dropped it and went searching in other directions.

5.7
The Role of Acrylonitrile in the Production of Nylons

Analogous to nylon 6,6, tetramethylene diamine together with adipic acid will form nylon 4,6 (see Fig. 5.3). Tetramethylene diamine is industrially produced in a surprising and interesting manner from a three-carbon industrial intermediate, acrylonitrile, by addition of HCN to the very reactive double bond followed by hydrogenation. And acrylonitrile, as we shall see, also yields a surprising route to hexamethylene diamine and therefore to nylon 6,6. But the importance of acrylonitrile hardly derives solely from its use in nylon production.

Acrylonitrile, which will be a central focus of our attention in Chapter 8 (Sections 8.10, 8.11), is a large-volume industrial intermediate that plays its own role in both the fiber, plastic and elastomer industries and is produced on a world scale of several *billions* of kilograms per year.

To understand how acrylonitrile allows production of both hexamethylene diamine and tetramethylene diamine – the essential intermediates for nylon 6,6 and nylon 4,6 – we need to look further into the nature of the cyano (nitrile) functional group and see how this leads to special kinds of reactivity for acrylonitrile.

While cyanide anion, CN^-, formed by the acid–base reaction of triethylamine with HCN, adds rapidly to acrylonitrile as shown in Fig. 5.12, the analogous addition to propylene would not occur. The difference is that the other end of the double bond has to bear the negative charge that is produced on addition of CN^-. In propylene such a negative

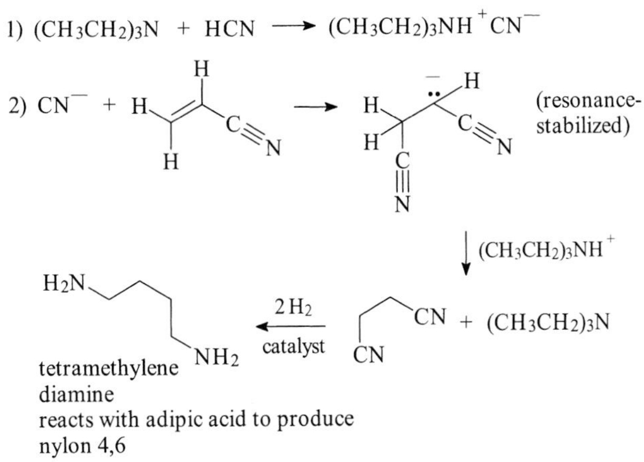

Fig. 5.12 Mechanism for the base-catalyzed addition of HCN to acrylonitrile, showing how succinonitrile is produced.

1) $(CH_3CH_2)_3N$ + HCN $\longrightarrow$ $(CH_3CH_2)_3NH^+CN^-$

2) CN^- + H₂C=CH-C≡N $\longrightarrow$ (resonance-stabilized anion with adjacent C≡N)

$\downarrow$ $(CH_3CH_2)_3NH^+$

H₂N~~~NH₂ (tetramethylene diamine) $\xleftarrow[\text{catalyst}]{2\,H_2}$ NC~~~CN + $(CH_3CH_2)_3N$

tetramethylene diamine reacts with adipic acid to produce nylon 4,6

charge has no means of stabilization and so would have a high energy. In acrylonitrile, on the other hand, the existing cyano group offers strong stabilization to a negative charge on the adjacent carbon (Fig. 5.12). Thus, the cyanide anion not only adds to the double bond of acrylonitrile but as well adds only to the unsubstituted end of that double bond (Fig. 5.12).

Stabilization of negative charge on carbon substituted with nitrile groups is a general phenomenon in organic chemistry, and is responsible for the surprising acidity of molecules such as acetonitrile, CH_3CN and dicyanomethane, $NC-CH_2-CN$. This can be seen by some comparisons. The acidity of methane is almost unmeasurable with a pK_a in the range of 50 to 60. Consider this in comparison to acetic acid with a pK_a of about 5. These are logarithmic relationships, meaning there is about 50 orders of magnitude difference between the acidity of these two molecules. But replacing a hydrogen on methane with a nitrile group makes an enormous difference lowering the pK_a to 31 and remarkably, replacing two hydrogens with cyano groups, $(CN)_2CH_2$, brings the pK_a to 11 so that the loss of the carbon-bound hydrogen as a proton, acting as a Brønstead acid, is almost in the range of phenol. Indeed going a step further to tricyanomethane, $(CN)_3CH$, brings the pK_a to a startling −5, meaning that this hydrocarbon is among the strongest acids known, stronger than sulfuric acid, for example.

The negatively charged intermediate following addition of the cyanide ion to the acrylonitrile as shown in Fig. 5.12 (so-called *conjugate addition*) produces just such a stabilized negative charge, which then accepts a proton from the quaternary ammonium salt formed from the triethylamine and HCN. Succinonitrile is formed and triethylamine is released to react with more HCN in the kind of efficient catalytic cycle essential to industrial processes. Just as addition of catalytic hydrogen to 1,4-dicyano-2-butene saturates the double bond and reduces the cyano groups to amine groups leading to hexamethylene diamine and therefore to nylon 6,6 (Fig. 5.11), catalytic addition of hydrogen to succinonitrile saturates the cyano groups and produces tetramethylene diamine leading to nylon 4,6 (see Fig. 5.3).

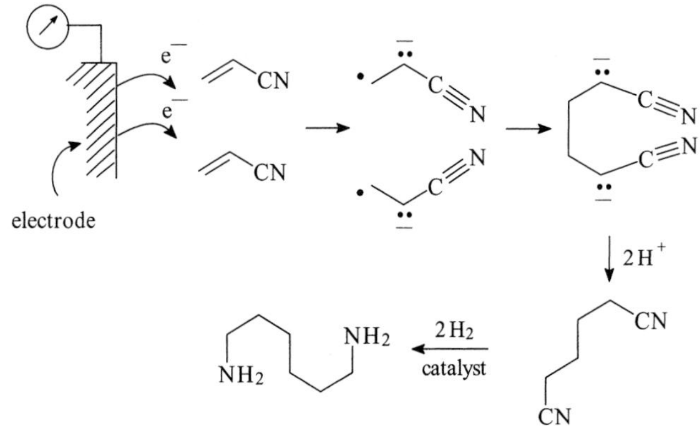

Fig. 5.13 Outline of the Monsanto electrochemical method for producing adiponitrile, a precursor of hexamethylene diamine.

But the story of acrylonitrile's role in nylon production is not yet finished. The capacity of the cyano group to stabilize a negative charge at an adjacent carbon has been used by Dr. Emmanuel Baizer, a famed electrochemist of the Monsanto Chemical Company, to produce also hexamethylene diamine, the precursor of nylon 6,6.

Fig. 5.13 shows the consequence of adding an electron to acrylonitrile – a process that can be accomplished electrochemically. A reactive intermediate is produced, which has both a free radical and a carbanion site, an anion radical. The radical site is constrained to be at the methylene carbon because the negative charge is most stable adjacent to the nitrile group.

In free radical chain mechanisms (see Chapter 1, Section 1.11; Chapter 2, Section 2.2; Chapter 3, Section 3.11; Chapter 4, Section 4.7), the formation of carbon–carbon bonds terminates chain processes and reduces the rate of formation of a desired product. In the radical anion produced by Dr. Baizer however, this propensity of carbon-based radicals to form a carbon–carbon bond produces adipontrile, which is the desired product (Fig. 5.13). Termination is what is sought. As we have seen in Fig. 5.11, adiponitrile is one hydrogenation step away from hexamethylene diamine on the route to nylon 6,6. The Monsanto invention in Fig. 5.13 is the only electrochemically based industrial method known that produces a high-volume organic chemical.

The electrochemical method for production of hexamethylene diamine was stimulated by the fact that Monsanto was the world's largest producer of acrylonitrile. This means that making hexamethylene diamine from acrylonitrile allowed for greater economy of scale in the acrylonitrile plant. Another reason for continued use of the electrochemical method is that the plants are all depreciated – a strong economic reason to continue the process. We will see more about the importance of plant depreciation in the competition among industrial processes in Chapter 6 relating to methyl methacrylate. Finally, DuPont will not license their process for production of hexamethylene diamine, shown below.

Workers at DuPont discovered during the late 1960s that butadiene can be reacted directly with HCN using zerovalent Ni or cuprous halides as catalysts. The process is complex with several isomerization steps, and we shall reserve discussion of it for Chapter 10

(Section 10.17). The economics of this process are outstanding compared to its competitors, and DuPont patents have helped the company to prevent others from using it. This is the basis for the new process displacing alternative methods. In industrial chemistry, one is playing in the major leagues!

5.8
From the Dicarboxylic Acid and the Diamine to Nylon

Neither adipic acid and hexamethylene diamine nor adipic acid and tetramethylene diamine can be mixed and heated to form a satisfactory polymer. This difficulty arises in spite of the well-known fact that carboxylic acids and amines on mixing easily form salts, RCO_2^- $^+NH_3R$, which on further heating eliminate water to form amides, $RCONH_2$. The problem is purity – a problem noted from the beginning by Carothers when he started the first experiments in the development of condensation polymerization. Carl (Speed) Marvel, of the University of Illinois, a friend of Carothers and DuPont consultant for sixty years, consistently emphasized that more time is wasted in polymer chemistry by impure monomers than for any other reason.

In the formation of an amide by the single reaction of an acid and an amine, the presence of impurities simply reduce the yield. But in a polymerization, impurities may stop a chain from continuing to grow. Consider, as an example, the presence of a monocarboxylic acid or monoamine, likely impurities in the mixture of adipic acid and the diamines used in the synthesis of nylon 6,6 or nylon 4,6 (Fig. 5.3). As shown in Fig. 5.14, whenever one of these impurity molecules reacts with a growing chain the chain growth must stop because the impurity, having only one reactive group, acts to "cap" the chain and block further reaction.

The impurity therefore causes far more damage to the desired result than its concentration would suggest. This growing chain may contain scores or even hundreds of units

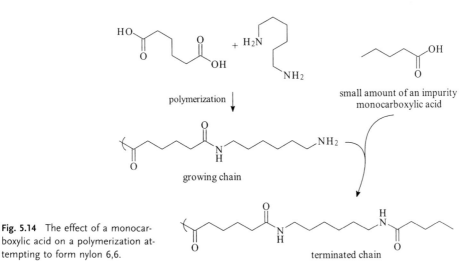

polymerization

small amount of an impurity monocarboxylic acid

growing chain

terminated chain

Fig. 5.14 The effect of a monocarboxylic acid on a polymerization attempting to form nylon 6,6.

when it encounters the chain-stopping impurity. The chain length at that degree of polymerization may be too small to yield a fiber-forming polymer, and all these units are therefore wasted in contributing to the desired result (Fig. 5.14). It is even worse than that because the chain with a low degree of polymerization may act to destabilize the fiber made by the longer chains. It is for this reason that every chemist working on polymers knows that monomer purity is far more important than it is in reactions that are not polymerizations.

Industrial chemists solved the problem of impure monomers in the formation of nylon 6,6 and nylon 4,6 in a clever manner. A salt was formed from adipic acid and the respective diamine, which was easy to purify by crystallization. The crystallization process is an excellent way to reduce impurities to levels that do not interfere with chain growth. Crystallization is generally the best route to high purity, although it is cumbersome and is used in industry only when other methods are not available. We have seen this in the purification of bisphenol A in Chapter 3 (Section 3.10).

Salt formation and crystallization in addition to guaranteeing purity, plays another important role. To make a high polymer, one needs not only exceptionally high purity for the monomers but also precisely equal proportions of the reacting monomers. Carothers also noted this requirement at the beginning of his work (Section 5.2). Because the crystals are formed from a salt of the diamine with the dicarboxylic acid, the necessary equal proportions of the diacid and diamine are automatically enforced. After all, the plus and minus charges have to be balanced.

After the necessary recrystallizations to bring the salt to high purity, it is heated to eliminate much of the water and the resulting concentrated solution of the salt in water allows formation of the amide linkages and the high polymer. The crystallization approach is the way nylon was synthesized from the beginning and still is, more than sixty years later (Fig. 5.15).

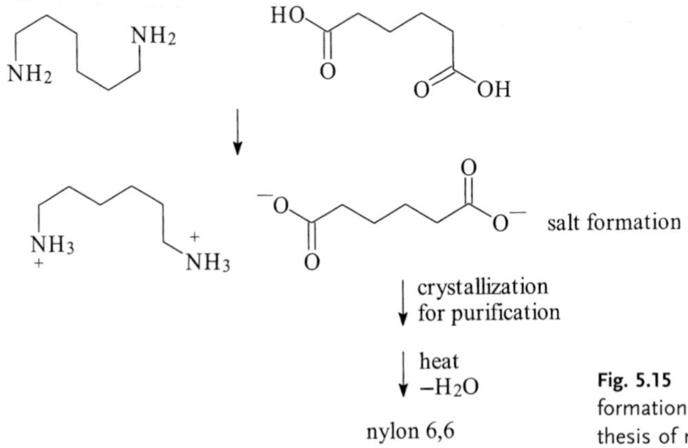

salt formation

crystallization for purification

heat −H_2O

nylon 6,6

Fig. 5.15 Purification process via salt formation and crystallization for synthesis of nylon 6,6.

5.9
Nylons made from a Single Monomer: Nylon 6

In the nomenclature of the nylons, a single number following the word nylon designates that the amino and carboxylic acid functions are in the same precursor monomer molecule. Rather than two molecules, a dicarboxylic acid and a diamine reacting together to form nylon 6,6, both nylon 6 and nylon 11 are formed from a single molecule, which contains both the amine and carboxylic acid functional groups. In the most important commercial nylon of this type, nylon 6, and in the less important nylon 11 (see Fig. 5.3), the amide functional groups must therefore be separated by a chain of 5 and 10 methylene units, respectively (Fig. 5.3). In nylon 6, the precursor molecule is cyclic, while in nylon 11 a surprising source is found in castor oil. Let's follow this chemistry starting with nylon 6. To get the whole picture we must trace a critical development back into the nineteenth century.

In the earliest years of the development of organic chemistry, long before spectroscopic and chromatographic methods eased the analysis of molecules, chemists sought to convert liquids to crystalline solids for two reasons. A solid usually meant that purification by crystallization was possible. This is invaluable compared to a liquid that had to be distilled, a far less selective purification process. Moreover, a crystalline solid has a precise melting point that could be used as an absolute means of identification and as a test of purity. This is judged by the narrowness of the melting point range and the change of melting point on mixing the material with standard samples. More experimentally complex procedures, boiling point determination and index of refraction, had to be used for liquids. Ketones and aldehydes were the earliest functional groups of importance in the development of organic chemistry and were often liquids. Thus, reagents that could convert such functionalized molecules to the crystalline state were of great value.

Among the several molecules that were used as reagents for this purpose was hydroxylamine, NH_2OH, which reacted with aldehydes and ketones to form what are called oximes. These derivatives were often crystalline even for molecules of low molecular weight. For example, the oxime of cyclohexanone (Fig. 5.16) melts at 90 °C, while cyclohexanone is a liquid at room temperature. But oximes were destined to play a much larger role. A distinguished German chemist, Ernst Beckmann found, in 1886, that oximes undergo a remarkable rearrangement. On treatment with strong acid the oxime of a ketone was converted to a molecule with an identical elemental composition but with an entirely different structure (Fig. 5.16).

This remarkable rearrangement (Fig. 5.16) mystified the chemists who first observed it, and stimulated many mechanistic studies to figure out what was driving this transformation. But these studies had no apparent practical use until nylons became of interest and a problem was encountered in the early DuPont work. When Carothers decided to turn his attention from the initially studied polyesters to the polyamides (Section 5.2), he focused on the monomer 6-aminocaproic acid (Fig. 5.17) in which the groups that would link the chain units together were in one molecule. But there is a route that is alternative to polymerization, which the molecule can take in forming the amide bond. Instead of the carboxylic acid and amine groups finding partners and forming amide groups among different molecules, the two groups within one molecule could react together and form a ring.

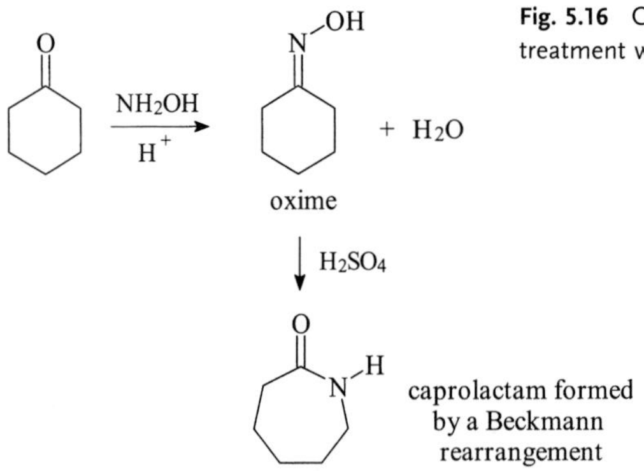

Fig. 5.16 Conversion of cyclohexanone to the oxime, and treatment with strong acid for conversion to caprolactam.

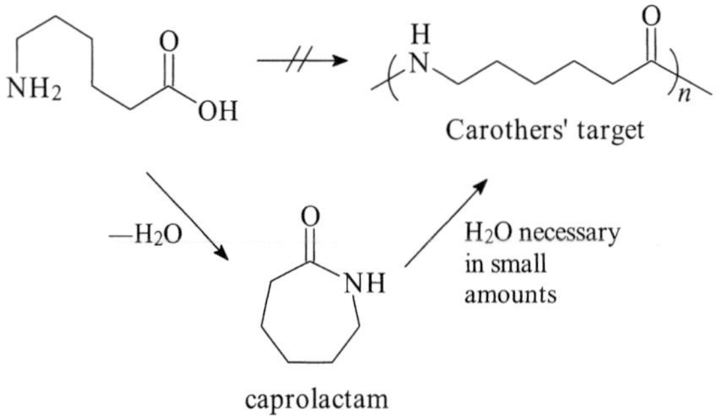

oxime

+ H₂O

| H₂SO₄

caprolactam formed by a Beckmann rearrangement

NH₂ ... OH

—//→

Carothers' target

—H₂O

caprolactam

H₂O necessary in small amounts

Fig. 5.17 Attempted polymerization of 6-aminocaproic acid yields caprolactam, which does yield nylon 6 on addition of small amounts of water.

On attempting polymerization of 6-aminocaproic acid, a waxy solid that appeared to be a low molecular-weight polymer was obtained by the DuPont workers in Carothers' laboratory. Such low molecular-weight polymers are called oligomers, and this one was mixed with a cyclic molecule. The cyclic molecule produced was caprolactam (Fig. 5.17), the identical molecule that Beckmann had observed in the rearrangement of the oxime of cyclohexanone (Fig. 5.16).

Carothers published a paper on his results in which he stated that this cyclic amide could not yield a high polymer. This report in the public literature was to cost DuPont a great deal of money. As reported by Matthew Hermes, when DuPont revealed nylon 6,6, German chemists at I. G. Farben in 1937 scoured the literature for every word on the path to its discovery. They soon discovered Carothers' statement about caprolactam and went on to find a way to polymerize this cyclic amide producing what is known as nylon 6 and gaining an exclusive license to a valuable nylon that DuPont might have claimed. The German success in polymerizing caprolactam rested on the addition of a small amount of water. A molecule of water spelled the difference between success and failure and you will have a chance to figure out why by doing problem 24 at the chapter's end.

The very molecule that Carothers knew had to be excluded in forming high molecular-weight polyesters, and the driving force for his use of molecular distillation (Section 5.2) was the molecule necessary to polymerize caprolactam. And so in this way the rearrangement Beckmann discovered in 1886, applied to the oxime from cyclohexanone, led to a major industrial product, nylon 6 (Figs 5.16 and 5.17).

5.10
Another Nylon made from a Single Monomer: Nylon 11

Let's look at nylon 11 and find out what role castor oil plays in the nylon saga. *Time* magazine, on September 14, 1931, wrote: "By heating castor oil with an alkali and mixing the result with motor antifreeze ... DuPont chemists produced an artificial silk fiber." The article goes on to report that it is too expensive to manufacture to make a commercial fiber. This is one of the earliest public reports that the DuPont Company was trying to better nature. This popular magazine article was based on a paper that Carothers had given at an American Chemical Society meeting in Buffalo, New York the previous month entitled "Castor Oil Silk." Let's see what led to this report, and find how it leads us to nylon 11.

Ricinoleic acid, having a hydroxyl group, is a curiosity among naturally occurring fatty acids and is found only as its triglyceride in castor oil, a natural product imported into the United States from India and Brazil. As for all triglycerides (see Chapter 4, Section 4.10), saponification with aqueous sodium hydroxide hydrolyzes the ester bonds yielding the salt of the fatty acids and glycerol (Chapter 4, Section 4.11). But in the case of ricinoleic acid, raising the temperature far higher than necessary for saponification and increasing the sodium hydroxide concentration to a high level breaks the fatty acid resulting from the saponification into two parts, 2-octanol and sodium sebacate (Fig. 5.18). The "castor oil silk" that Carothers described resulted from the polymerization of sebacic acid with ethylene glycol (antifreeze) to form the polyester.

Although DuPont allowed this research to be publicly disclosed because it was not commercially viable, in the years to follow, castor oil did prove to be a starting material for nylon 11. The story begins with the French seeking to industrialize North Africa and realizing that the castor oil bean could grow in this part of the world and by the realization that nylon 11, which could be synthesized from chemicals derived from castor oil, would resist absorption of water because of the large ratio of methylene units to amide bonds (Fig. 5.3).

The essential intermediate necessary to synthesize nylon 11 could be produced not by heating ricinoleic acid with base in water, but rather heating the acid to a very high temperature in the absence of solvent. This technique is called *dry distillation*. Two compounds distill, heptaldehyde and undecylenic acid (Fig. 5.19). The zinc salt of the latter compound from this source is used today as a fungicide against athlete's foot with the name of "Desenex." But undecylenic acid can also be converted to 11-aminoundecanoic acid, which can be polymerized to produce nylon 11.

The path to the monomer for nylon 11 requires placing an amino group at the terminal position of the double bond in undecylenic acid (Fig. 5.19). This reaction is accomplished by an unusual mode of addition of HBr to the double bond. If HBr were added to undecylenic acid in the normal manner, the usual polar addition would occur with the first addition of the proton to the CH_2 terminal end of the double bond placing therefore

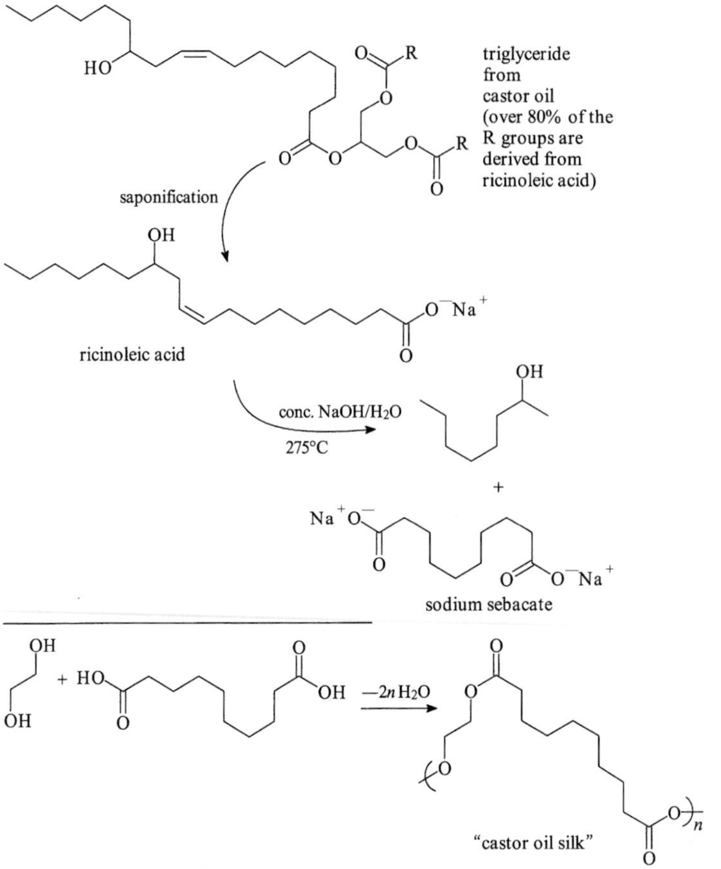

Fig. 5.18 Triglyceride of castor oil undergoing hydrolysis and conditions for conversion to sodium sebacate, followed by reaction with ethylene glycol to form the polyester ("castor oil silk").

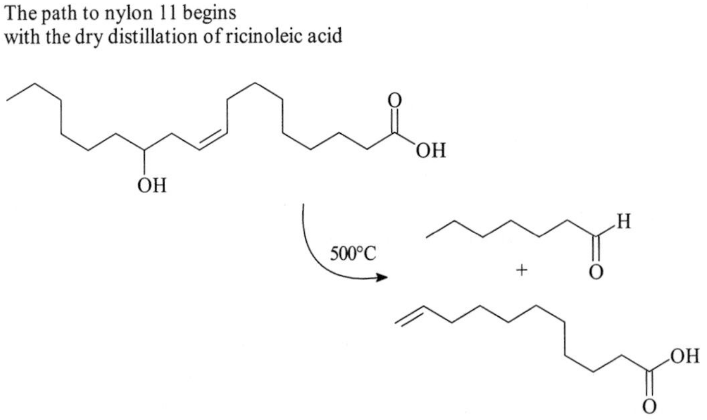

The path to nylon 11 begins
with the dry distillation of ricinoleic acid

Fig. 5.19 Dry distillation of ricinoleic acid to undecylenic acid and heptaldehyde.

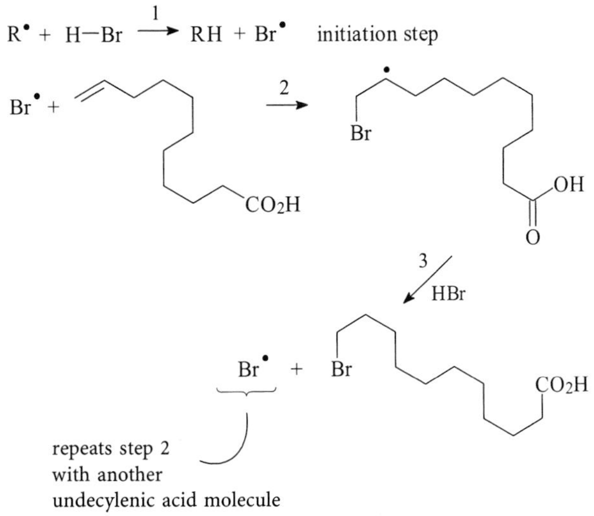

Fig. 5.20 Free radical chain reaction for addition of HBr to undecylenic acid.

$$R^\bullet + H{-}Br \xrightarrow{\ 1\ } RH + Br^\bullet \quad \text{initiation step}$$

the positive charge on the more substituted end. It follows that the site of capture by the bromide anion will be the more substituted carbon, which is simply the familiar Markovnikov addition.

But because the site of the bromine must be on the terminal carbon for preparation of the intermediate to nylon 11, the reaction is conducted in an entirely different manner. HBr is mixed with the undecylenic acid in the presence of a source of free radicals such as peroxides. The radicals abstract hydrogen atoms from HBr releasing Br$^\bullet$, which then adds to the double bond. Production of the most stable radical arising from this addition of the bromine radical requires adding the bromine radical to the terminal CH$_2$ (Fig. 5.20). The overall process occurs as a free radical chain mechanism with all the familiar steps, initiation, propagation and termination seen for this mechanistic category (Chapter 2, Section 2.2). In each addition step the bromine is placed at the same desired terminal position producing therefore 11-bromoundecanoic acid (Fig 5.20), which is readily separated from the impurities produced by the termination steps in the mechanism.

A bromine atom attached to a primary carbon is a set-up for nucleophilic displacement via the S$_N$2 mechanism by which bromide anion is pushed out and substituted for by the nucleophile, which in this industrial production of nylon 11 is ammonia. In this way, 11-aminoundecanoic acid is produced, which forms amide bonds between amino groups of one molecule and carboxylic acid groups of another and therefore forms, nylon 11 (Figs 5.3 and 5.21).

In this formation of nylon 11 it is apparent that the formation of the cyclic amide, that is, the lactam, does not take place. This behavior is in contrast to Carothers' observation on attempted polymerization of 6-aminocaproic acid and is a demonstration of one of the principles of organic chemical reactivity. Mutually reactive functional groups on the same molecule may be able to react with each other in principle if the structure is flexible but

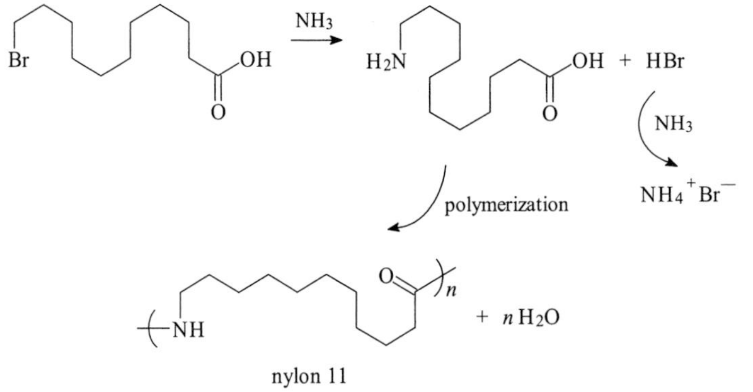

nylon 11

Fig. 5.21 Nucleophilic displacement of bromide ion by ammonia, and formation of nylon 11.

forming large rings involves too many restrictions on conformation and therefore creates an entropic barrier to reaction.

More will be said about the difficulty of forming large rings in Chapter 7 (Section 7.7). The difficulty of forming the ring leaves only the intermolecular path for amide formation and therefore the formation of the polymer, nylon 11. This principle of intramolecular reactivity worked against Carothers when 6-aminocaproic acid cyclized too readily and blocked his attempt to form nylon 6, while the difficulty of forming the ring from 11-aminoundecanoic acid favors the formation of nylon 11.

5.11
Summary

We see here the desire of a brilliant young man to apply a different kind of synthesis to the formation of polymers. He wanted to use synthetic methods based on known chemical reactions instead of the ill-understood free radical-based methods in use at that time. This effort led to the development of condensation polymerization. The work, based on the formation of amides and esters, gave rise to the first fully synthetic fiber, nylon 6,6, a material that transformed industrial chemistry and generated great wealth and prominence for one of America's largest corporations. We learn how the structure of nylon has parallels to the structure of silk and how hydrogen bonding plays an essential role in the nature of the fiber-forming character of both substances.

The first of the commercial nylons arose from a chemistry based on benzene, a starting material available from coal tar and later available in great abundance from petroleum processing. In this chemistry we find how the aromatic character of benzene can be overcome in saturating the benzene ring and how, unexpectedly, aliphatic compounds of precise functionalization can be produced. Here reductions and oxidations play the starring roles.

But all industrial products and especially those produced in large amounts must become subjects for improvement in the procedures to make them. Small changes in efficiency can make a large difference in profit. Yet one of the monomers for nylon, adipic acid, is still made the same way as sixty years ago, while the production of the other, the diamine monomer, is now based on new chemistry. This chemistry based on acrylonitrile, in which we observe the importance of conjugate addition to double bonds and the carbanion stabilizing character of the cyano group, lead to a new nylon product, nylon 4,6. We see how the diamine necessary for nylon 6,6, hexamethylene diamine, can be made from intermediates related to the route to the synthetic rubber, neoprene, and how this chemistry confronts industry with the conflict between thermodynamic versus kinetic control of chemical reactivity. We discover how the fundamental nature of free radicals and carbanions allows an electrochemical method to yield an important nylon intermediate.

We also see how the original investigators lost the opportunity to capture the market for another type of nylon, nylon 6, which would not yield to the newly minted polycondensation methods but instead became accessible to a version of polymerization never considered before, ring opening polymerization. And this chemistry became possible because certain derivatives of carbonyl compounds, oximes, undergo entirely unexpected rearrangements to yield cyclic amides, lactams. And finally we discover how unusual fatty acids derived from castor oil can undergo fragmentation reactions that yield intermediates, which on using anti-Markovnikov addition to double bonds can form less important but still useful nylons.

Some of the subjects treated in this chapter are listed below. These are key words and terms that act as reminders of the chapter's contents and should become a valuable part of your chemical vocabulary.

- Esters and polyesters
- Amides and nylon 6 and 4,6, and 11
- Amines and carboxylic acids
- Polycondensation
- Hydrophilicity
- Hydrophobicity
- Silk
- Hydrogen bonding
- Reduction of benzene
- Oxidation of hydrocarbons
- Nitriles and acidity
- Conjugate addition
- Thermodynamic versus kinetic control of reactivity
- 1,2 versus 1,4 addition to butadiene
- Hydrolysis of nitriles
- Acrylonitrile
- Electrochemical synthesis
- Polymerization and monomer purity
- Ring-closing reactions
- Ricinoleic acid
- Free radical chain reaction
- Anti-Markovnikov addition to double bonds

Study Guide Problems for Chapter 5

1. Bakelite, the crosslinked resin shown in Fig. 5.1, is made by via an electrophilic aromatic substitution reaction between formaldehyde and phenol. A hint at the driving forces and mechanism comes from the reaction between phenol and acetone on the route to bisphenol A described in Chapter 3. Write mechanistic steps to explain the structure shown in Fig. 5.1.

2. Years before Carothers joined DuPont, Herman Staudinger carried out an experiment in which he reacted formaldehyde with an anion to form a polymer. What do you think the structure of this polymer might be? Staudinger then adjusted the conditions of this polymerization to obtain oligomers of various degrees of polymerization and studied how the properties of the oligomers changed as the degree of polymerization increased. Why do you think he made these measurements?

3. Describe the structure of a polymer that could be formed from adipic acid and 1,3-propanediol. What functional group links together the units of the chain? Do the same for a polymer formed from succinic acid and 1,3-diaminopropane.

4. What would happen in the polymerization of nylon 6,6 if adipic acid contained butyric acid as an impurity.

5. Show all mechanistic steps in the formation of a single ester bond in the reaction of the dicarboxylic acid and diols in Fig. 5.2, demonstrating how water can reverse the esterification.

6. How do the following points relate to the observation that water does not have to be rigorously removed in the formation of nylon 6,6? The amide bond tends to be planar and far less flexible than the ester bonds. Proteins, which exist in an aqueous environment, are polyamino acids with amide groups linking the amino acid units together.

7. Natural amino acids have the (S) configuration at the stereocenter bearing the pendent group. What are the configurations of the amino acid units in the segment of silk shown in Fig. 5.4?

8. Cobaltic ion, that is, Co^{+3}, is capable of oxidizing a saturated hydrocarbon to a ketone or an aldehyde in the presence of oxygen. What products of a single oxidation step could be produced by this oxidizing agent with *n*-hexane, with 2,3-dimethylbutane, with cyclohexane, with methylcyclohexane?

9. The reaction of vinylacetylene with HCl (Fig. 5.8) demonstrates something about the relative reactivity of double and triple bonds and as well about Markovnikov addition. Expand on this statement by rationalizing in mechanistic terms the observed difference in reactivity.

10. Use the information in Fig. 5.9 to draw a segment of the chain in both neoprene and natural rubber including several units.

11. Notice that the double bonds along the chain in both neoprene and natural rubbers are *cis*. Draw the structure of the polymer chain for several segments as in Question 10 with the *trans* double bond. Although we shall discuss this difference in Chapter 7, can you suggest how this change in the double bond stereochemistry might affect the properties of these polymers?

12. The molecular weight of a rubbery polymer is important to its properties. How might ozone damage the properties of natural rubber?

13. Offer a mechanistic explanation for why elimination of HCl from the 1,2-addition product of chlorine to butadiene in Fig. 5.10 produces 2-chloro-1,3-butadiene rather than 1-chloro-1,3-butadiene.

14. Addition to conjugated dienes as in addition to 1,3-butadiene in Fig. 5.10 proceeds via 1,4 and 1,2 addition. Explain how heat of hydrogenation measurements are used to judge the relative stability of double bonds and what could be expected for such a measurement for comparing the 1,2 versus the 1,4-addition products shown in Fig. 5.10.

15. Use a reaction coordinate diagram to present the 1,2 versus the 1,4-addition products of chlorine to 1,3-butadiene.

16. The cyanide (nitrile) functional group can react with water under acid or base catalysis to produce amides and carboxylic acids. Outline a series of mechanistic steps to show how these interconversions take place.

17. Outline a synthesis from malononitrile to butyric acid using as starting materials, malonic acid and ethyl bromide.

18. Consider the polymerization of adipic acid and hexamethylene diamine. Why would unequal proportions of these two monomers act to limit the molecular weight of the resulting polymer?

19. Several of the crystalline derivatives of ketones and aldehydes that are used for characterization contain NH_2 groups as for example, hydroxylamine or 2,4-dinitrophenylhydrazine. These molecules form imines on reaction with carbonyl compounds. These derivatives are sometimes known as Schiff bases. Outline the structure of such derivatives and the mechanism by which they are formed.

20. Molecules containing two hydroxyl groups such as ethylene glycol form derivatives with ketones and aldehydes, that is, ketals and acetals, which can serve important purposes as protecting groups in synthesis. Although water is a byproduct produced in the formation of both the amine derivatives in Question 19 and in the formation of acetals and ketals, it is necessary to remove the water as it is formed only in the formation of acetals and ketals. Why?

21. The rearrangement shown in Fig. 5.16, which produces the cyclic amide, caprolactam from the oxime of cyclohexanone is driven by protonation of the hydroxyl group. This first step leads to loss of water. Try to complete a reasonable mechanistic picture for this rearrangement.

22. The formation of caprolactam from 6-aminocaproic acid in Fig. 5.17 could be expected because of what is known now about ring-closing reactions. Support this statement with as many examples as you can. Offer examples of molecules containing both amino and carboxylic acid functional group that would behave differently.

23. The branches in low-density polyethylene are formed by a rearrangement (see Chapter 2, Section 2.5). How is the fact that these branches are mostly four carbons long relat to the formation of caprolactam from 6-aminocaproic acid (Question 22)?

24. The mechanism for formation of nylon 6 from caprolactam involves addition of water to caprolactam to form 6-aminocaproic acid. The 6-aminocaproic acid then acts as an acid catalyst. The carboxylic acid group of this molecule protonates the carbonyl oxygen of the caprolactam. The amine group of the 6-aminocaproic acid then attacks the

carbonyl carbon to open the caprolactam. Show this mechanism in structural detail to produce several units of the nylon 6 polymer chain.

25. It is not easy to come up with a reasonable mechanism for the transformation from ricinoleic acid to 2-octanol and sebacic acid as shown in Fig. 5.18. Try it.

26. The path from ricinoleic acid to heptaldehyde and undecylenic acid (Fig. 5.19) takes place at 500 °C, which is more than enough energy to break carbon–carbon bonds homolytically. The weakest carbon–carbon bond in ricinoleic acid is between C_{11} and C_{12}. Propose a series of mechanistic steps leading to the observed products of the dry distillation.

27. Compare the mechanistic steps in the free radical addition versus the polar addition of HBr to undecylenic acid.

6

Competition for the Best Industrial Synthesis of Methyl Methacrylate

6.1
Economic and Environmental Factors are Driving Forces for Industrial Innovation

Small changes in production cost per kilogram for chemicals that are produced in the range of many billions of kilos per year make a large difference in profit. When the chemical is a large-volume product like the ethylene or propylene that results from the cracking of petroleum fractions, changes in production costs are difficult to acheive; there is not much leeway in how these intermediates are produced. Focus, however, on a compound such as methyl methacrylate (Fig. 6.1), which on addition polymerization produces poly(methyl methacrylate), better known as Plexiglas^{TM}, and we shall discover that there may be room for improvement via the organic chemist's ingenuity. In particular, there is impetus for improvement if the synthesis makes use of a noxious chemical such as HCN and produces a useless byproduct like $NaHSO_4$. But before we discuss the problems with the industrial production of methyl methacrylate, examine the candidates for improvement, and learn how the principles of organic chemistry are involved in the economic and chemical factors intertwined with this decision, let's look first at why methyl methacrylate is important.

6.2
Plexiglas^{TM}

In 1936, the Rohm and Haas Company sent a young chemist, Dr. Donald Frederick, to its parent company, Rohm AG in Germany to learn about an exciting new transparent plastic with excellent optical properties that later was to be called Plexiglas^{TM}. He did indeed learn all about the chemistry and engineering of the polymer, but not about the military use the Germans planned for it – a use the Americans were later able to find out for themselves. Certainly, he was not shown how this marvelous plastic could be molded into windshields and nose cones for fighter planes and bombers. This application was important because it was no longer feasible to have the open cockpits found in airplanes used in World War I. The new generation of aircraft simply flew too fast and too high for open cockpits. Plexiglas, aptly named, had the multiple advantages of being moldable and lighter than glass and, remarkably, also shatterproof even when pierced by a bullet or shrapnel.

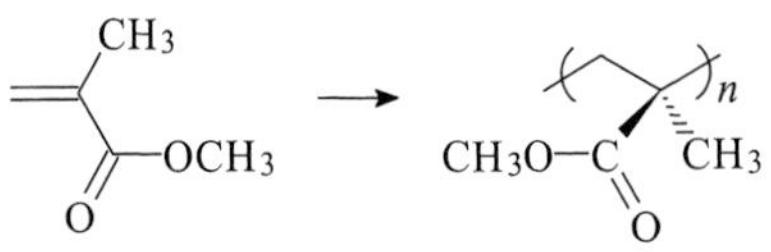

Fig. 6.1 Methyl methacrylate and its polymerization.

Nevertheless, even without seeing the entire picture, this young American visitor saw enough to stimulate his company to begin production of poly(methyl methacrylate). That same year, Rohm and Haas produced and sold Plexiglas as sheets in dimensions of up to 36 by 48 inches and thickness from 0.06 to 0.24 inches. Although total sales that year came to only $ 13,000, the company decided to invest heavily in the production of this plastic. This decision turned out to be a good one. Within a few years war arrived and the need for Plexiglas surged.

By the middle years of World War II the United States was producing a large number of warplanes (86,000 in 1943 alone), and they all used Plexiglas. The Rohm and Haas Company were selling an enormous amount of Plexiglas, 22 million dollars worth in 1944. A relatively small company had seen the value of Plexiglas through the vision of its founder, Otto Haas, and invested heavily and with some risk in its production. By this means there was supplied a critical war material and Rohm and Haas was transformed into a major American corporation.

Poly(methyl methacrylate), Plexiglas[TM], as it is called by Rohm and Haas, or Lucite[TM] as it was termed by DuPont, or Perspex[TM] as ICI called it in England, found many other commercial uses. This is easily attested to by the large amount of methyl methacrylate produced, several billion kilograms per year, demonstrating the important economic impact of this material, most of which is converted to poly(methyl methacrylate) or its co-polymers.

Higher alkyl esters of methacrylic acid can be produced in the same process used to produce methyl methacrylate by replacing methanol with a longer alkyl chain alcohol. As a matter of curiosity, these long-chain esters with as many as 18 carbon atoms were first investigated by Herman Bruson at Rohm and Haas in the 1930s. The poly(alkyl methacrylate) produced from these esters was found to dissolve in motor oils and to impart the interesting characteristic of keeping the fluidity high, even at very low temperatures. The discovery was patented and put aside, since at the time there seemed to be no commercial use for this property. But then came World War II. The United States government had a standard procedure of searching through old patents to see if any might be of use to the war effort. This old patent was pulled out since it could contribute to the successful use of vehicles at extreme low temperatures.

Astonishing as it may seem, this discovery turned out to be important at a turning point in World War II. The information about these long-chain esters was shared with the Russians in the midst of their life-and-death struggle with Germany. The polymeric esters were added to the motor oil of the Russian tanks, allowing them to move in the extreme cold of the Russian winter during the tank battle at Stalingrad. The German tanks with their frozen motor oil were of little value, and the Russians won a decisive victory turning the tide of the war for the first time against Germany. Not a single German tank from this battle returned to Germany.

6.3
The Classical Route to Methyl Methacrylate involves the Essential Role of Cyanohydrins, which can be Easily Converted to Unsaturated Carboxylic Acids

During the 1920s, Otto Rohm, the other chemist after whom Rohm and Haas is named, and the founder of both the German and the United States companies, was working on chemistry extending the studies conducted in his doctoral thesis presented in 1901 at Tuebingen. He worked in the area of acrylates and their esters, and his synthetic approach involved chemical intermediates called cyanohydrins – molecules with both a cyano (nitrile) and a hydroxyl functional group. We have discussed the cyano group in Chapter 5 (Section 5.7), where its primary role was reductive conversion to a primary amine. Now we will discover the importance of this functional group in an entirely unrelated area involved with a different aspect of its chemical reactivity.

The connection between cyanohydrins and acrylates is that cyanohydrins are ideal for creating unsaturated carboxylic acids and their esters. The fact that the carbon of the cyano group is in the same oxidation state as the carbon in the carboxylic acid group CO_2H (see Chapter 5, Fig. 5.7), allows transformation of cyanide to carboxyl by appropriately catalyzed addition of water. No oxidation or reduction step is necessary. In addition, the hydroxyl function of the cyanohydrin is the basis for formation of a double bond via loss of water. Carboxylic acid and alkene functions are the fundamental structural features of all acrylates, including methyl methacrylate (Figures 6.1 and 6.2).

Cyanohydrins are formed from carbonyl compounds by addition of HCN in the presence of base. Sodium hydroxide is used in the industrial route, which forms the salt, sodium cyanide, and therefore cyanide anion.

$$HCN + NaOH \Rightarrow NaCN + H_2O$$

The addition of cyanide anion to acetone is a classic example of nucleophilic addition to carbonyl carbon. The electropositive character of the carbonyl carbon and the nucleophilicity of the attacking group are the necessary prerequisites in this reaction, as it is for all nucleophilic reactions at carbonyl carbon.

It is instructive to compare the consequences of attack of cyanide anion on acetone with the nucleophilic attack of bisphenol A on phosgene in the route to polycarbonate

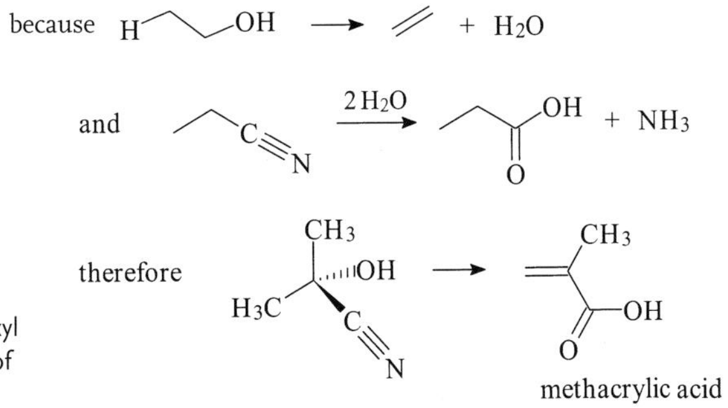

Fig. 6.2 Cyano groups are easily transformed to carboxylic acids, while hydroxyl groups lead to alkenes via elimination of water.

studied in Chapter 4 (Section 4.20). The initial steps of both reactions are identically driven as noted above, but lead to different final results, addition with acetone and substitution with phosgene (Fig. 6.3). The reason is straightforward: two methyl groups flank the carbonyl carbon in acetone, while two chlorine atoms occupy the equivalent positions in phosgene. It is impossible for a methyl group to be driven out as the methyl anion while it is not only possible but in fact favorable for the chlorine to be driven out as chloride. The difference is that CH_3^- is the conjugate base of CH_4, as weak an acid as can be imagined, while Cl^- is the conjugate base of a very strong acid, HCl. In general, nucleophilic attack at carbonyl carbon in ketones and aldehydes leads to addition. If there is no group to leave then there is no possibility of substitution. On the contrary, nucleophilic attack at an acyl carbon, as in esters, anhydrides, acid chlorides and even amides under certain circumstances, as we shall see below (Figures 6.4 and 6.5) lead to substitution (for examples see: Chapters 4, 7, 10, Sections 4.10, 4.20, 7.8, 10.13)

As seen in Fig. 6.3, the resulting negatively charged oxygen resulting from the addition of the cyanide anion to the carbonyl group of acetone, grabs a proton from water reforming the hydroxide anion catalyst that had been used to form the cyanide anion in the first step. The overall result is that the HCN, catalyzed by sodium hydroxide, has added across the π-carbon-oxygen double bond. The product is the cyanohydrin.

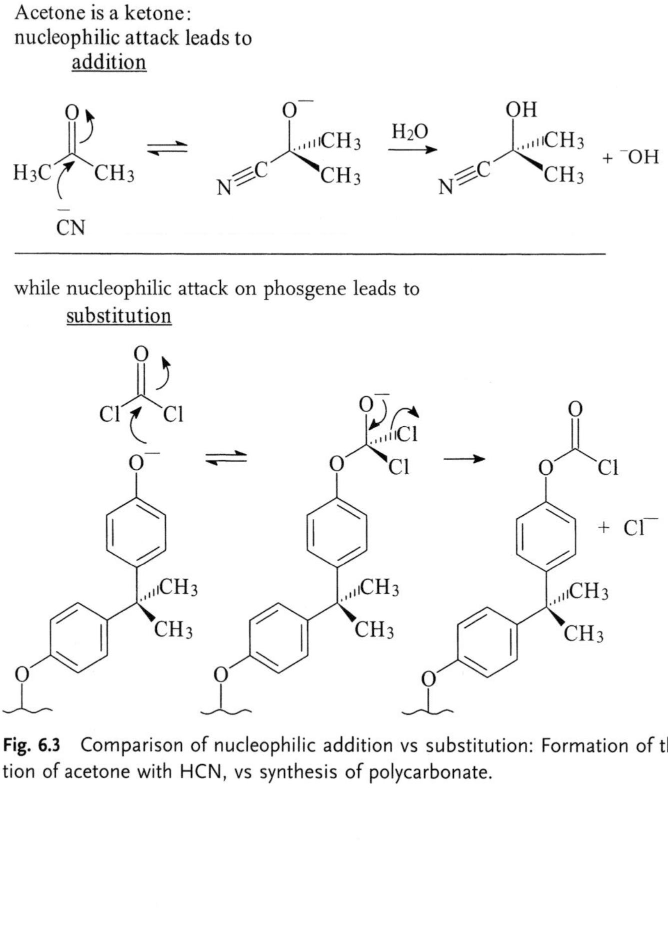

Fig. 6.3 Comparison of nucleophilic addition vs substitution: Formation of the cyanohydrin from reaction of acetone with HCN, vs synthesis of polycarbonate.

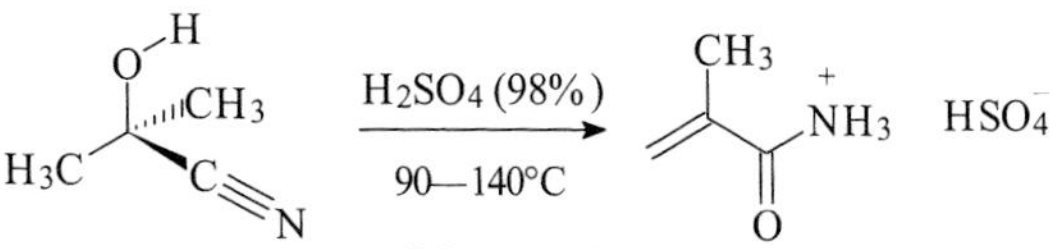

Fig. 6.4 Conversion of the cyanohydrin of acetone to the amide of methacrylic acid.

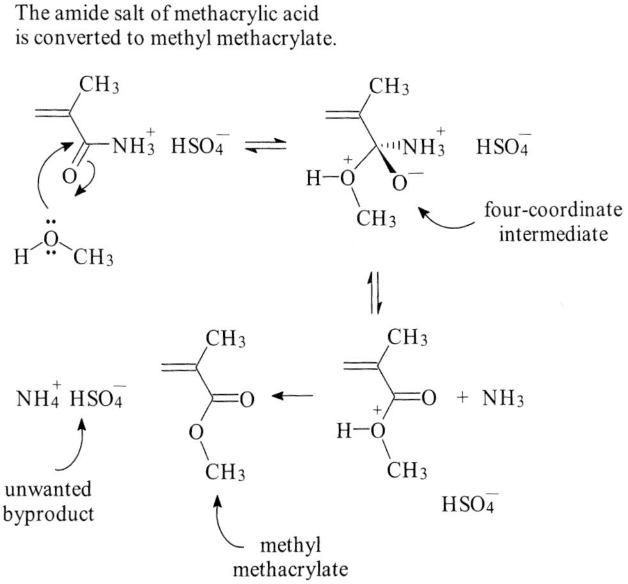

The amide salt of methacrylic acid is converted to methyl methacrylate.

Fig. 6.5 The four-coordinate intermediate formed by addition of methanol to the protonated amide and the formation of methyl methacrylate. Ammonia is converted to the byproduct, ammonium bisulfate.

From an industrial process point of view, the conversion of the cyanohydrin of acetone to methyl methacrylate is not complicated, and this simplicity certainly accounts for the longevity of this process. Sulfuric acid plays two roles in the transformation of the cyanohydrin. It is necessary for conversion of the hydroxyl group of the cyanohydrin to the double bond via protonation and loss of water. It is also necessary, via acid-catalyzed hydrolysis, for transformation of the cyano group to the amide salt. The end result of both of these reaction paths is shown in Fig. 6.4.

Let's look closely at the conversion of the cyano-derived amide group (Fig. 6.4) to the final product, the methyl ester (Fig. 6.5). All carboxylic acid derivatives undergo reaction with nucleophiles in the same manner. The addition of a nucleophile to the carbonyl carbon forms a four-coordinate intermediate. This is followed by reformation of the carbon-oxygen double bond to return to the trigonal structure, and this creates two possibilities. The incoming nucleophile may be rejected or, alternatively, a group already present on the carbonyl carbon may be ejected, thus forming a new compound. If the incoming nucleophile is the conjugate base of a weaker acid than the negatively charged moiety that would form from the group already present, then the reaction will proceed to substitution. This is the path seen in Fig. 6.3 for the reaction of bisphenol A with phosgene, where the incoming phenoxide group substitutes for chloride. Chloride is the conjugate base of the stronger acid, HCl, compared to phenol.

Addition of methanol to the salt of the amide (Fig. 6.5) plays by the rules outlined above. If ammonia is ejected from the four-coordinate intermediate formed following addition of methanol, then methyl methacrylate will be formed. However, methanol is the conjugate base of protonated methanol, $CH_3OH_2^+$, which is a stronger acid than the ammonium ion, NH_4^+, which means that the reaction would tend not to go on to product.

Ammonia is the less-favored leaving group, and therefore the equilibrium will favor rejecting the incoming methanol from the four-coordinate intermediate (Fig. 6.5). However, each loss of ammonia, when it occasionally occurs, will be irreversible. The reason for this is that ammonia, as shown in Fig. 6.5, will be quickly protonated after being ejected from the four-coordinate intermediate. The ammonium ion, NH_4^+, once formed, is then no longer available to add back again since there is no longer a lone electron pair on the nitrogen. This removal of one of the components of the equilibrium, as is the situation for all equilibria, will inevitably displace the equilibrium to produce more of the removed component. The result here is transformation of the protonated amide to the desired methyl ester (Fig. 6.5). However, the dark side of this protonation of the ejected ammonia is that this ammonium ion forms the unwanted ammonium bisulfate byproduct, which is the Achilles' heel of the classical synthesis.

6.4
Problems in the Classical Approach to Synthesis of Methyl Methacrylate

In essence the conversion, therefore, from the cyanohydrin to methyl methacrylate involves heat, sulfuric acid, and methanol. This process is simple to engineer, and most of the manufacture of methyl methacrylate in the world takes place in this way. It is nevertheless remarkable that this process, simple though it may be, has achieved commercial status. The major negative factor is the need for HCN, which is probably the most toxic chemical used industrially. HCN was the infamous Zyklon B used by the German Nazis in World War II to exterminate millions of people. And indeed there have been the inevitable industrial accidents and deaths associated with its use. This danger, coupled with the desire to eliminate NH_4HSO_4 formation and therefore the higher manufacturing costs associated with its disposal, provided the motivation for the development of new processes.

The ammonium bisulfate produced as a byproduct in the industrial synthesis of methacrylic acid on the route to methyl methacrylate is too acidic to use as a fertilizer, and its disposal presents difficulties. In fact, the only thing to do is heat it to high temperatures and convert it to sulfuric acid. To produce one tonne of methyl methacrylate, more than one tonne of sulfuric acid is required. All of this is converted to NH_4HSO_4. Thus, for one tonne of methyl methacrylate nearly 1.5 tonnes of ammonium bisulfate are produced.

So why does this method of synthesizing methyl methacrylate continue? The answer is that many of the industrial plants dedicated to this outdated method are still capable of functioning and any new process must involve building a new facility engineered to the new synthesis. As long as these old plants work this puts a competitive burden on new processes since doing it a new way involves spending a great deal of money to engineer and build a new plant. But slowly all the problems associated with the aging of these plants and the greater environmental acceptability of a new approach counters the economy of the old method so that new processes become reasonable.

6.5
What New Possibilities exist for Replacing the Old Process? Can the Ammonium Bisulfate Disposal Problem be Solved?

One possibility, long considered, would be to deal more effectively with the ammonium bisulfate. If this problem could be solved, the HCN-acetone route would become more economical. How can this be done? There have been two approaches. In one, Imperial Chemical Industries (ICI) in England, developed a high-temperature process for conversion of the ammonium bisulfate to produce ammonia and sulfur trioxide, the anhydride of sulfuric acid. The ammonia is further oxidized under these oxidative conditions to nitrogen and water and the water added to the sulfur trioxide produces sulfuric acid, which can be recycled back into the methyl methacrylate process (Fig. 6.6).

A second potential commercial solution to the problem has come not from the traditional manufacturers of methyl methacrylate, but from Japan where the byproduct problem and the difficulty of the need for large amounts of HCN are especially acute.

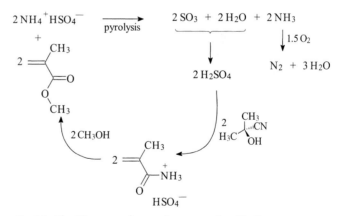

Fig. 6.6 The ICI process for recycling ammonium bisulfate.

6.6
The Mitsubishi Gas Chemical Company Approach to improving the Synthesis of Methyl Methacrylate

Instead of discarding the nitrogen of the cyano group, which then becomes the nitrogen of ammonium bisulfate, and therefore having to use a new molecule of HCN for each molecule of methyl methacrylate produced, the approach taken by the Mitsubishi Gas Chemical chemists is to reuse the nitrogen of the cyano group by recycling this atom to produce more HCN. In this way, the cost of the classical HCN/acetone process is reduced in two ways: (1) it is no longer necessary to pay for an HCN molecule for each methyl methacrylate molecule produced; and (2) there will no longer be any unwanted and expensive-to-discard ammonium bisulfate. But, as we shall see, these savings are not enough to make the process truly economical.

In this approach (Fig. 6.7), the first step, HCN addition to acetone, is retained to produce the cyanohydrin (Figures 6.2–6.3). The hydrolysis of the cyano group to the amide is carried out not with sulfuric acid as in the classical route but catalytically with manganese dioxide as a slurry with water and acetone. Sulfuric acid is not involved. But what is most important – the key to the advantage of this synthesis – is that the conversion of the amide to the methyl ester uses methyl formate, HCO_2CH_3, which undergoes an exchange reaction with the amide of the *a*-hydroxyisobutyric acid to produce the methyl ester. What the reaction accomplishes is to exchange the $-NH_2$ of the amide group with the $-OCH_3$ of the ester group (Fig. 6.7). This produces the desired methyl ester of *a*-hydroxyisobutyric acid, one step removed from methyl methacrylate via loss of water to form the double bond.

How does the reaction between methyl formate and *a*-hydroxyisobutyramide occur? One possible, although somewhat unsatisfactory, answer is seen in the mechanism

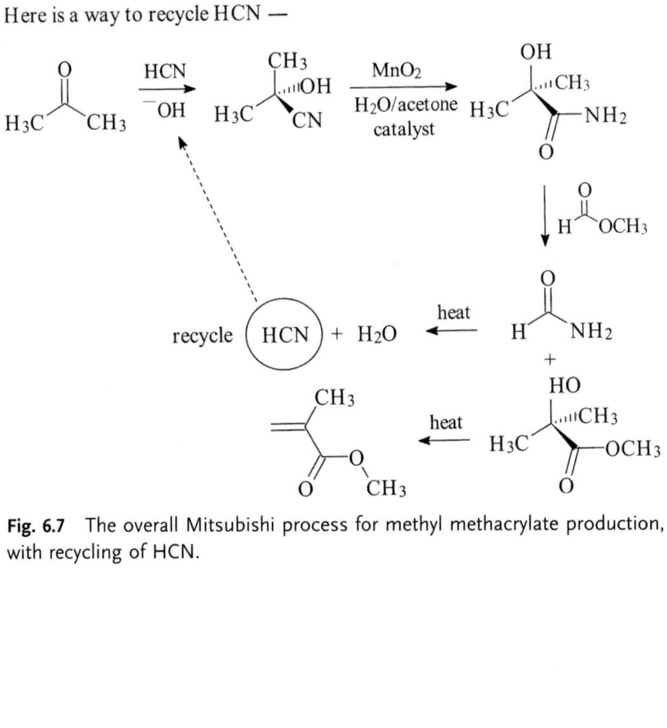

Here is a way to recycle HCN —

Fig. 6.7 The overall Mitsubishi process for methyl methacrylate production, with recycling of HCN.

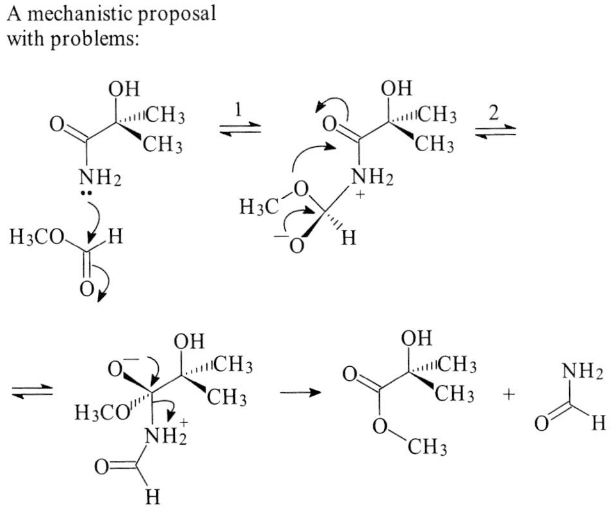

A mechanistic proposal
with problems:

Fig. 6.8 Mechanistic possibility for reaction of α-hydroxy isobutyra-
mide with methyl formate.

drawn in Fig. 6.8. Elements of the mechanism are certainly reasonable. There is the involvement of the same driving forces seen in all acyl substitutions as discussed in Section 6.3, that is, a series of nucleophilic additions followed by reformation of the carbonyl carbon. However, there are some problems such as the transfer of the methoxy group in step 2, which is shown here as occurring via an "unlikely" four-membered ring transition state. And there is the amide nitrogen in step 1 playing a role that amide groups are usually not suited for. But certainly the overall changes do take place. Perhaps the catalyst is involved to help the atoms find their way. We'll leave it up to the reader to look further and perhaps find a better mechanism or maybe justification for the mechanism shown.

The advantage of the Mitsubishi process is capture of the nitrogen of the originally used HCN – not in ammonium bisulfate, where it is unavailable for further use, and must be discarded – but rather in a molecule, formamide, $HCONH_2$, that allows that nitrogen to be released again as HCN for recycle. This conversion occurs because formamide has the identical elemental composition as H_2O and HCN and on heating to high temperatures breaks down to these molecules (Fig. 6.7). The decomposition of formamide is driven by a large positive entropy change; two molecules, HCN and H_2O, are produced from one, $HCONH_2$ (Fig. 6.7). In addition, high temperatures speed the rate, accounting for conducting the process at 500 °C.

Any extra HCN needed is made up by reacting methyl formate with ammonia producing more formamide, which is then heated high enough to form HCN and water. However, there remains one more step to get to our goal, methyl methacrylate.

6.7
The Double Bond still has to be Introduced to form the Final Methyl Methacrylate Product

But what about the introduction of the double bond in this Mitsubishi process? In the classical route (Figures 6.2 and 6.4), loss of water occurs from the cyanohydrin in an acid-catalyzed process allowing the double bond to form under relatively mild conditions. These mild conditions are essential since the molecule, with the amide group, cannot stand the very high temperatures and conditions necessary to eliminate water from an alcohol directly.

By contrast, in the Mitsubishi Gas Chemical route (Fig. 6.7) the methyl ester can be heated to a higher temperature because the ester functional group is far more resistant to chemical change. Any alcohol capable of forming an alkene will do so if heated to a high enough temperature, and especially in the presence of water-absorbing materials such as zeolites. In addition, loss of water in this situation is facilitated by the conjugation of the resulting double bond with the ester group. The overlapping π-bonds of the carbonyl group of the ester and that of the developing double bond are a stabilizing interaction, a resonance interaction, which favors loss of water. We shall see this same resonance stabilizing effect play a large role in the loss of water step in the aldol chemistry to be studied in Chapter 9 (Section 9.3). A similar effect, but one involving the stabilization of a double bond conjugated with an aromatic ring, is also at work in the elimination of H_2 from ethylbenzene to produce styrene (see Chapter 3, Fig. 3.4). The eliminated molecule and the details of the conjugation may differ, but the principle is identical.

The loss of water occurs directly from the a-hydroxymethylbutyrate in the vapor state at 250–300 °C over a water-absorbing zeolite to form the double bond and, therefore, the target methyl methacrylate (Fig. 6.9).

By these clever manipulations of the same cyanohydrin intermediate produced by the reaction of HCN and acetone in the classical route, the Mitsubishi chemists avoid the ammonium bisulfate with all its disposal problems, and also recycle the nitrogen to form new HCN. Beautiful! However, we will find that the engineering costs of the complex plant necessary to do this will affect the economics adversely. Let's hold that calculation (Section 6.10) until we look for further ideas for improvement.

6.8
Can Things still be Improved Further?

The problem with the ICI method for dealing with the ammonium bisulfate, and also with the Mitsubishi Gas Chemical process for recycling the HCN and eliminating ammonium bisulfate entirely, is that both retain the original cyanohydrin approach. The win-

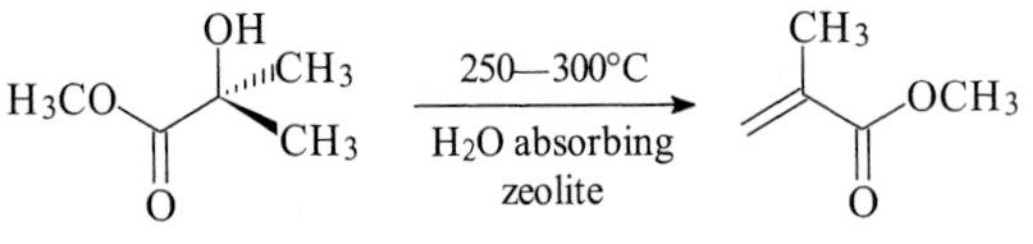

Fig. 6.9 Loss of water from a-hydroxymethylbutyrate.

ner in this competition will have to find an entirely new source for the carbon atoms in methyl methacrylate that arise not from acetone and HCN but from molecules that are closer to those produced directly from the cracking of petroleum fractions. This is a way to be both economical and environmentally sensitive. Isobutene is an excellent candidate since it comes directly from the cracking of petroleum fractions or can be easily produced from isobutane, a product of steam cracking. This molecule has all four carbons as well as the double bond necessary for methacrylic acid. Therefore, all that would be necessary is to convert one of the methyl groups of isobutene into a carboxylic acid group and esterify with methanol to obtain methyl methacrylate.

Alternatively, a synthesis of methacrylic acid starting from ethylene or propylene would have the advantage of an economical starting material but would require finding the carbon source to add two carbons or one carbon respectively. As we shall see below, for ethylene these needed carbons arise from carbon monoxide and formaldehyde. For propylene, the extra carbon is found in carbon monoxide. Another approach is found in a minor product of steam cracking of petroleum, methyl acetylene where the additional carbon comes from carbon monoxide. In each of these syntheses the source of the ester group in methyl methacrylate is methanol.

Let's look in some detail at these various candidates for replacement of the classical HCN/acetone approach and the modifications of this process we have just studied (Sections 6.5 and 6.6). Let's begin with the apparently best approach, isobutene.

6.9
From Isobutene to Methyl Methacrylate: Mitsubishi Rayon versus Asahi

Surprising as it may seem, in order to understand the conversion necessary for isobutene to methacrylic acid we must return to the chemistry that blocked the conversion of propylene to polypropylene. This makes it necessary to introduce a subject to be covered in greater depth in Chapter 8.

In Chapter 2 (Section 2.3) we discovered that propylene could not be polymerized to polypropylene using free radical initiators, although such initiators worked very well for ethylene. The reason was the ease of breaking of a carbon-hydrogen bond in the methyl group of propylene arising from resonance stabilization of the allylic radical.

This resonance-enhanced reactivity of a methyl group next to a double bond, such as the methyl group of propylene, is not necessarily a disadvantage. It can be turned to advantage to achieve a desired reactivity important to industry – a method of adding considerable value to propylene and by analogy, to isobutene. Isobutene and propylene have identical relationships of a methyl group to a double bond.

Synthesis of methacrolein and further oxidation to methacrylic acid from isobutene is carried out using the same molybdenum, vanadium, and bismuth catalysts used for oxidation and ammoxidation of propylene described in some detail in Chapter 8 (Section 8.14). Only one more step is then required – formation of the methyl ester – to produce the target methyl methacrylate (Fig. 6.10). This is the Mitsubishi route. But watch out for innovation. Another Japanese company, Asahi, has found a catalyst, the properties of which are not being fully revealed, that will convert the intermediate isobutraldehyde di-

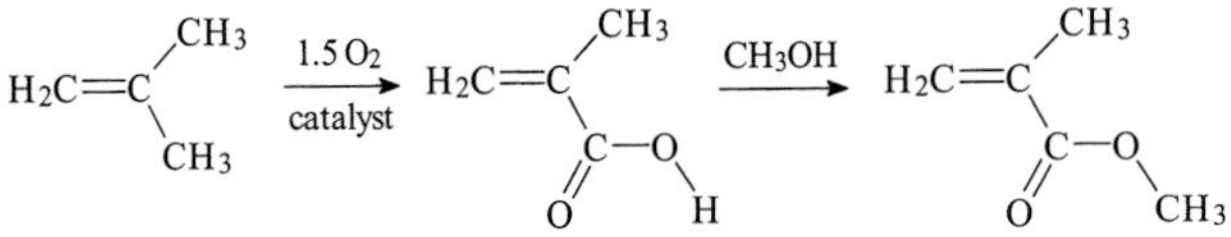

Fig. 6.10 The isobutene route to methyl methacrylate.

rectly to methyl methacrylate, saving one step and therefore considerable engineering costs. Plants have been built following both the Mitsubishi and Asahi routes, thus setting up an interesting competition. But whichever one wins, the necessary raw material will be isobutene. Therefore, only one thing could stand in the way of such ideal routes to methyl methacrylate – the availability of isobutene. Are there other important uses for this chemical? If so, this could raise the price of isobutene and therefore make the methyl methacrylate synthesis less competitive. As we will see, there could have been a problem, but it disappeared for an interesting reason associated with the interplay of environmental considerations and the octane rating of gasoline.

6.10
How Environmental Reasons stopped the Use of Tetraethyllead as an Octane Improver in Gasoline leading to its Replacement with Methyl *Tertiary* Butyl Ether (MTBE). But MTBE is synthesized from Isobutene, which *could have* blocked the Supply of Isobutene for Production of Methyl Methacrylate. But Environmental Concerns about MTBE have caused it to lose Favor as an Octane Improver in Gasoline therefore releasing Isobutene for Production of Methyl Methacrylate. A Story of the Ups and Downs of the Chemical Industry – What a Ride!

Tetraethyllead's ($(C_2H_5)_4Pb$) property of enhancing the octane number of gasoline was first discovered in the 1930s, and it has been an additive to gasoline since that time. This valuable property arises from the very weak bond between the lead and the four hydrocarbon moieties attached to it. The explosive combustion of gasoline, when sparked in the cylinder of the internal combustion engine, is a free radical chain reaction. The $CH_3CH_2^{\bullet}$ radicals formed by breaking the four lead-to-carbon bonds moderate this reaction to reduce "knocking" and therefore add smoothness to the explosive power and therefore value to the gasoline.

However, if the lead is left over in the engine cylinder after the four ethyl groups are broken off the lead would be oxidized to lead oxide with eventual fatal consequences to the life of the engine. Therefore, ethylene dibromide, $BrCH_2CH_2Br$, was added together with the tetraethyl lead so that lead bromide would be formed, which is volatile enough to be swept out of the engine with the exhaust. However, after many years it became apparent that the lead swept out with the exhaust as a byproduct of the valuable property of tetraethyllead was an environmental hazard. This is especially true for children, so the use of tetraethyllead had to be reduced or eliminated. The basic problem is the same one that has forced the elimination of lead-based paints.

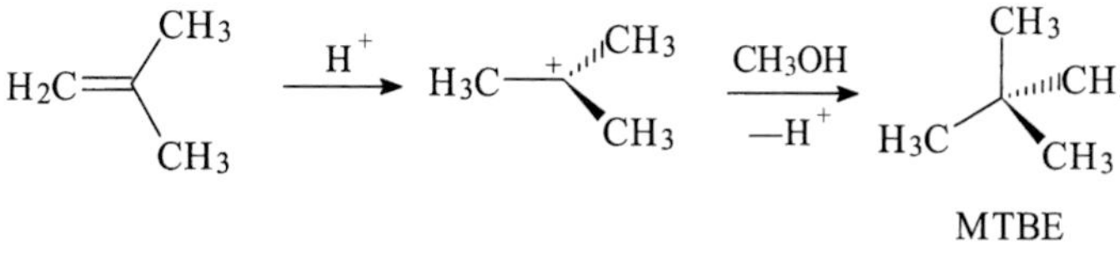

Fig. 6.11 Formation of MTBE from isobutene and methanol.

The discontinuation of tetraethyllead almost put a large American corporation, Ethyl Corporation, out of business and reduced the United States production of ethyl chloride by 80% between 1975 and 1991. But interestingly enough, Ethyl Corporation found a way to turn its organometallic chemical technology to a new use and became a prime supplier of the pain killer, ibuprofen. But that is another story.

But what does all this have to do with isobutene? Discontinuing use of tetraethyllead caused a search for replacement octane improvers. Among the possibilities were the so-called oxygenates. And among the best of these were sterically hindered ethers and partic-ularly methyl *tertiary* butyl ether, MTBE, which is widely used – as you can see by inspection of the list of components on almost all gasoline pumps in gasoline stations in the United States. As shown in Fig. 6.11, the best synthesis of MTBE is the reaction of isobutene with methanol under acid catalysis. Addition of a proton to isobutene forms the tertiary carboca-tion, which will readily react with methanol to form the ether. How simple.

But the production of huge amounts of MTBE consumed much of the world's supply of isobutene. The use of isobutene for production of methyl methacrylate has, in MTBE, a competitor. Any intermediate such as isobutene that is competitively used for gasoline (MTBE) and a chemical product (methyl methacrylate) will be used for the chemical, which adds more value. The four carbon atoms in isobutene will find themselves in a more expensive product per atom in methyl methacrylate and the derived Plexiglas[TM] than in an additive to gasoline. The chemical industry always takes precedence over the gasoline industry in the call on raw materials. However, the competition would raise the price of the isobutene therefore potentially making the isobutene route to methyl methac-rylate less competitive to other processes. This is not good news for the Mitsubishi Rayon or Asahi processes.

Now comes the twist in the story. The competition is turning out not to be an issue after all, since it has recently been discovered that the small water-solubility of MTBE has caused it to find its way into potable water derived from ground water sources. And this water solubility is combined with an unpleasant odor. When this is added to the fact that MTBE is accused of affecting human health – even though this is not proved – it be-comes clear that MTBE in the early 2000s may be on its way out, with great benefit to the use of isobutene for production of methyl methacrylate.

In fact, the use of isobutene for production of methyl methacrylate via the oxidation to methacrylic acid followed by formation of the methyl ester with methanol (Mitsubishi) (Fig. 6.10) or via the direct conversion of isobutraldehyde to the methyl methacrylate (Asahi), are both generally agreed (but see Section 6.11 below) in the chemical industry to be superior to the classical route. Several plants have been built. A story with an en-tirely different ending might have arisen if MTBE had not had environmental problems.

It is interesting to look at the economics of the two isobutene oxidation process to methyl methacrylate (Section 6.9) versus the best that can be done with the original HCN-acetone route, that is, the Mitsubishi route involving recycling of the HCN (see Fig. 6.7). In terms of raw material costs, the two isobutene processes, Mitsubishi Rayon and Asahi are nearly identical in the range of 12 cents per kilogram. But the advantage is in the engineering simplicity of the Asahi isobutene route, which saves the conversion to the carboxylic acid and the subsequent esterification (Fig. 6.10).

The original HCN-acetone route, on the contrary, with its large consumption of HCN (see Figures 6.4 and 6.5) is left far behind with raw material costs of 15 cents per kilogram and engineering that is far more complex than that for either isobutene route.

The future production of methyl methacrylate will depend on one of the isobutene routes, probably the Asahi process because of its fewer steps. The main reason that keeps the original HCN-acetone process going is that the plants are still in use and do not have to be built, and also that the old process has been optimized to high yield over the many years of its use. But as the plants wear out, this route will disappear as a source of methyl methacrylate production.

6.11
A Competitive Process for the Synthesis of Methyl Methacrylate based on Ethylene

There is no question that war, for all its horrors, drives innovation – and one of these innovations of war occurred in the hands of a German chemist, Otto Roelen who was working at Ruhrchemie at the time of World War II. Looking for ways to make carbon–carbon bonds from simple starting materials during those times of scarcity of raw materials, Roelen invented what is called the *oxo process* or *hydroformylation*. He found that carbon monoxide and hydrogen could be added to alkenes via catalysis with cobalt complexes to form aldehydes. We are going to look more closely at this important reaction in Chapter 9 (Section 9.9) and find out how it played a role in markedly decreasing acetaldehyde production by the chemical industry. However, our attention now is how in recent years the German chemical giant, BASF, has used the oxo reaction to produce methyl methacrylate (Fig. 6.12). The reaction of ethylene with carbon monoxide and hydrogen to produce propanal is the first step.

The step following propanal production in the BASF process (Fig. 6.12) is an aldol reaction, which we shall also look at in detail in a later chapter (see Chapter 9, Section 9.3). The example of the aldol reaction in Fig. 6.12 follows the classic pattern: The methylene carbon adjacent to the aldehyde group in propanal reacts with base to lose a proton and form the carbanion. This carbanion then attacks the carbonyl group of another aldehyde, formaldehyde. The product, containing an *ald*ehyde and an alco*hol* group, an aldol, then loses water to form an unsaturated aldehyde, methacrolein (Fig. 6.12).

Aldehydes are one oxidation level below carboxylic acids and can take this step up in oxidation level so rapidly in the presence of air that aldehydes are difficult to isolate in high yield without taking special precautions to avoid further oxidation. In the BASF procedure (Fig. 6.12) this ease of reaction is a favorable characteristic of aldehydes because oxidation is the desired path for producing methacrylic acid. Esterification with methanol produces the

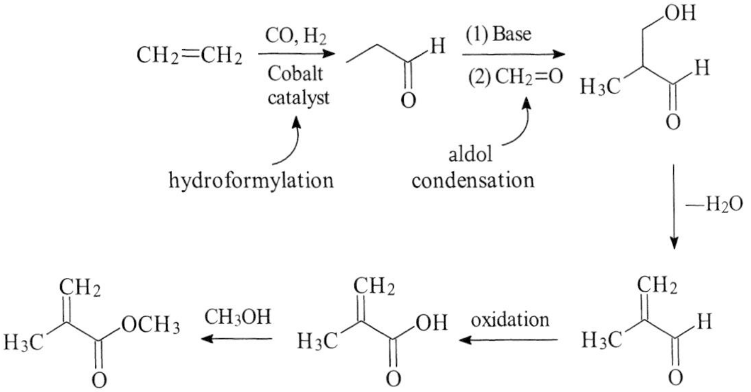

Fig. 6.12 BASF ethylene-based process for methyl methacrylate. An ethylene-based route to methyl methacrylate uses two important reactions: hydroformylation and the aldol condensation. Both reactions will be discussed in detail in Chapter 9.

target methyl methacrylate. The BASF process (Fig. 6.12) is competitive enough to be in commercial production and competes directly with the classical HCN/acetone route.

6.12
A Competitive Process for Synthesis of Methyl Methacrylate based on Propylene

Let's look at another approach for the synthesis of methyl methacrylate, which is based on a Markovnikov addition of carbon monoxide to the double bond of propylene. A representation of the bonding structure of carbon monoxide, CO, places a pair of electrons on carbon with molecular orbital theory supporting a triple bond between carbon and oxygen so that the carbon bears a negative charge. This is the source of the nucleophilic characteristics of carbon monoxide, accounting for strong complexation with transition metals such as cobalt (Fig. 6.12) and for that matter the iron in hemoglobin, which is then responsible for the deadly characteristics of carbon monoxide. The nucleophilic character of carbon monoxide is also the basis of a classic synthesis of carboxylic acids from olefins, the Koch reaction, which has been used for the synthesis of methyl methacrylate (Fig. 6.13).

Addition of HF to propylene produces a carbocation. Carbon monoxide, with the nucleophilic character described above, captures this carbocation as shown in Fig. 6.13, producing an acylium ion intermediate. Participation of water then yields isobutyric acid. In a step that finds analogy to the formation of styrene from ethylbenzene, the formation of a double bond that is driven by conjugation (see Chapter 3, Fig. 3.4). High temperature eliminates H_2 from isobutyric acid to yield methacrylic acid. Now, a simple esterification with methanol yields our target, methyl methacrylate (Fig. 6.13).

In principle, this process could be an excellent competitor. However, the reaction yield – especially for the first step – is not high enough to proceed to a commercial level. By-products are also formed. The chemistry is clever, but it is not clean enough.

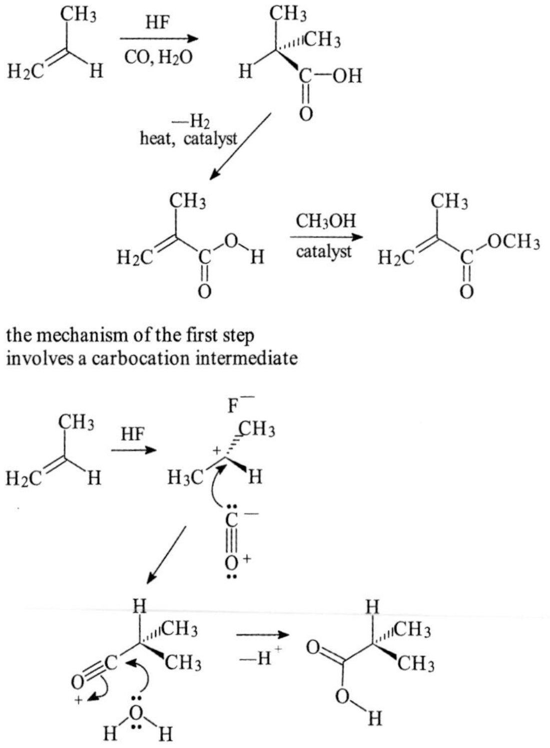

Fig. 6.13 Process for synthesis of methyl methacrylate using propylene, HF and carbon monoxide.

the mechanism of the first step
involves a carbocation intermediate

6.13
A Possible Commercial Synthesis of Methyl Methacrylate starting from Methylacetylene

Ineos Acrylics has developed a process for methyl methacrylate synthesis, which they bought from ICI, who had bought the process from Shell. A close-knit world, industrial chemistry! This process uses a palladium-catalyzed addition of carbon monoxide to methylacetylene. Palladium is an important catalytic transition metal, and we are going to study the mechanism of its action in detail in the Wacker process that produces acetaldehyde from ethylene (Chapter 9, Section 9.8).

In the process in focus here, the first step (Fig. 6.14) produces methacrylic acid directly so that the entire process takes place with only two steps counting methanol esterification to the target methyl methacrylate. Starting with methylacetylene rather than propylene avoids the high-temperature cracking necessary to eliminate hydrogen to obtain methacrylic acid from isobutyric acid (Fig. 6.13).

The difficulty in the methylacetylene approach is the starting material. Methylacetylene is only a minor product of steam cracking of petroleum fractions. To feed a world-scale methyl methacrylate plant it must be obtained from multiple steam crackers. If this messy supply problem can be overcome, the process may become an important commercial route to methyl methacrylate.

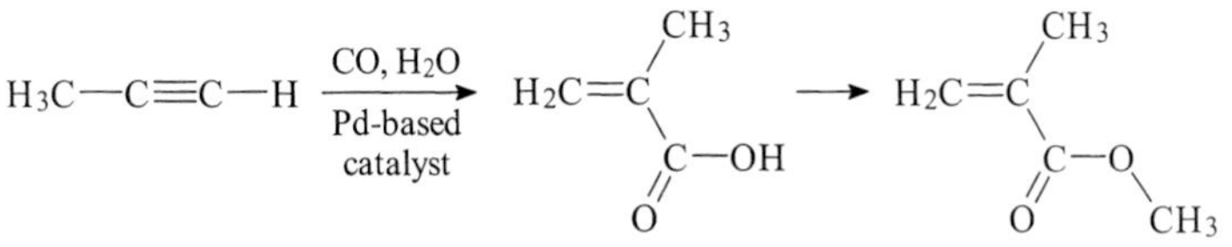

Fig. 6.14 Methyl acetylene approach to methyl methacrylate.

6.14
Summary

Poly(methyl methacrylate), one of the oldest and best-known plastics, is produced by free radical polymerization of methyl methacrylate, which therefore must be produced by the chemical industry in very large amounts. The long-used manufacture of this monomer is, however, based on a flawed process, which uses the exceptionally dangerous chemical HCN, and produces very large amounts of a useless byproduct, ammonium bisulfate. In this classical process, acetone and HCN react together to form a cyanohydrin, which undergoes chemical transformations converting the nitrile group to the carboxylic acid function and the hydroxyl group to the double bond of methacrylic acid. Esterification with methanol produces the target monomer, which then is polymerized to the commercial materials we know so well. We learn why this outdated process continues to be used: The plants for the classic process are already built meaning that any new process must bear the costs of engineering and building new plants.

As time passes, however, new synthetic ideas, which can overcome the economic handicap are put on the table to try to replace this aging method. First, they offer ways to deal with ammonium bisulfate byproduct by converting it to nitrogen and sulfuric acid. Then, new ideas find a way to use the cyanohydrin intermediate in a manner that does not form ammonium bisulfate at all and recycles the nitrogen by recycling HCN. That comes closer, but cannot compete with the apparently best innovation of all – oxidation of isobutene directly to methacrylic acid.

Oxidation of isobutene is possible because of the weak carbon–hydrogen bonds of the allylic methyl groups, a consequence of the resonance stabilization of the intermediates formed on breaking these bonds. This connects us to the story of the conversion of propylene to value-added chemicals derived by oxidation of its methyl group (a subject to be studied in Chapter 8), and to the inability to polymerize propylene via free radical methods (a subject studied in Chapter 2).

At the time of this writing, the Japan-based isobutene approach to methyl methacrylate appears to hold the greatest promise, and several plants are in operation, though this promise can only come to fruition if enough isobutene is available. To evaluate this aspect we have to take a side trip into the problems of replacing tetraethyllead in gasoline with methyl tertiary butyl ether (MTBE) – an innovation that would require large amounts of isobutene in competition with its use for methyl methacrylate manufacture. But the suspected environmental hazard of using MTBE has dashed prospects for its use and promises to release the world production of isobutene for methyl methacrylate. There will be more than enough, because vastly more isobutene is needed for MTBE than for methyl methacrylate.

But the competition is not settled. In a transition metal-catalyzed approach, ethylene can be converted in multiple steps to methyl methacrylate via initial addition of carbon monoxide to the double bond. This approach uses the cobalt-based hydroformylation reaction (oxo chemistry) developed by BASF in Germany, followed by an aldol reaction with formaldehyde to gain the four carbons in methylacrylic acid – one step via esterification with methanol from the target. This process is now in commercial production.

Propylene has also entered the competition in a process that uses a classic reaction to produce carboxylic acids from olefins, the Koch reaction. Propylene and HF produce the carbocation, which adds carbon monoxide and water to yield isobutyric acid – two steps from the target methyl methacrylate. This original approach, however, appears to be losing out because of low yields in the first step.

However, another three-carbon starting material derived in small amounts from steam cracking of petroleum fractions, methylacetylene, has entered the competition. Here, Ineos Acrylics is involved with catalysts based on palladium that allow the addition of carbon monoxide and water to form methacrylic acid in one step, and therefore only an esterification with methanol is needed to produce the target monomer. But methylacetylene is hardly available, requiring many suppliers in order to provide the volume needed for a world-class methyl methacrylate plant. If this problem can be overcome, the process appears to be an excellent candidate.

And given the ingenuity of the chemical industry there may be more to come.

Some of the subjects treated in this chapter are listed below. These are key words and terms that act as reminders of the chapter's contents and should become a valuable part of your chemical vocabulary.

- Cyanohydrins
- Methyl methacrylate
- Poly(methyl methacrylate), PMMA
- Carboxylic acids from cyano groups (nitriles)
- Acid-catalyzed dehydration
- Acetone reactivity
- HCN
- Nucleophilic substitution at carbonyl carbon
- Leaving groups
- Nucleophilic addition at carbonyl carbon
- Amides
- Esters
- Ammonium bisulfate
- Ester–amide exchange chemistry
- Resonance-enhanced reactivity of methyl groups adjacent to double bonds
- Oxidation chemistry of propylene
- Oxidation chemistry of isobutene
- MTBE
- Gasoline combustion
- Tetraethyllead
- Oxo chemistry

- Aldol chemistry
- Carbon monoxide as a nucleophile adding to carbocations
- Plexiglas™ and its history
- World War II and the battle of Stalingrad

Study Guide Problems for Chapter 6

1. Methyl methacrylate is polymerized successfully to poly(methyl methacrylate) using free radical initiation. Offer a reason why the allylic methyl group in propylene blocks formation of poly(propylene) (Chapter 2, Section 2.3) while the similarly situated methyl group in methyl methacrylate allows a high polymer to form.

2. The transformation from the cyanohydrin in Fig. 6.4 to the alkene intermediate begins with proton addition to the oxygen of the hydroxyl group. Try to continue from this beginning to trace the mechanistic path with the hint that the oxygen of the amide group might be derived from a molecule of water that was lost.

3. Amides are resistant to acyl attack by nucleophiles, in contrast to many other derivatives of carboxylic acids such as acid chlorides, acid anhydrides, and esters. In addition, amides tend to be planar, a factor that plays a key role in the conformations of proteins such as silk (see Chapter 5). Offer an explanation that connects the reactivity resistance of amides to the planar characteristics of this functional group.

4. Heating phenyl acetate with an aliphatic alcohol such as methanol, catalyzed by acid in a sealed container, will yield predominantly phenol and methyl acetate. Heating methyl acetate with phenol in a sealed container will yield the same mixture as the previous experiment. Conducting this experiment in an open container will increase substantially the yield of phenyl acetate. Offer a mechanistic explanation for these experimental results.

5. A new approach for the synthesis of polycarbonate that avoids use of phosgene involves the catalyzed reaction of phenyl carbonate with bisphenol A under high temperature. Reduced pressure would help the polymerization. Show how these reactants can yield polycarbonate. What kind of reaction mechanism is involved?

6. Using the principle of loss of the conjugate base of the strongest acid as the best leaving group from the four-coordinate intermediate arising from nucleophilic attack at acyl carbonyl, evaluate and criticize the mechanism in Fig. 6.8. Why does the reaction proceed to produce the desired ester?

7. Hydroperoxides, such as the intermediate in the formation of phenol and acetone in Chapter 3 (Section 3.11) can be reduced to alcohols. Propose a synthesis of styrene involving an intermediate hydroperoxide.

8. The German chemist, F.A. Paneth carried out experiments in the 1920s in which tetramethyllead was heated in a tube with a flowing inert gas to carry whatever was formed toward a film of lead further down the tube. The lead film disappeared and was converted to tetramethyllead. Offer an explanation for the results of this experiment.

9. Addition of tetraethyllead to gasoline served for many years as an octane improver – that is, to reduce engine knocking. Knocking is thought to arise from reactive radicals forming too quickly and in high concentration during combustion. Offer an explana-

tion of how tetraethyllead works. Is this related to your answer to Question 8? Does your answer have any relationship to the three characteristic steps in a free radical chain reaction?

10. The aldol reaction of propanal and formaldehyde in Fig. 6.12 is not the only aldol reaction possible from the reaction mixture. What other possible aldol reaction could occur?

11. One of the important uses of acetaldehyde in the chemical industry is conversion to *n*-butanol. Outline a synthesis of this alcohol using the aldol condensation of acetaldehyde as one of the steps.

12. Loss of a proton from a carboxylic acid and loss of a proton from the *α*-carbon of an aldehyde produce a negative charge in identical relationships to a carbonyl group. Yet one is stabilized more by resonance than the other. Offer an explanation for the relative stability.

13. What are the common mechanistic elements of addition of cyanide anion to an aldehyde and the aldol reaction of the same aldehyde?

14. In the process for production of methyl methacrylate in Fig. 6.13, the first step involves reaction of HF with propylene to produce the carbocation shown. Although reaction of this carbocation with water and carbon monoxide leads to the desired product, the process is bedeviled by byproducts. These byproducts can be controlled to some extent with the propylene pressure. Why?

15. What do the terms Markovnikov and anti-Markovnikov have to do with byproduct formation in the route to methyl methacrylate shown in Fig. 6.14?

7
Natural Rubber and Other Elastomers

7.1
Introduction to Rubber

The essential feature of a rubbery substance is that it can be stretched several fold and then return to its original shape when the stretching force is withdrawn. It took a very long time after the first observation of this elastic property to understand the fundamental basis of this behavior.

Columbus, in his first voyage to the Caribbean, reported seeing native people playing with elastic balls, and it is reported that a group of Aztecs were taken to Spain to demonstrate the game. Later, a French explorer actually observed the tapping of rubber trees, the source of this mysterious substance. But it took more than two centuries after these observations, and not until physical chemistry became a rigid discipline, to gain a fundamental understanding of rubber-like elasticity. And this understanding explained how the organic chemical structure of elastomeric substances, of which natural rubber is one, accounts for this property.

The first clues come from physical observations nearly 200 years ago that hinted at the fundamental forces at work. If rubber is stretched, heat is released. Try it yourself. Take a thick rubber band or piece of rubber tubing and touch it to your lips, stretching it rapidly and you will feel heat. Why should heat be released? What we see here is one of the beautiful aspects of polymer chemistry. What is observed macroscopically reflects what is happening at the molecular level. When the rubber is stretched, the polymer chain molecules that make up this material are stretched and, as we shall see, it is this stretching of the polymer chains that accounts for the release of heat.

This characteristic of rubber is of great practical importance. When a tire on an automobile rotates, the rubber in contact with the pavement flattens out, it stretches, and heat is released. When this part of the tire then leaves the road, the rubber contracts and cools, but the process is not perfectly reversible, and so the tire becomes hot. All tire manufactures have to take this heat release into account because it affects the lifetime and failure potential of the tires. With only one exception, natural rubber is superior to synthetic rubber in this regard, and for this reason airplane tires are nearly 100% natural rubber. When an airplane hits the tarmac, the tires generate a tremendous amount of heat, the dissipation of which is vitally important to the tires' performance.

7.2
Why are Some Materials Rubbery?

As with many polymers, the polymer chains in natural rubber are very long and exist in the relaxed state in a randomly coiled shape. The best way to imagine a random coil is to think of tracing a random path in three dimensions. This will be a very disorderly array, which has often been compared to a plate of cooked spaghetti. When the rubber is stretched, the individual polymer chains are forced to change to a much more extended shape. For a polymer to be an elastomer it must have a structure that allows facile and reversible change of its shape. But for heat to be released on stretching the elastomer, the extension of the polymer must take place with little change in energy.

If the extended state and the relaxed state of the polymers are energetically similar, thermodynamics teaches us that the work expended in the stretching process must cause the temperature to rise. But what kind of chemical structure allows a polymer to be extended without much change in energy? The answer to this question must reside in what happens to the conformations about the numerous bonds in a polymer when it changes its shape.

7.3
The Conformational Basis of Elasticity

If a polymer chain changes its overall shape, this must mean that the conformations of some of the thousands of bonds that make up a long-chain polymer must also change. But we have just learned above that the coiled and extended state of the polymer must have nearly the same energy. This energy equivalence allows for two possibilities: (1) The local conformations about these bonds in the relaxed and extended state cannot differ very much in energy; and (2) if there is a significant energy difference between the local conformations about these bonds in the relaxed and extended states, then very few of these local conformations would have to change to make a big difference on the dimension of the whole chain. In many elastomers, both conditions, (1) and (2), are met. The energy differences between the conformations are not large, and very few conformational changes cause large changes in the overall shape of the chain.

Let's consider a polymer with the kinds of single bonds we are familiar with from the study of simple molecules such as ethane or butane, that is, *anti* and *gauche* conformations. Imagine a polymer with many linkages that are subject to similar conformations as, for example, in the carbon–carbon single bonds of the commercially important elastomer, ethylene propylene rubber. Here, one has numerous numbers of linked $-CH_2-$ groups, as shown in Fig. 7.1. This figure graphically demonstrates that a large portion of the chain can go between an extended and a coiled state by changing relatively few of these bonds between *anti* and *gauche* conformations. Therefore, extending the polymer chains – which must take place when the elastomer is stretched – does not cost a great deal of energy.

But there is a thermodynamic price for stretching the polymer chains. The tangle of chains in the relaxed state of a rubber is very disorderly, and at the same time there is the disorder of the many types of conformations along the backbone of the chain. This disorder is greatly reduced when the chains are stretched. The stretched state of rubber is more orderly than the coiled or relaxed state since the chains are organized, in a general way, in the same direction, the direction of the stretching force. This is shown in Fig. 7.2, which gives a representation of a rubbery polymer. In thermodynamic terms, the stretched state is therefore of lower entropy than the coiled or relaxed state. As we shall see below, understanding the entropy change associated with stretching rubber provides understanding of one of the longstanding mysteries about rubber elasticity.

We noted above that one of the characteristics of rubber is that heat is released when it is stretched. The work expended in the stretching process does not raise the energy of the rubber but rather appears as heat. The other side of the coin of this observation, discovered nearly 200 years ago, was that the restoring force to return the stretched state to the relaxed state was greater at higher temperatures. In other words, natural rubber became stiffer as temperature increases. This was astonishing to these early investigators of rubber elasticity and still seems amazing today. We are used to materials getting softer on heating, not stiffer. Is this not your normal experience? Why do elastomers behave in a way that is counter to our intuition?

When the polymer chain is extended, the number of different conformations along the chain from one end to the other decreases. Fewer different kinds of conformations correspond to a higher order, less disorder, and therefore to a decrease in entropy. Looking at the situation from the other direction, when the chain returns to its coiled state there is an increase in the number of different kinds of conformations and therefore an increase

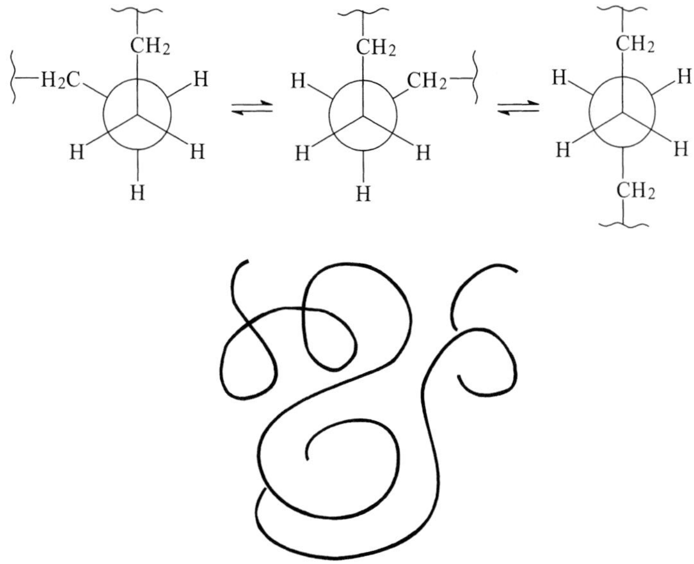

Fig. 7.1 Ethylene–propylene elastomer chain changes between coil and extended states via *gauche* and *anti* conformational change around only a few bonds.

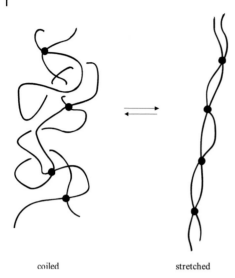

Fig. 7.2 Expanded and relaxed state of a cross-linked rubber.

coiled   stretched

in entropy. Since the enthalpy, ΔH, of the chain changes little between the extended and contracted states, as we have pointed out above the change in free energy, ΔG = ΔH–TΔS, depends almost entirely on the –TΔS term. Since stretching the rubber decreases entropy, ΔS is negative and the –T(–ΔS) term is therefore positive, making ΔG positive and therefore disfavoring the stretching process. Increasing temperature will increase the magnitude of the –T(–ΔS) term even further, therefore further favoring the unstretched state. The rubber will become stiffer as is experimentally observed. How beautiful!

7.4
How does the Structure of Natural Rubber fit into the Theoretical Picture of Elasticity drawn above?

The polymer shown in Fig. 7.3 is natural rubber derived from the tree *Hevea brasiliensis*, and it has a distinguishing feature in the *cis* stereochemistry of the double bond in each unit. This is a critical feature, since a polymer made of the identical units – called *gutta percha* – can also be found in nature with a *trans* arrangement about each of the same double bonds (Fig. 7.4). *Gutta percha* is the opposite of an elastomer, and in fact an early use in 1850 was to coat with it a copper wire cable under the English Channel between Dover and Calais. This application was the first underwater telegraph connection. The *gutta percha*, also extracted from trees, as is natural rubber, could be heated to a state in which it could be formed into a coating. But on cooling it hardened into a substance that was so impenetrable that it protected the copper cable from the corrosive effect of salt water, a property that natural rubber was incapable of achieving.

Why should *Hevea* rubber and *gutta percha* have such enormously different properties? The chains are stereoisomers, in this case diastereomers, so they should be different. But it is very difficult to understand the magnitude of this difference based simply on the *cis*

Fig. 7.3 Structure of a portion of a chain of natural rubber derived from the *Hevea brasiliensis* tree.

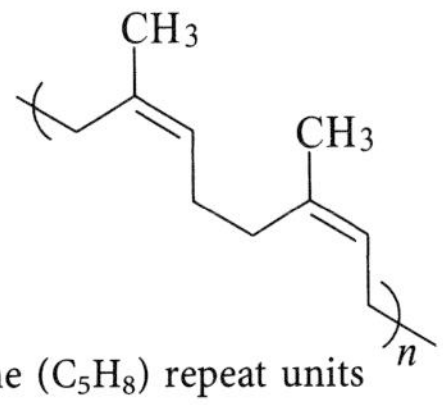

two isoprene (C_5H_8) repeat units

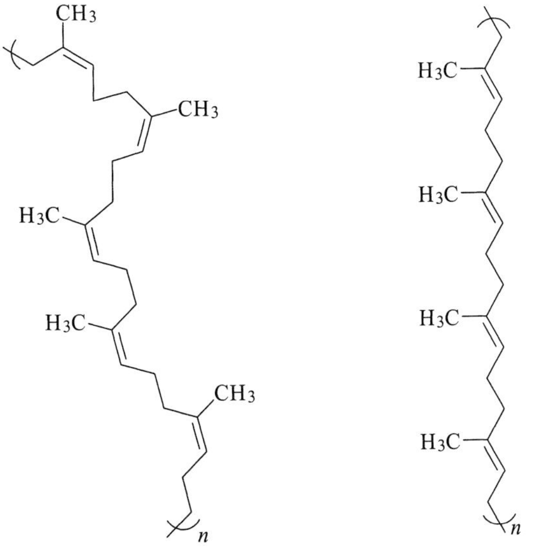

Fig. 7.4 How the *cis* and *trans* double bonds in *Hevea* rubber and *gutta percha* affect chains approaching each other and, therefore, crystallinity.

Hevea rubber

Gutta percha

versus *trans* configurations about the double bond. The answer is not reasonably found in the properties of single chains. We must examine the relationships among chains to find the answer, and in doing so we encounter a property that is able to preclude elastomeric characteristics.

If a polymer chain is capable of a regular structure, the chains may approach each other closely. Portions of the chains may therefore associate and form crystalline regions. Although only very small crystals are formed in polymers with only relatively small portions of the chains participating, these crystalline regions are hard and inflexible. This is true for all crystals whether they are sodium chloride or the small crystals formed by *gutta percha*. However, the necessary chain-chain approach for crystallization is impeded by the *cis* double bonds in *Hevea* rubber.

We have studied crosslinks arising from covalent bonds in Chapter 4 and their effect on epoxy resins and polyurethanes and the fleeting crosslinks arising from hydrogen bonding in water and glycerol. The crystalline regions in *gutta percha* act as crosslinks for an entirely different reason. The crosslinks arise from crystallization involving participation of different chains. Unless the temperature reaches a level that melts these crystalline regions, allowing the chains to regain individual freedom of motion, the crosslinks are as permanent as those from covalent bonds. In *gutta percha*, the density of such physical crosslinks, which act to restrict chain motion greatly, is so high that the necessary flexibility for elastomeric character is lost. But why should *gutta percha* crystallize so much easier than *Hevea* rubber?

The answer is that the *trans* double bond in *gutta percha* provides a chain structure that allows the chains to approach each other closely and to form crystals. In contrast, the bend necessitated by the *cis* double bond in *Hevea* rubber interferes with the chains achieving the molecular nearness necessary to form the crystalline regions. The unstretched state of *Hevea* rubber therefore is entirely free of crystallization.

But on stretching *Hevea* rubber the chains are extended, and the number of conformations is greatly reduced. This more orderly arrangement of the stretched chains should then be more consistent with crystal formation and in fact experimental observations show that highly extended *Hevea* rubber does crystallize to some extent. Scientists who work on rubber elasticity are greatly interested in what is called strain-induced crystallization, but detailed discussion is beyond the scope of this book.

7.5
Let's take a Short Diversion from Elastomers

The difference between *Hevea* rubber and *gutta percha* has a parallel in the nature of the fatty acids in cell membranes. It has been observed that cold-blooded animals have a far higher proportion of unsaturated fatty acids than warm-blooded animals, and that the double bonds of the unsaturated fatty acids are predominantly *cis*. The *cis* double bonds of the fatty acid chains, as in *Hevea* rubber, do not allow close approach of the fatty acid chains and therefore give more flexibility to the cell membranes at lower temperatures (Fig. 7.5). This is necessary for cold-blooded animals. The saturated fatty acids allow close chain packing and thus require higher temperatures to achieve the fluidity necessary for them to function in a membrane, a property which is consistent with warm-blooded animals. Moreover, as far as shape is concerned, a *trans* double bond is similar to the *anti* conformation about a carbon–carbon single bond, and in that way we see *gutta percha* as analogous to a fully saturated chain (Fig. 7.5).

From this insight into the difference between *Hevea* rubber and *gutta percha* compounds produced by different kinds of trees, we discover that the intermolecular relationships between polymer chains – that is, molecular nearness – can play a critical role in

portion of the chain of a

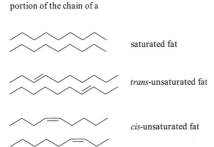

saturated fat

trans-unsaturated fat

cis-unsaturated fat

Fig. 7.5 Structures of unsaturated and saturated portions of fatty acids and relationship of unsaturation to packing.

preventing elastomeric characteristics. Not only must the chains be inherently flexible but also the interactions among different polymer chains must not act to restrict that flexibility. Later in the chapter (Sections 7.11–7.1●) we shall see that industry uses this interaction among different chains to invent new kinds of elastomers by controlling and limiting the chain–chain interactions and using these as the limited number of cross-links necessary for elastomers to be commercially useful.

7.6
Elastomers require Essentially Complete Recoverability from the Stretched State.
The Story of Vulcanization and How Sulfur supplies this Characteristic to *Hevea* Rubber

We have learned one necessary condition for elastomeric behavior – a flexible, mobile polymer with a high deformability, a polymer in which the local bonds have different accessible conformational states that can be interconverted rapidly. But this is not sufficient. Rubber-like elasticity requires complete recovery from the stretched state. This means that the force causing the stretching cannot cause a permanent deformation. Therefore, the polymer chains must not be able to slip by each other. They must be anchored, and this requires that the chains be crosslinked, but not too crosslinked, not anywhere as crosslinked as in the epoxy and polyurethane resins discussed in Chapter 4.

Perhaps you were curious about the black dots in Fig. 7.2? These represent the necessary crosslinks, and in an elastomer they occur about once in a hundred units along the main flexible chain. As we will see shortly, this is what vulcanization of rubber is about. For a contrasting situation, make a connection in your mind to the crosslinks in cured epoxy resin discussed in Chapter 4, where the large number of crosslinks greatly restricts flexibility. Added to this is the stiffness of the main chain contributed by the aromatic character of the bisphenol A, which confers rigidity to the epoxy resin. Thus, we have the antithesis of an elastomer, which requires highly flexible long polymer chains with a flexible network structure.

Before there was any understanding of rubber elasticity, even before the concept of a polymer was understood, there were a series of events that led to rubber turning from a curious material with little practical value to a commodity of great economic importance. This event involved the discovery of how to crosslink rubber with sulfur.

How in the world did anyone conceive of a relationship between natural rubber and sulfur? The story is well told by Morawetz. In the 1830s, natural rubber – known as gum elastic – had many deficiencies. It became very hard in cold weather, but sticky and soft in warm weather. The superintendent at a plant in Massachusetts dealing with rubber, Nathaniel Hayward, had first exposed rubber-coated fabrics to sulfur and sunlight with the idea that the sulfur would act as bleach and would also remove the stickiness of rubber. Indeed, it did remove the stickiness, but it only affected the surface of the rubber. Charles Goodyear, who became aware of this dusting process on meeting Hayward, made a chance observation that exposure of the sulfur-treated rubber to the high heat of a stove led to the kind of charring encountered in leather rather than the usual melting seen on heating rubber. This was the initial observation that led to the development called vulcanization, which provided the elastic character of rubber we know today. Somehow the necessary crosslinks were formed.

One could argue that the term vulcanization, named after the Roman god of destructive fire, as in the fire from volcanoes, although appropriate in reference to the heat used in the initial discovery, is inappropriate in other respects. Vulcanization of rubber, after all, leads to good results.

7.7
What happens When Sulfur and Natural Rubber are mixed and heated?

We should begin by admitting that the precise mechanism of vulcanization of rubber is not fully understood. There are several types of reactions proceeding at the same time when elemental sulfur is heated with natural rubber, but the series of reactions shown in the following discussion are widely accepted as important and have pedagogic value. In commercial processes accelerators are often used leading to complex chemical reactions, which we will not describe here but which are reviewed by odian.

Elemental sulfur is a highly reactive substance, which can be linear and may also exist as rings. Under conditions of treating rubber, sulfur substantially exists as eight-membered rings. This form of sulfur has solubility properties that cause it to mix easily with natural rubber meaning that the sulfur molecules can come close to the double bonds along the chain. Under these conditions a familiar reaction can take place.

Sulfur–sulfur bonds are weak and easily cleaved. This identical statement can be made about other weak elemental bonds such as those in chlorine and bromine, which undergo 1,2-addition to double bonds via a polar mechanism. We could therefore expect the same kind of reactivity for sulfur. But there is something special about polar addition of sulfur to a double bond. The anion formed is remote but still attached to the positive end that has reacted with the double bond as shown in Fig. 7.6. This means that the negatively charged sulfur produced by the addition as shown in Fig. 7.6 has to close a ten-membered ring in order to add to the other side of the double bond.

However, the formation of large rings such as a ten-membered ring is improbable. We have come across this before in the discussion of why the monomer for nylon 11 does not cyclize (see Chapter 5, Section 5.10). With ten atoms intervening to close the ring necessary completion of the 1,2-addition to the double bond following the first step shown in Fig. 7.6 is not likely. This is expressed by a small pre-exponential factor, A, in the Arrhenius equation, $k = Ae^{-E/RT}$ for the ring closure. The small A term reflects the entropy of activation, which is unfavorable for closing such a large ring. The rate constant for this reaction, k, is reduced in spite of the fact that the energy of activation, E, may be low.

The discussion of the difficulty of closing large rings is related to the reaction of sulfur with the double bonds in rubber. But ironically, it happens that this difficulty of closing the large ring shown in Fig. 7.6 is precisely responsible for the success of vulcanization and therefore essential to transforming natural rubber to one of the most important materials of the twentieth century. How is this true?

The addition of sulfur to a double bond in natural rubber produces a pair of intermediates in which the resulting positive charge is traded between the sulfur and the tertiary

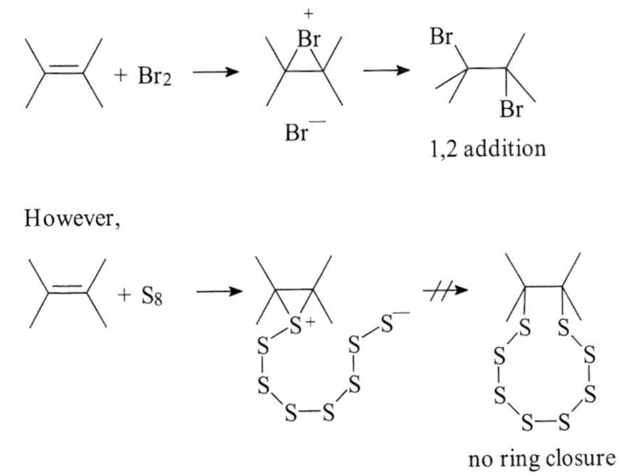

Fig. 7.6 The contrast between the addition of Br$_2$ versus S$_8$ to a double bond.

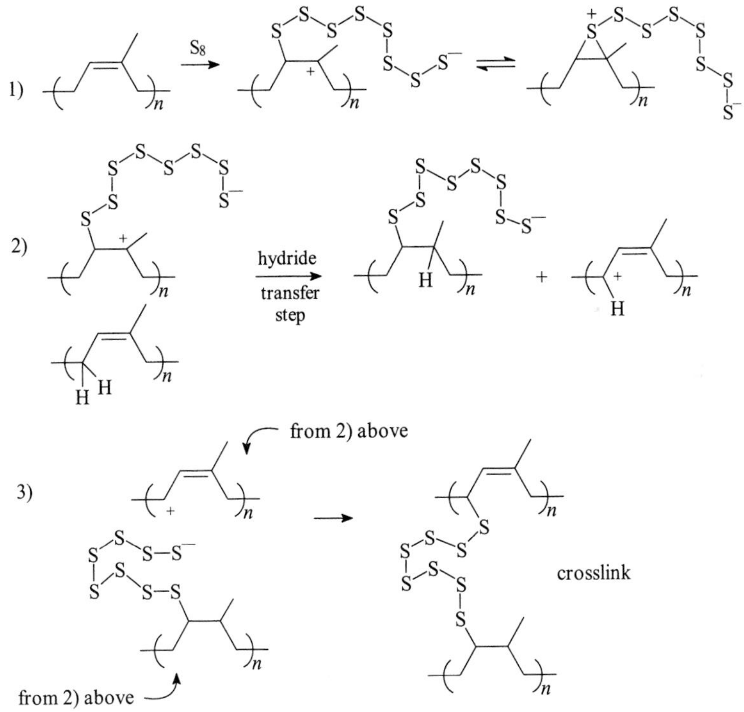

Fig. 7.7 Initiation and termination steps in the vulcanization of natural rubber.

carbon (step 1 in Fig. 7.7). It is the stability of the tertiary carbocation that drives the opening of the charged three-membered ring containing two carbons and one sulfur atom. In the coiled mass of the rubber polymers at the elevated temperatures of vulcani-

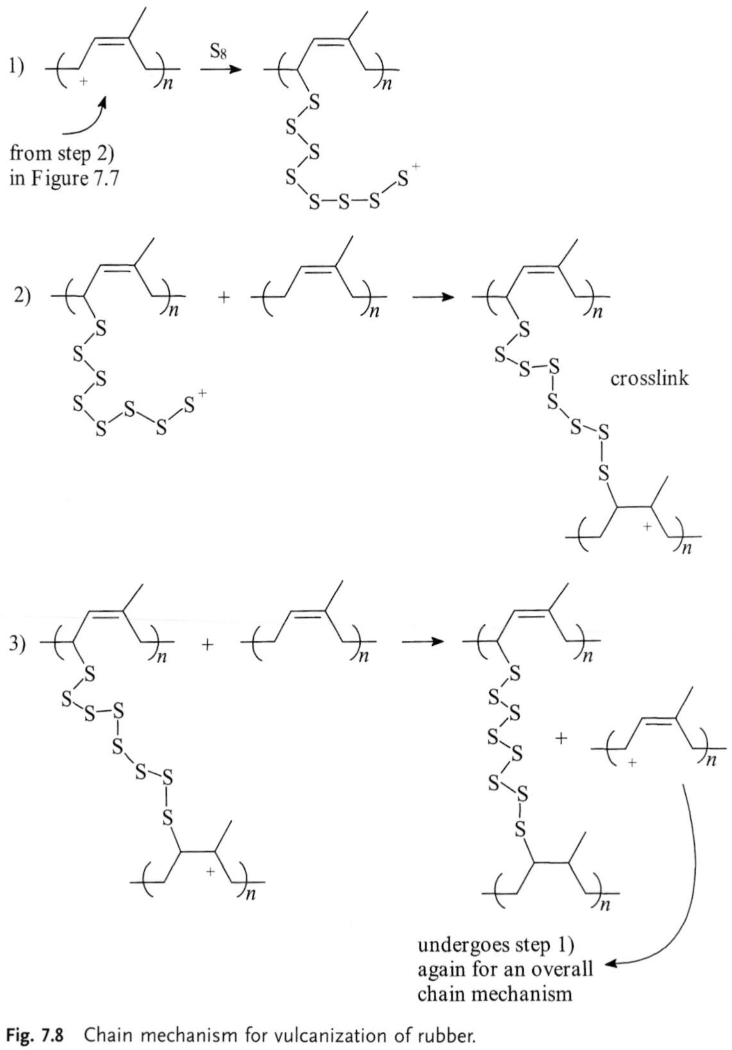

Fig. 7.8 Chain mechanism for vulcanization of rubber.

zation, other polymer chains come close to the polymer chain to which sulfur has added. This proximity allows a hydride to be transferred from a methylene group adjacent to the double bond in the proximate polymer chain to the carbocation (step 2 in Fig. 7.7). There is an analogy here to the hydride transfers occurring in the carbocations present in the catalytic cracking of petroleum fractions (see Chapter 1, Fig. 1.12).

In Fig. 7.7, we see that the hydride transfer in step 2 produces an even more stable carbocation, which is both tertiary and resonance-stabilized. This transfer step in Fig. 7.7 can lead to a crosslink, as shown in step 3 in Fig. 7.7, but step 2 can also initiate a chain mechanism leading to crosslinking (Fig. 7.8) as discussed below.

Until now, the only chain mechanisms we have described have involved free radicals such as, for one example, the steam cracking of petroleum fractions (see Chapter 1, Section 1.11). But in vulcanization of rubber we encounter a chain mechanism involving charged entities. Consider the polymeric carbocation produced in Fig. 7.7 by the hydride transfer, step 2. This carbocation could act to open one of the eight-membered sulfur rings, forming a single bond between sulfur and the polymer chain and a positively charged sulfur as shown in step 1 of Fig. 7.8. Reaction of the positively charged sulfur intermediate formed in step 1 of Fig. 7.8 with another rubber polymer chain then forms a crosslink and generates a new carbocation as shown in step 2 of Fig. 7.8. The approach of another rubber polymer chain then allows a hydride transfer (as shown in step 3 of Fig. 7.8) that neutralizes the crosslinked entity that was formed in step 2 of Fig. 7.8 and also forms a resonance-stabilized carbocation, which is the carbocation intermediate that started the process (step 1 in Fig. 7.8). And so in this manner we repeat the process again and again – a chain mechanism – which will be terminated occasionally by step 3 in Fig. 7.7. But every time sulfur adds to a rubber polymer double bond (step 1 in Fig. 7.7), a new chain (Fig. 7.8) starts again.

Notice that in the mechanistic steps outlined in Figs 7.7 and 7.8, the crosslinked rubber polymer can form in two different ways – one as a step that terminates the chain mechanism (step 3 in Fig. 7.7) and one that propagates the chain mechanism (steps 2 and 3 in Fig. 7.8).

If a great deal of sulfur is used, in the range of 30–50%, then a hard rubber results that is hardly elastic at all. This product was formerly used to make battery cases, which are now made from easily moldable and less expensive polypropylene. The use of about 5% of sulfur for vulcanization of natural rubber yields the elastic rubber necessary for tires. The difference is in the number of crosslinks. There must be enough crosslinks to prevent the polymer chains from slipping by each other so that the stretched state can recover to the relaxed state, but not so many as to retard the flexibility of the polymer chains. This is understandable based on the discussion in this chapter of the physical basis for elastic behavior (see Section 7.00 ◉).

7.8
Hypalon: an Elastomer without Double Bonds that can be Crosslinked

Polyethylene has a highly flexible chain backbone, which arises from the rapid interconversion of the *trans* and *gauche* conformational states accessible to the single bonds along the chain backbone (see Fig. 7.1). However, the crystalline character of polyethylene interferes with the inherent flexible character of the individual chains in analogy to the same phenomenon in *gutta percha* (Fig. 7.4), and therefore blocks elastomeric behavior. Why attempt to develop an elastomer based on polyethylene?

The reason is that the absence of double bonds gives polyethylene a great advantage as an elastomer. Double bonds are reactive, and in particular they react with the ozone that is found in the air in many cities around the world. The reaction of ozone with double bonds is a classical method used in synthesis to cleave double bonds and form carbonyl compounds, which means that if the double bond is in a polymer chain, ozone will act

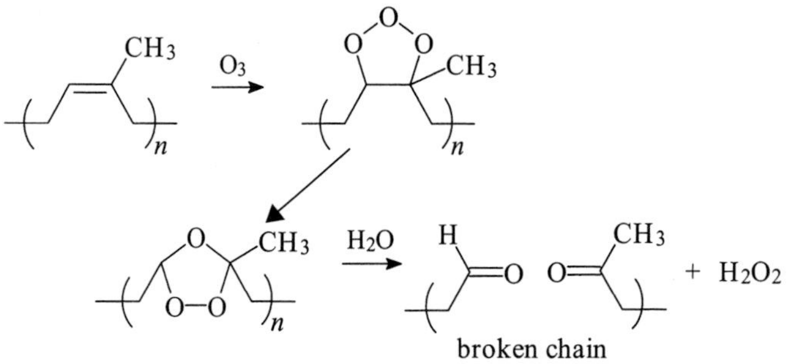

Fig. 7.9 Ozone forms an ozonide from the double bonds in natural rubber causing chain cleavage.

to break the chain into smaller segments. Such a process severely weakens the properties of any elastomer and is a well-known problem for vulcanized natural rubber used for automobile tires (Fig. 7.9).

Any resistance to ozone is valued in polymers used outdoors, and although in a saturated polymer such as polyethylene the resistance is found in the absence of double bonds, we have seen ozone resistance arise in a different manner from the presence of a chlorine-substituted double bond in neoprene (see Chapter 5, Section 5.6).

If one could reduce the crystallinity of the polyethylene and at the same time put in the required crosslinks for elastomeric properties without the presence of unsaturation in the chain, a commercially important elastomeric material could be formed with a prized property, ozone resistance. DuPont has in fact accomplished this in an interesting manner in an industrial product called Hypalon.

To start with polymers where crystallization is minimal, the DuPont chemists developed the product at first from LDPE (see Chapter 2, Section 2.5), which has the lowest level of crystallinity of all the polyethylenes. Nevertheless, the identical chemistry described below can also be applied to HDPE and to LLDPE (Chapter 2, Section 2.5). Let's see how Hypalon is made.

To overcome the absence of a crosslinkable functional group, polyethylene was reacted with a mixture of sulfur dioxide and chlorine gases, leading to substitution of carbon–hydrogen bonds along the chain with the sulfonyl chloride group.

In the first step, heating breaks the Cl_2 bond to yield chlorine radicals in the initiation step of this free radical chain mechanism. In the propagation step that follows, the chlorine radicals abstract hydrogen atoms from the polymer chain forming carbon-centered radicals. At this point SO_2 participates in the chain mechanism. The carbon radicals along the chain backbone react with SO_2 to form sulfur to carbon bonds with a radical site on sulfur (Fig. 7.10). This sulfur radical can then react with Cl_2 to form the sulfonyl chloride.

The reactivity of the sulfonyl chloride group is ideally suited for crosslinking the polyethylene chain, for one example, with amines as shown in Fig. 7.11. This reaction yields a sulfonamide in a direct parallel to the formation of an amide from a carboxylic acid chloride and an amine. The beauty of this chemistry is that the sulfonyl substituent along the backbone of the polyethylene and the crosslinks they form also play another es-

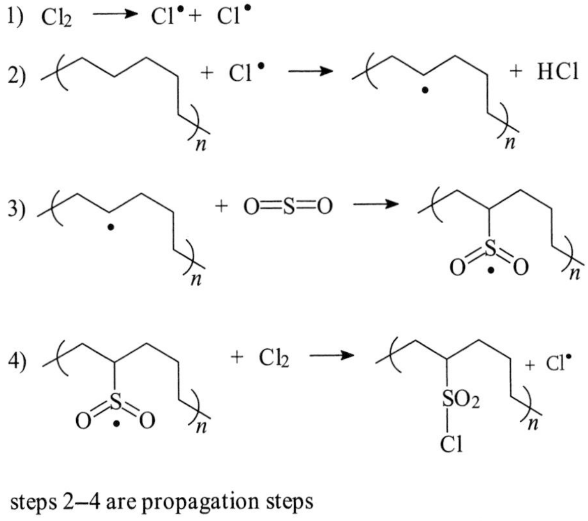

1) $Cl_2 \longrightarrow Cl^{\bullet} + Cl^{\bullet}$

2) [polyethylene chain] $+ Cl^{\bullet} \longrightarrow$ [polyethylene chain radical] $+ HCl$

3) [polyethylene radical] $+ O=S=O \longrightarrow$ [chain with SO_2 radical]

4) [chain with SO_2 radical] $+ Cl_2 \longrightarrow$ [chain with SO_2—Cl] $+ Cl^{\bullet}$

steps 2–4 are propagation steps

Fig. 7.10 Free radical chain mechanism forming sulfonyl chloride groups along a polyethylene chain.

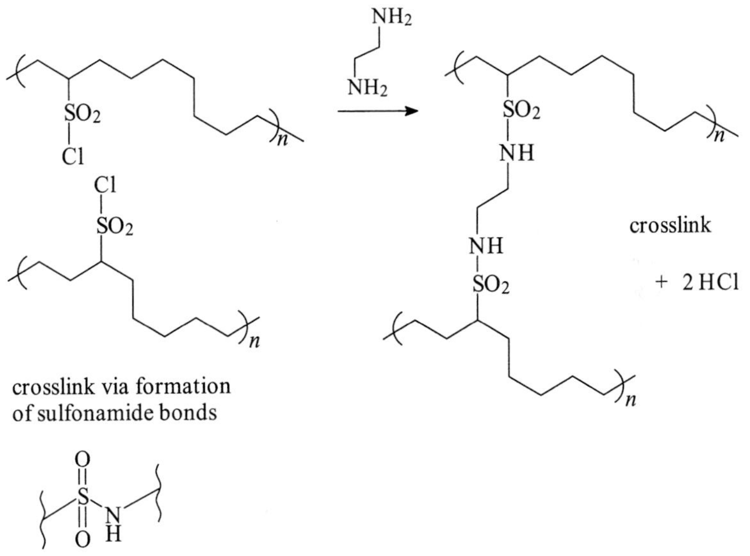

crosslink via formation
of sulfonamide bonds

crosslink

$+ 2\,HCl$

Fig. 7.11 Crosslinking to form Hypalon from sulfonyl chloride-substituted polyethylene.

sential role. They greatly interfere with the tendency of the unsubstituted polyethylene chains to form crystalline regions, which was one of the problems blocking the use of polyethylene as an elastomer. The crosslink arising from the sulfonyl chloride group on reaction with the diamine, which is essential for elastomeric behavior, precludes the crosslink, that is, crystallization, which blocked such behavior. Neat!

Hypalon is useful, for example, as a roofing material, where the absence of degradable double bonds is critical to prevent oxidative degradation. The ozone resistance of neoprene (see above) made it valuable for the same commercial use (see Chapter 5, Section 5.6). Hypalon is also useful for strip coatings of automobile doors and windows, for oil-resistant hoses and under the car hood belts, as well as for coatings for electrical wiring. Hypalon is also used as an elastomeric coating for tarpaulins and awnings where exposure to air and sunlight easily breaks down elastomers with double bonds.

Produced in the range of 100,000 tonnes per year, Hypalon sells for a premium of several dollars per kilogram. The difference between this price and the much lower price for a commodity such as styrene butadiene rubber (SBR), which will be discussed below (Section 7.9), can be accounted for by the fact that one product is a commodity made in large volumes and this allows for economy of scale. Hypalon is made in much smaller amounts and is sold with a great deal of technical service. There is no doubt that Hypalon is considerably more profitable.

Now we have seen two very different examples of elastomers based on covalent crosslinking, one derived from natural rubber and one from polyethylene. They are very different in detail but are identical in the basic principles involved. About one in one hundred units are sulfur crosslinked in vulcanized natural rubber. Almost the same number of crosslinks form from the reaction of amines with the sulfonyl chloride groups in the polyethylene-based rubber.

There are many more elastomers based on these same fundamental principles. But while the polymer used must invariably have the necessary flexibility to allow easy extension of the chain, the nature of the crosslinks vary as we have seen in the difference between Hypalon and vulcanized rubber.

7.9
Many Synthetic Elastomers are produced by the Chemical Industry. In Every Case the Physical Principles are Identical to those at Work in Natural Rubber. The Essential Characteristic of an Elastomer must be present, that is, a Crosslinked Flexible Polymer Chain

Vulcanization is a word used in industry as a synonym for covalent crosslinking of elastomers, whether or not sulfur is involved. Vulcanization with sulfur, however, requires the presence of double bonds in the elastomeric chain to undergo the chemical reactions discussed above (see Figs 7.6–7.8).

In addition to natural rubber, many synthetic rubbers containing double bonds are crosslinked with sulfur such as: SBR, polybutadiene rubber; butyl rubber containing a small percentage of isoprene; synthetic natural rubber, that is, *cis*-1,4-polyisoprene; chloroprene (Neoprene); ethylene-propylene-diene monomer rubber and nitrile rubber. Fig. 7.12 shows the structures of these rubbers, the monomers they are made from, and the type of polymerization used. In addition are listed three other types of rubbers that are not crosslinked with sulfur: the class of rubbers known as thermoplastic elastomers to be discussed later in this; ethylene-propylene rubber; and Thiokol. Let's discuss these materials as well as some other elastomers that are not major elastomers in current use, and let's add some of the interesting stories associated with their invention and production. The

Major elastomers

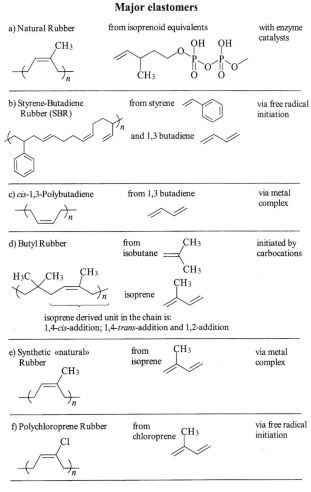

Fig. 7.12 Structures of synthetic rubbers crosslinked industrially and the structures of the monomers from which they are made.

first synthetic rubbers and the story of their invention are related to the history of the twentieth century, and especially the two world wars.

In the years from the beginning of the century to the time just preceding World War I, which began in 1914, laboratories in both Russia and Germany had experimented with polymerization of at least three different conjugated unsaturated molecules, isoprene (Fig. 7.12e), 2,3-dimethyl-1,3-butadiene and 1,3-butadiene (Fig. 7.12c) to produce materials that could be vulcanized with sulfur to obtain rubber characteristics. Some patents were filed and it was felt by several prominent chemists that a substitute for natural rubber could be at hand given some effort to work out details necessary for industrial production. But little was done until Germany felt the effect of the British blockade, which cut off Germany's supply of natural rubber. This led to the production of poly(2,3-di-

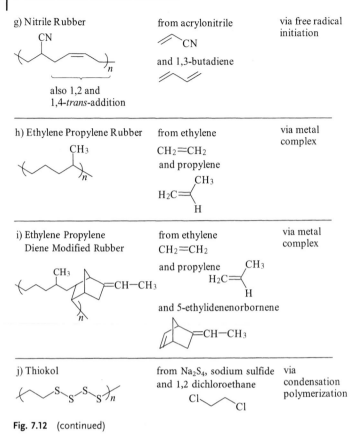

Fig. 7.12 (continued)

methyl-1,3-butadiene) by a process using oxygen to initiate the polymerization. The low rate of reaction with the oxygen required several months to produce a batch of polymer. At that time, the free radical mechanism of polymerization was not understood, and it was not clear what was causing the transformation of the liquid monomer to the rubbery mass. Even with this inefficient process, between 1916 and 1918 Germany produced 2500 tons of two grades of rubber based on this monomer, which were used for a variety of military purposes.

After World War I, important progress was made, particularly with the production of a copolymer of styrene and 1,3-butadiene (Fig. 7.12b). Sodium was used to initiate the process, and this rubber was named Buna S, Bu for butadiene, na for sodium (natrium in German) and S for styrene. This led to an interesting connection to the second world war effort in the United States in the 1940s. Prior to World War II, Standard Oil of New Jersey had entered into a joint research agreement, called the Joint American Study Company, with the German chemical giant, I.G. Farben Industrie. When it became apparent that the United States would inevitably participate in the war, Standard Oil was criticized for collaborating with the enemy. But in fact the collaboration was invaluable because Standard Oil was able to make the German patents on Buna S available to the U.S. Government. These patents served as a starting point for a Rubber Reserve Corporation research program initiated in 1940 to develop methods to manufacture large amounts of synthetic rubber.

This information from I.G. Farben, which was the basis for the research in the United States, was critically important to the American effort since rubber was soon realized to be the most important strategic material in short supply. The seriousness of the shortage was demonstrated by gasoline rationing at that time. Rationing was instituted not because gasoline was in such short supply, but rather because with lesser supplies of gasoline, cars would be driven less and the rubber in the tires would be preserved. The development of the synthetic rubber project by industry, academia and the government led eventually to a maximum production of 719,000 tons of synthetic rubber during the last year of the war.

Until the development of Buna S, which led to what is now known as SBR rubber (Fig. 7.12 b), only two synthetic rubbers had been produced in the United States, neither of which was appropriate for tires. One of these, neoprene (Fig. 7.12 f) was derived from efforts in Carothers' group at DuPont based on acetylene chemistry as discussed in Chapters 5 and 10 (Section 5.6) while the other, Thiokol® (Fig. 7.12 j) had been known since 1929 and was based on the reaction of ethylene dichloride with sodium polysulfide. Thiokol® has a use as a sealant for windows and seams of airplane cabins, and for liners to make oil-resistant O-rings and gaskets. The latter characteristic makes it important as a binder for rocket fuel. It has excellent oil and solvent resistance and is highly impermeable to gases.

Another elastomer used as a component in tires, particularly to enhance abrasion resistance, is poly(*cis*-1,3-butadiene) (Fig. 7.12 c), which remarkably is superior in its elastomeric characteristics to natural rubber. However, as a result of this "superior" characteristic, a tire cannot be made solely from poly(*cis*-1,3-butadiene), since it is so resilient that the car practically bounces as it is driven, and the turning of corners is almost impossible. The *cis* double bond in poly(1,3-butadiene) is essential to the elastomeric character for the same reason as for natural rubber (see Figs 7.3 and 7.4) as discussed above. But whereas the *cis* double bond in natural rubber arises from the stereospecificity of an enzymatic reaction in the *Hevea* rubber tree, the *cis* double bond in poly(1,3-butadiene) arises via Ziegler–Natta catalysts – the same class of catalysts that permitted polymerization of propylene to isotactic polypropylene as described in Chapter 2 (Sections 2.6 to 2.10).

Related catalysts based on the Ziegler–Natta process could also be applied to isoprene to synthesize what is called the oxymoron, synthetic natural rubber (Fig. 7.12 e). The synthetic material, however, is not an exact duplicate of natural rubber. The *cis* content of the synthetic material is lower, while the molecular weight and groups at the ends of the polymer chains, which change the conditions necessary for its commercial processing, also differ between the synthetic and natural materials. Synthetic natural rubber is usually more expensive than natural rubber, so commerce focuses primarily on the rubber produced by nature, although there is one plant in the United States producing synthetic natural rubber. But the synthetic material may be used in the future if the supply of natural rubber does not meet the growing demand.

Another rubber used in tires is butyl rubber (Fig. 7.12 d). Although butyl rubber has poor resilience and therefore cannot be used in tire treads, it is highly gas impermeable so that it served well for inner tubes to prevent escape of air. Today, it is used to line tires for the same purpose, to prevent the escape of air. Butyl rubber is made from isobutylene with a small percentage of isoprene. The isoprene supplies the double bonds for vulcanization.

Before World War II, chemists at I.G. Farben Industrie in Germany developed nitrile rubber (Fig. 7.12 g), which was named Buna N, by substituting acrylonitrile (see Chapter 8, Section 8.10) for styrene. Whereas Buna S, that is, SBR (Fig. 7.12 b) is competitive in its properties with natural rubber and therefore was virtually discontinued when the price of natural rubber came down after World War II, Buna N has unique properties with excellent resistance to oils, fuels and solvents so that is it is used to line fuel tanks, underground storage tanks and fuel hoses. These properties have justified a higher price than for natural rubber.

Although synthetic rubbers are one of the inventive wonders of the chemical industry of the twentieth century, they are not a growth industry today. One of the main problems arises from the success of tire technology, which means that tire life has greatly increased requiring therefore less rubber production.

One type of elastomer that is growing in use, although it is not used in tires, is based on a copolymer of ethylene and propylene (Fig. 7.12 h). This is an interesting story because it demonstrates the way tradition works in the chemical industry. Industrial methods were designed using the Ziegler–Natta catalyst systems to form copolymers of about 60 parts of ethylene and 40 parts of propylene. These Ziegler–Natta catalysts were the same ones used to produce isotactic polypropylene as well as both high-density and linear low-density polyethylene (see Chapter 2, Section 2.5).

The resulting copolymer of ethylene and propylene has the polymer chain characteristics of an elastomer since the many carbon–carbon single bonds along the chain backbone provide flexibility. In addition, the ethylene and propylene-derived units follow each other along the chain backbone in a random manner. Crystallization is virtually impossible with such a random structure, which means that there will be no interchain restrictions to the inherent flexibility of the polymer chains.

Just as the randomly pendant sulfonyl groups block excessive crosslinking due to crystallization in the polyethylene backbone used to synthesize Hypalon (Section 7.8), the random arrangement of the ethylene and propylene-derived units accomplish the same purpose. But as we well understand, some crosslinking is necessary to achieve a commercially important elastomer. How can this goal be accomplished? There are no double bonds to react with sulfur and therefore vulcanization is not possible.

An ingenious solution was designed in which peroxides were added to the ethylene–propylene (EP) copolymer. The peroxides, on heating, produced free radicals which abstracted hydrogen atoms along the chain producing carbon free radicals, which can couple with each other as in a termination reaction to produce crosslinks as shown in Fig. 7.13. The problem was solved, and an acceptable elastomer resulted. But there was one glitch. The elastomers industry is set-up for vulcanization with sulfur, and therefore the invention of EP rubber and its peroxide-based crosslinking method was not well accepted. The inventors solved the problem by adding molecules to the copolymerization mixture that contained two double bonds. This is the source of the name diene modified or DM. Only one of the double bonds entered into the polymer backbone and the other was pendant to the chain. The latter double bond was then available for vulcanization in the traditional manner with sulfur (see Figs 7.6–7.8) and everyone was satisfied. This polymer is called EPDM (Fig. 7.12 i).

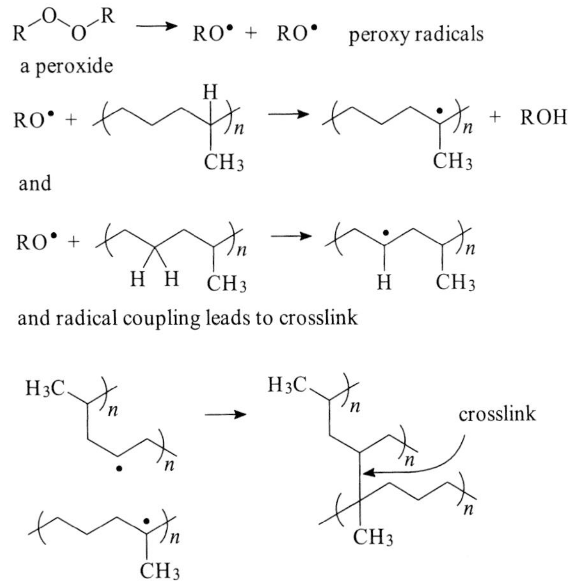

Fig. 7.13 The peroxide crosslinking mechanism for ethylene-propylene (EP) rubber and the structure of ethylene–propylene diene monomer (EPDM) rubber.

In the industrial synthesis of EPDM rubber, the proportion of diene monomer is kept to a minimum and most of the double bonds are used up in reaction with sulfur in the vulcanization step. In contrast, therefore to the many elastomers in which every unit is unsaturated with only a small proportion reacting with sulfur, as in natural rubber, EPDM rubber after vulcanization has fewer double bonds remaining. The small number of of double bonds is the factor that leads to good heat, oxygen, ozone (Fig. 7.9), and general weathering resistance and makes EPDM rubber excellent for roofing as well as for parts of automobiles that need to stand up to extreme conditions such as bumpers, radiators, heater hoses, weather strips, seals and mats.

EPDM rubber is also excellent for wire and cable insulation, and is even used to coat fabrics. Wherever an elastomer must resist tough conditions including the weather, EPDM rubber is a good choice. And to reiterate an important point, the reason derives from the absence of unsaturated groups in the vulcanized product. However, peroxide-crosslinked EP rubber is also free of unsaturation and has therefore also found a commercial niche for itself.

Peroxide crosslinking of elastomers was invented long before EP rubber. Silicone rubbers were first produced using peroxides and this procedure continues today. The story begins during World War II at the General Electric Company as part of the focus of researchers on silicone-based polymers and is told in Herman Liebhafsky's book.

Chemists at General Electric had developed a synthesis for dimethyldichlorosilane. On hydrolysis under certain conditions, this monomer gave rise to high molecular-weight poly(dimethylsiloxane) (Fig. 7.14), which had the kind of flexible chain characteristics necessary to form an elastomer. The researchers at General Electric realized that production of a commercial elastomer required crosslinking but knew that vulcanization with sulfur was not possible because the polymer contains no double bonds.

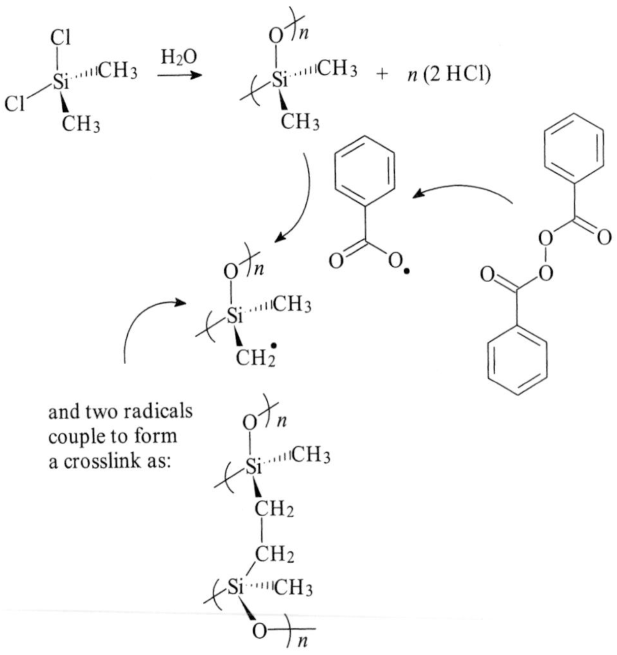

Fig. 7.14 Hydrolysis of dimethyldichlorosilane to produce poly(dimethylsiloxane) followed by peroxide crosslinking.

An unexpected breakthrough came from General Electric's experience with the development of alkyd resins in the 1930s. Alkyd resins are crosslinked by using free radical coupling of the double bonds in the unsaturated fatty acid chains (see Chapter 4, Section 4.10). This did not seem related at all to the problem of crosslinking the poly(dimethylsiloxane), which had no double bonds. But Gilbert Wright of the General Electric laboratories decided to add benzoyl peroxide to the silicone-based polymer (Fig. 7.14). On heating this mixture to 150 °C, the sample was transformed into what appeared to be a useful rubber. This material was the basis of further developments by General Electric to produce what is marketed today as silicone or siloxane rubber. Just as shown in Fig. 7.13 for the peroxide crosslinking of EP copolymer, the peroxide-derived radical abstracted hydrogen atoms from methyl groups in the poly(dimethylsiloxane), which then coupled with other radicals formed on other chains to form the crosslink (Fig. 7.14).

This unexpected discovery was of immediate use to the United States military. Shortly after the summer of 1943, in the midst of World War II, the peroxide crosslinked silicone rubber (Fig. 7.14) was used to make gaskets that became a critical component of aircraft engines, where previous gaskets made entirely of organic materials had not been able to withstand the high temperatures and pressure from the heavy metal components.

Standing up to high temperatures also made the silicone gaskets essential as components of the intense searchlights used on battleships. These same battleships fired powerful canons, which gave rise to intense heat and concussion. Only the silicone-based gaskets were able to withstand these stressful conditions. To this day, the outstanding properties of silicone rubbers make them useful for gaskets and sealing rings in jet engines, for vibration damping and for cable insulation for ships. Wonderful stuff!

There is a special quality of silicone-rubbers worth mentioning here. They maintain their flexibility at very low temperatures, whereas other elastomers such as natural rubber stiffen, losing their resiliency. The ease of conformational motion of the silicon oxygen bonds in the backbone is responsible for this low temperature flexibility. Polymers with backbones constructed of silicon–oxygen bonds are the *most* flexible polymers known. Such a characteristic raises the question: What are the *least* flexible polymers? What kind of polymer is the antithesis of flexibility? Are there polymers that could be called anti-elastomers? In seeing the opposite behavior, we can better appreciate the structural characteristics necessary for elastomeric behavior.

7.10
What Kinds of Polymer Properties will preclude Elastomeric Behavior?
What Kinds of Polymers could be called Anti-elastic?

The essential requirement of a rubber is a flexible polymer chain, and although there are many polymers that fit this requirement, there are also many that do not. Many polymers have backbone structures that resist conformational change. If the polymer cannot readily change shape, then elastomeric behavior becomes impossible. In this way we discover that elastomeric characteristics can be blocked in a flexible chain polymer not only from too many crosslinks, either from covalent bonds or crystallization, but also, in the absence of crosslinks, if the polymer backbone structure is too stiff.

Stiff or rigid polymers have many important properties arising from the material strength that often accompanies such backbone properties. We have seen this in Chapter 4 in the polycarbonates formed from bisphenol A. Here, the stiffness is prized as the foundation of an engineering plastic, one that can replace metals in many applications. In polycarbonate, the stiffness, as discussed in Chapter 4 (Section 4.19), arises from the high proportion of benzene rings in the chain and the resistance of the aromatic ring to conformational change. As a contrast to the type of chain with the flexibility for elastomeric behavior, Fig. 7.15 shows the structures of several industrially important stiff chains. In every structure shown in Fig. 7.15, just as for polycarbonate, there are aromatic rings in the backbone of the polymer chain.

7.11
Physical Interactions among Polymer Chains can be used to form Elastomers with Unique Properties. How the Polymeric Glassy State can act as a Physical Crosslink

As an introduction to the title of this section we have to look first at the nature of crystallization of polymers to understand the glassy state of polymers.

Almost all types of small molecules will crystallize when cooled from the melt state. This result is reasonable, since small molecules are free to move about quickly and therefore to find their way into proper relationships with each other to assume the regular arrangements necessary for crystallization. Polymers, and especially flexible polymers, on the other hand,

Examples of several "anti-elastomers", that is, stiff chain polymers

Poly(ether imide)

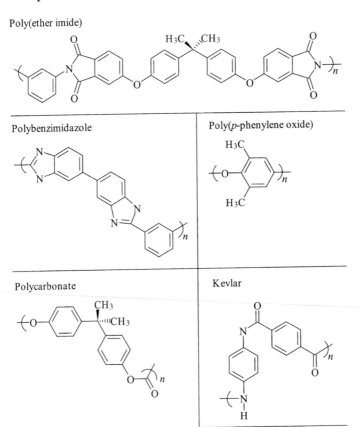

Polybenzimidazole	Poly(*p*-phenylene oxide)
Polycarbonate	Kevlar

Fig. 7.15 Structures of stiff-chain polymer structures.

Polyetherimides: Printed circuit boards, microwaveable containers, electrical connectors, sterilizable medical equipment.

Polybenzimidizoles: asbestos replacement (like Kevlar), high temperature filter-cloth and other textiles uses that require its high-temperature resistance, solvent resistant, etc. properties. Proposed as a component of fuel cells.

Poly(2,6-dimethyl-1,4-phenylene oxide): Usually called poly(phenylene oxide) or poly (phenylene ether): Electrical connections; food packaging for microwave heating; business machines; automo-

biles such as wheel covers, grills, mirror housings, instrument panels; internal radio and television parts; motor housings.

Polycarbonate: automotive in combination with poly(butylene terephthalate) or ABS for doors and side panels of cars; for compact discs. There are many others, for example, glazing in schools and stores in areas prone to rioting; enclosure of a squash court; telephone dials.

Kevlar: As an asbestos replacement. This is its main use. Many others: cloth to reinforce epoxy resins in body armor, canoes (a 6 meter-long canoe weighs only 20 kilograms), protection for firefighters, for sails for sail boats.

resist forming crystals. If crystals are formed, only portions of chains will participate. This result is also reasonable. These very large molecules can move only slowly and are entangled with each other, which restricts the ability of the polymer chains to come together in the regular manner necessary to crystallize even a small portion of the chain.

There is another reason for the resistance of many polymers to form crystals. In a sample of small molecules purification is possible, because every molecule is identical. This is essential for crystal formation, which requires regular arrangements only possible where all molecules are indeed identical. But in polymers the individual chains in a sample may differ from each other. Although one of the ways the chains differ from each other, that is, molecular weight, does *not* contribute to the difficulty of crystallization, other differences are critical. The backbones may differ from chain to chain. For example, in low-density polyethylene, LDPE, branches are formed whenever there is a statistical possibility for the backbiting reaction from the radical at the chain end (see Chapter 2, Section 2.4).

These branches will occur along each chain in an irregular manner so that polymer crystallization involves only sections of the chains, between branch points. The same thing is true of the stereochemical arrangements of the methyl groups along the backbone of atactic polypropylene. No two chains will be the same. This is not a situation that is suited to the formation of crystals and certainly not large crystals. Even if some regions of differing chains, by chance, are arranged in the same way, and by chance can find each other in the tangled mass, these regions will be limited in length and number and therefore the size of the crystals that can form will be limited. Sometimes structural irregularity of a polymer is so high that no crystals can form at all no matter how small the crystals might be.

Even if crystallization is possible because there exist regular portions of the chains, the slow motions of the chains, arising from the high viscosity associated with polymers, may preclude crystallization because the structurally regular parts of the backbone may not be able to encounter each other in time before the rapidly dropping temperature "freezes" everything in place. The formation of a glass stops the necessary motion for the crystal to form. Many polymers are entirely amorphous on cooling; crystallization does not occur.

But what happens in the absence of crystallization as a polymer sample is cooled? Polymers, or portions of polymer chains, unable to crystallize for the reasons given above, enter a glassy state. All motion associated with flow stops, the center of mass of the material is unable to change. A glass forms because as the temperature decreases the chains are crowded closer and closer together as the density increases until any movement that displaces the position of each chain becomes impossible. Thus the overall array of polymer chains becomes fixed in place in an entangled mass. The glassy state has the isotropic character of a liquid and the mobility properties of a solid. Poly(methyl methacrylate), discussed in Chapter 6, is an excellent example of a polymer forming such a glassy state.

The glassy state of a polymer is reached by lowering the temperature from the melt state. But at what temperature does the polymer exhibit glassy properties? At what temperature does the polymeric material become brittle and hard? Each polymer has a unique temperature at which this occurs, the so-called glass transition temperature, T_g. How T_g is related to the structure of a polymer is a subject of great interest because

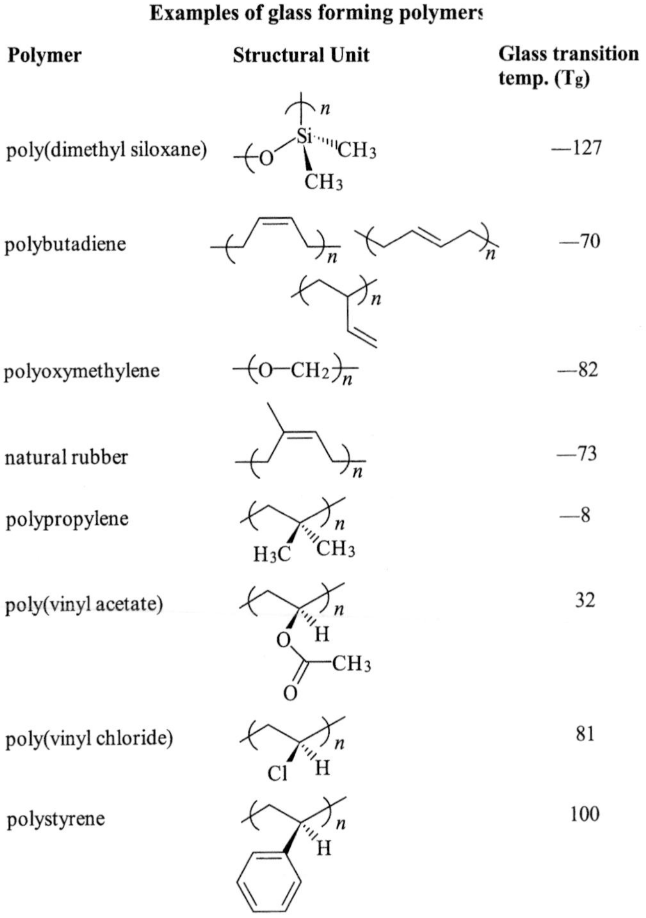

Examples of glass forming polymers

Polymer	Structural Unit	Glass transition temp. (Tg)
poly(dimethyl siloxane)		−127
polybutadiene		−70
polyoxymethylene		−82
natural rubber		−73
polypropylene		−8
poly(vinyl acetate)		32
poly(vinyl chloride)		81
polystyrene		100

Fig. 7.16 Examples of glass-forming polymers, and glass transition temperature (T_g) as a function of polymer structure.

many polymers are used commercially in their glassy state. Understanding the glass state is one of the unsolved mysteries of the science of materials and has attracted much experimental and theoretical interest.

It is known that the more restrictions there are to motion of the polymer backbone, the higher the temperature below which the polymer enters the glassy state. For example, poly 1,3-butadiene exists in the glassy state at all temperatures below about −70 °C, while polystyrene exists in the glassy state at all temperatures below 100 °C. Polystyrene with a benzene ring pendant to the chain on every third carbon atom is far more conformationally restricted then is poly (1,3-butadiene). The polymer with the higher T_g, polystyrene in this example, will therefore exist as a glass over a wider temperature range. It will be a glass from that high temperature, +100 °C, to all temperatures below that point. Fig. 7.16 shows the T_g values of a variety of polymers where a general correlation be-

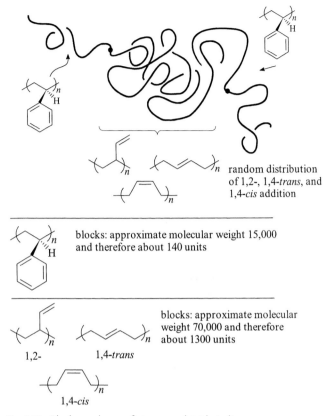

random distribution of 1,2-, 1,4-*trans*, and 1,4-*cis* addition

blocks: approximate molecular weight 15,000 and therefore about 140 units

blocks: approximate molecular weight 70,000 and therefore about 1300 units

1,2-   1,4-*trans*

1,4-*cis*

Fig. 7.17 Block copolymer of styrene and 1,3-butadiene.

tween conformational flexibility and T_g can be observed. With this general introduction to the polymeric glass state we are now prepared to understand how the glass state can serve as a crosslink in forming an elastomer. Let us see how.

Consider the polymer structure shown in Fig. 7.17. This arrangement is called a block copolymer. In such a structure, the chain will have a long section made of one kind of monomer unit followed by and covalently linked (the dark spots represent the connection) to another section of the chain made of a different monomer unit. In Fig. 7.17 you see blocks of units polymerized from 1,3-butadiene and blocks of units polymerized from styrene. Each polymer chain has three blocks, a 1,3-butadiene-derived block flanked by blocks derived from styrene.

Most polymers are immiscible with each other; they do not mix together to form a homogeneous phase, and polystyrene and poly 1,3-butadiene are typical polymers in this regard. If one were to mix a sample of separate chains of polystyrene and poly 1,3-butadiene the result would be two phases, a polystyrene and a poly 1,3-butadiene phase. This behavior is analogous to what one would find for two immiscible liquids such as hexane and water. But the immiscibility of polymers is remarkable. Even polymers of closely sim-

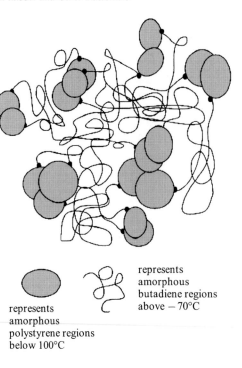

represents
amorphous
polystyrene regions
below 100°C

represents
amorphous
butadiene regions
above − 70°C

ilar structure are most often immiscible. The reason is associated with the small entropy of mixing because the molecules are so large, that is, they are macromolecules. Whatever may be the details of the reason for immiscibility of polymers, this immiscibility of the separate polymer chains is also the situation for the blocks in the copolymers, shown in Fig. 7.17. But since the blocks are linked covalently, they must respond to this incompatibility in a special way. The structurally different blocks cannot separate from each other as if they were in separate chains. This result is shown in Fig. 7.18.

When the block copolymer (Fig. 7.17) is at a temperature such that both blocks are in the melt state, that is, above T_g for both blocks as for example substantially above 100°C, the chains are free to move. There will be a driving force for the different kinds of blocks to separate to satisfy their immiscibility and the freedom to move allows "phase separation." The styrene-derived blocks within a single chain and among different chains will congregate, as will the 1,3-butadiene-derived blocks. The material will evolve toward a mixture of regions (domains) of different block structure within the limits of the covalent linkages between the blocks as is shown in Fig. 7.18.

In this segregated structure, the T_g of the aggregate of the blocks made of 1,3-butadiene will remain flexible at all temperatures above about −70°C. But the aggregate of blocks made from styrene units will be glassy at all temperatures below about +100°C (Figs 7.16–7.18). This means that the overall block copolymeric material at temperatures between above about −70°C and below about +100°C consists of a mixture of flexible and glassy regions. The aggregated blocks of styrene units among various chains will be below their T_g and therefore will be glassy, whereas the aggregate of blocks of 1,3-butadiene units in various chains will be above their T_g and will be what is called rubbery. What is the overall material like between these temperatures?

The glassy regions of the material (Fig. 7.18) will act in the same way as the covalent crosslinks arising, as example, from vulcanization since the glassy segments are regions of the material in which the chains are immobilized. But the aggregates made of 1,3-butadiene units will remain flexible and can stretch and change shape. How clever! At any temperature between about $-70\,°C$ and $+100\,°C$ this material will act as an elastomer, although there are no covalent crosslinks, only aggregated blocks that serve the same purpose, albeit they are physical not covalent bonds. And what we have described is the commercial thermoplastic elastomer "Kraton" developed by Shell and sold to the extent of many hundreds of million kilograms per year globally.

Kraton costs in the range of 1US$ per kilogram and is used in rubber footwear and shoe soles as well as for hot melt adhesives and for asphalt flexibilization, which is necessary if the asphalt is to be used to coat a surface that is below grade. And if Kraton is heated substantially above the T_g of the styrene-derived blocks, that is, above about $100\,°C$, the crosslinks change from rigid glassy regions to flowable melt regions, as described above, and the entire material flows and therefore can be cast, molded, or extruded into any desired form. On cooling, this new form resumes its elastomeric character. This is the reason such a material is called a thermoplastic elastomer.

There are other variations of thermoplastic elastomers that are based on the same principle, blocks of flexible and glassy polymers. Depending on the relative T_g's of the blocks used, and especially the structure of the flexible block, the application and use temperature will vary and therefore the materials will have different markets. The lower the T_g of the flexible block, the lower the temperature the material will be elastomeric, whereas the higher the T_g of the rigid block, the higher the temperature at which it remains an elastomer. Below the lowest T_g the material becomes rigid (all glass), while above the highest T_g the material will flow because no crosslinks remain.

Major advantages of thermoplastic elastomers arise from the ability to process these materials by means such as injection molding normally used to process common plastics such as polystyrene or poly(methyl methacrylate). As long as the temperature is high enough, the glassy crosslinks will melt away and the chains can move about and the material can flow. Many thermoplastic elastomers also have excellent low-temperature properties for use in cold environments where retention of flexibility is important. This is true, as we have seen above for Kraton, which can act as an elastomer to nearly $-70\,°C$.

7.12
Variations on the Block Theme produce Thermoplastic Polyurethane Elastomers including Spandex (Lycra™), the Elastic Fiber. Here, the Crosslinks involve a Kind of Physical Interaction, which is Different from Glass Formation

This concept of polymers with mutually incompatible segments in which one segment type is flexible and the other is not flexible over a wide temperature range has been extended by the chemical industry to structures in which flexible segments are connected to segments that crystallize. In this case, the temperature range of use of the thermoplastic elastomer varies from the T_g of the flexible segment, which will be very low, to the melting temperature of the crystalline segment, which will be high. Here, we are using the crystallinity that blocked traditional elastomeric behavior (Sections 7.4 and 7.8) to develop a new kind of elastomeric behavior.

Spandex, known also in the clothing trade as Lycra^{TM} or as Elastane^{TM}, is a well-known example of this concept. And well known it is indeed. Joseph C. Shivers, who led the ten-year research effort at DuPont that culminated in 1959 with the invention of spandex hardly needed to be awarded the prestigious Olney Medal for Achievement in Textile Chemistry to testify to the power of the spandex concept.

The world-wide web, more than forty years after the invention of spandex, under the search words spandex and Lycra^{TM} yields more than half a million sites and demonstrates the amazing impact that spandex has had on the world. The impact is not only as a fiber that is added to enhance fit and appearance of garments made with cotton, wool, silk, linen and nylon, but as a concept that goes far beyond the chemistry to be discussed here, a concept that has been called "hug your body elasticity." Although invented many years ago, spandex fiber is still finding new uses all the time. A headline several years ago in *Chemical and Engineering News* asked the question: "What's That Stuff?" The structural answer for one kind of spandex is found in Fig. 7.19. We can focus on this one example because no matter what is the variety of spandex the structural characteristics yield the same results in parallel ways.

As shown in Fig. 7.19, spandex is a linear polymer with both urethane and urea bonds. We have looked at polyurethanes in detail in Chapter 4 (Sections 4.15–4.18) and indicated there that industry uses the word *polyurethane* for polymers with both urethane and urea recurring groups arising from the reaction of isocyanates with hydroxyl or amino groups respectively. Fig. 7.19 quickly reveals spandex's secret. Large segments in the chain arise from recurring groups of $-O-CH_2-CH_2-CH_2-CH_2-$. In commercial spandex these recurring oxytetramethylene units form a chain in the molecular weight range of no less than 500 and up to several thousand. This means that there will be between seven and fifty C_4H_8O units until a urethane bond appears along the chain to initiate a segment with aromatic rings and urea bonds, a stiff segment.

The carbon–oxygen single bonds in the oxytetramethylene units are highly flexible and allow the chain to take on a variety of shapes. This makes up the flexible part of the spandex structure, allowing the change of polymer shape necessary for elastomeric character. The aromatic urea-bonding sites, on the other hand, provide the part of the segmented structure (Fig. 7.19) for crosslinking, which are the other necessary component for long-range elasticity. And long-range elasticity it is! Spandex is famed for its ability to stretch to a range of 500 to 600% over and over again and return to its original shape. These crosslinks must exert a powerful restriction on the chains slipping past each other, for if such slippage occurred the original shape could not be regained.

The crosslinks associated with the urea linkages in Fig. 7.19 play the identical role as the crosslinks that arise from the vulcanization process in rubber (see Fig. 7.8) or the crosslinks in Hypalon^{TM} (see Fig. 7.11) or from the glassy regions from the aggregation of the styrene-derived blocks in Kraton^{TM} (Fig. 7.18). Spandex, as we have noted, is a thermoplastic elastomer and belongs to the same class of materials as Kraton^{TM} (Figs 7.17 and 7.18). But in spandex we are dealing with a different kind of physical

Fig. 7.19 The structure of spandex.

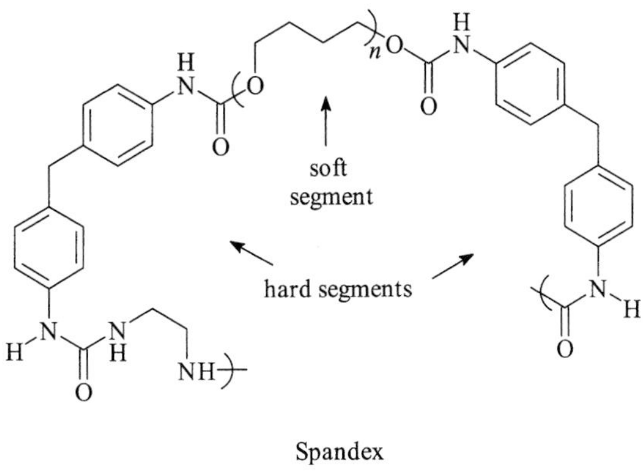

Spandex

crosslink, not a physical crosslink derived from glass formation as in Kraton™, but rather from physical forces derived from crystallization and hydrogen bonding.

Just as the blocks derived from 1,3-butadiene and from styrene in Kraton™ do not mix, the soft and hard segments in spandex (Fig. 7.19) behave in the same way. When the temperature is high enough to allow flow of all segments of the chain, the material segregates into the soft oxymethylene-dominated regions and the hard diphenylmethane urea regions, as designated in Fig. 7.19. Although the structures of spandex and Kraton™ are entirely different, the representation pictured in Fig. 7.18 for Kraton™ would apply equally well to spandex. When the temperature is lowered, the hard regions readily crystallize since they have a regular structure derived from both aromatic rings and the strong hydrogen bonding (Chapter 5, Section 5.4) from the urea groups.

It is the formation of these crystalline regions from the hard segments that immobilizes the chains, form the physical crosslinks and prohibit slippage. The remarkable structure (Fig. 7.19) allows the soft segments to be greatly stretched without disrupting the crystalline connections among the chains. And this property is maintained over a wide temperature range since the T_g of the soft regions is very low and the melting point of the hard regions is high. This chemistry answers the mystery behind "hugs your body elasticity," and brings up the question: how could this remarkable material be made?

7.13
Spandex: A Possible Synthesis

The structure in Fig. 7.19 appears quite complicated, but in fact spandex is synthesized from simple components using familiar reactions. It all begins with tetrahydrofuran (THF) which, as shown in Fig. 7.20, may react with fluorosulfonic acid to place a positive charge on the ring oxygen. The bond between this positively charged oxygen and the methylene group it is bonded to is weakened. Under the right conditions, THF rings that have not reacted with the fluorosulfonic acid can act as nucleophiles since the lone pair electrons on the oxygen are available. In this way unreacted THF rings act to open THF rings that have formed salts with fluorosulfonic acid. This is another example of the importance of an S_N2 reaction in the chemical industry, which we have seen at work several times in Chapter 4, and to the use of ring opening of cyclic ethers, which we have seen in both ethylene oxide and propylene oxide in other industrial processes in Chapter 8.

The conditions are perfect for the S_N2 mechanism. The site of nucleophilic attack is a methylene group. The leaving group arises by reaction of the THF oxygen with fluorosulfonic acid forming a positive charge and greatly weakening the methylene-to-oxygen bond. Quenching with water at the right time and adjusting the conditions of this ring-opening reaction (Fig. 7.20) can yield oligo(tetramethylene ether) in the molecular weight range desired for the spandex production.

The product of the ring opening of the THF is a polyether with terminal hydroxyl groups (Fig. 7.20). The next step, involving a reaction of this diol with a diisocyanate, is directly analogous to the formation of polyurethanes discussed in Chapter 4 (Section 4.15) except for one critical difference. A two-to-one excess of the diisocyanate is used so that instead of forming a high polymer, the reaction stops at the stage where a molecule is formed with an isocyanate group at each end (Fig. 7.21).

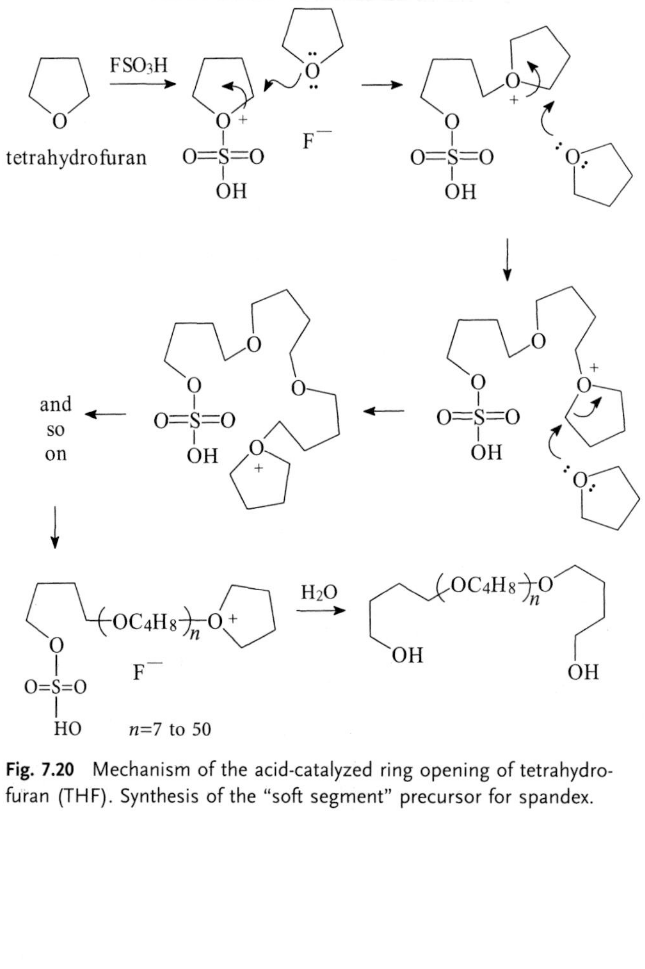

Fig. 7.20 Mechanism of the acid-catalyzed ring opening of tetrahydrofuran (THF). Synthesis of the "soft segment" precursor for spandex.

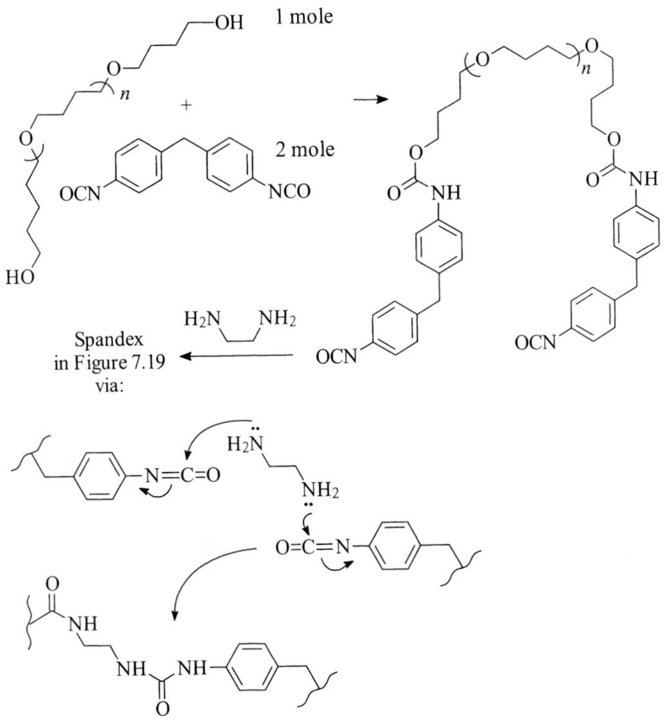

Fig. 7.21 Formation of the prepolymer for spandex synthesis. How spandex is "put together."

The diisocyanate in Fig. 7.21 can be called a prepolymer or a macromonomer. It is certainly a large molecule, and has the recurring units of the polyether derived from THF (Fig. 7.20). Reaction with an equal number of moles of diaminoethane (ethylene diamine) will yield a high polymer, spandex (Fig. 7.19), by forming urea bonds. Alternatively, reaction with ethylene glycol will yield another version of spandex by forming urethane bonds.

Spandex certainly is an application of the basic chemistry used to form polyurethanes (Chapter 4, Section 4.15). In fact Bayer, which was the chemical company that played an important role in the development of polyurethane chemistry, introduced its own version of spandex in the early 1960s in Germany and has opened production of the fiber within the last decade in Bushy Park, South Carolina. Although admitting that DuPont was in the spandex business first, a marketing executive at Bayer, presumably trying to stake out Bayer's important role and their claim to discovery, recently pointed out that, "Bayer has been involved in polyurethane chemistry longer than anyone else."

Now we have seen elastomeric behavior with covalent crosslinks as in vulcanized rubber, with physical crosslinks derived from glassy behavior as in Kraton^{TM}, and also from crystalline character and hydrogen bonding as in spandex. Is there still another physical interaction that can give rise to crosslinking and what different kinds of properties might that lead to?

7.14
Ionomers: Yet Another Approach to Reversible Cosslinking

When a carboxylic acid is neutralized in a nonpolar environment, the carboxylate is close-ly associated with its counterion. These ion pairs have a large dipole moment and strongly attract each other. Industry has used these interactions as the basis for a revers-ible crosslinking, allowing development of tough materials, some of which may be elasto-meric. Such materials are called ionomers.

DuPont has developed a thermoplastic material based on this attractive force. Surlyn®, which although not an elastomer, has a toughness that can be varied allowing for a wide range of uses. In Surlyn®, however, there is no necessity to synthesize a block copoly-mer, as was necessary for Kraton™ and spandex fiber (Figs 7.17–7.19). Let's see why.

The best way to understand why a block copolymer is not necessary for the reversible crosslinking in ionomers is to consider forming a random copolymer from 1,3-butadiene and styrene instead of the block copolymer (Fig. 7.17). The two different units would be randomly distributed along the chain. Since there are no segments of differing units – that is, no blocks – the polymer property does not depend on a flexible portion and a glass-forming portion. The random copolymer demonstrates no particular pattern in the arrangement of the two units and exhibits behavior that is the average of the two proper-ties of the units that make it up. It is impossible to make a thermoplastic elastomer from the random copolymer. Only the block copolymer will do.

But with a random copolymer of ethylene and an alkene containing an ionized car-boxylic acid group that is neutralized with a counterion (Fig. 7.22) average properties are not found. The properties of the carboxylic acid salt and those of the ethylene-derived units are simply too different to mix, even when placed next to each other along the chain. The powerful repulsion between these units is so strong that forming them into individual blocks is not necessary.

So how does the chain containing these units deal with the conflict? If the tempera-ture is high enough to allow motion of the chains, the carboxylate-counterion ion-pairs in the same chain or in different chains form aggregates. And these aggregates are ar-ranged in a form that minimizes the contact to the ethylene-derived units along the chain. This organization is made easier by the flexibility of the backbone of the chain, which allows the motions necessary to accomplish the separation.

These aggregates of the ion-pairs are highly hydrophilic. They are so repelled by the hydrophobic, non-polar ethylene-derived units that the chains twist and turn so as to seg-regate these groups (Fig. 7.22). In a manner of speaking, this is an example of the fact that oil and water do not mix. In fact, moisture is attracted to the ionic aggregates and away from the hydrophobic regions so that microscopic "puddles" are formed. The mate-rial separates into regions comparable to the situation for the block copolymer in Fig. 7.18, but for an entirely different reason.

The hydrophilic regions are in the range of several nanometers in which the chains cannot move without disrupting the attractive force of the ion dipoles. If chains could slip by each other they would drag some of the charged groups out of the hydrophilic re-gion into the hydrophobic region. Horror! But this does not happen. Thus, the regions in which the charged groups congregate are crosslinks. The flexible hydrophobic units form

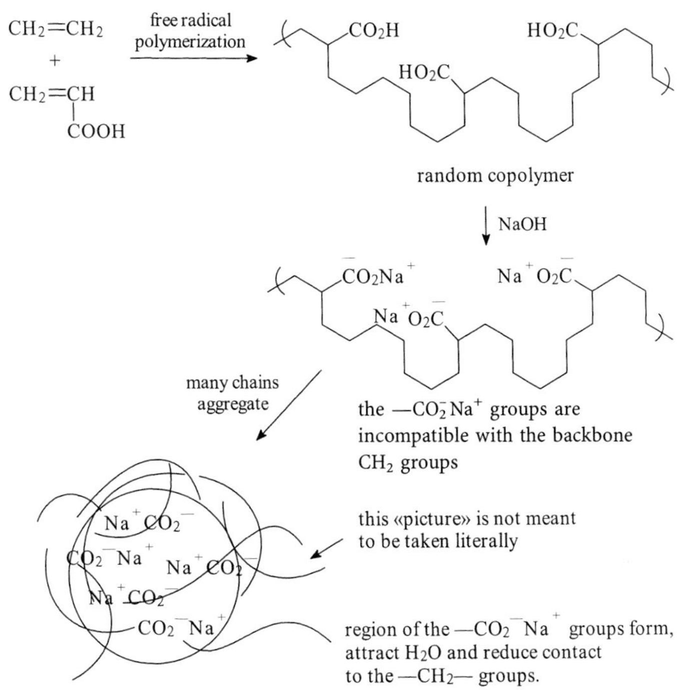

Fig. 7.22 The structure and synthesis of the ionomer, Surlyn®.

their own regions but their motion is restricted by the crosslinking derived from the hydrophilic groups. The dipole attractions in the hydrophilic regions are reversible if enough heat is supplied. In this way Surlyn® and the other commercial ionomers can be molded at higher temperatures just as for the glass-derived crosslinks of Kraton™ or the crystalline-derived regions of spandex, and then cooled in the form necessary for their use.

There are many uses for this unique kind of thermoplastic material, which was created in the early 1960s. A search of the web under the name Surlyn® comes up with many. A few of these are: golf ball covers; bowling pins; dog chew toys; ski boots; floor tiles; rodeo vests; hockey helmets; door handles and tennis shoes.

The ionomeric concept can also be used to form thermoplastic elastomers in addition to tough thermoplastic materials such as Surlyn®. This can be accomplished by copolymerizing hydrophobic monomers that yield more flexible chain structures, such as 1,3-butadiene, with a small proportion of neutralized acidic monomers such as methacrylic acid.

Polymer science is a very complicated subject. Here, we touch it lightly but enough to see how beautiful the subject is, and what exotic materials it is capable of producing. And all of that without even mentioning the fact that nature has used polymer science as the most essential component in the construction of life, consider proteins and DNA. There is so much to know, and so much to discover.

7.15
Summary

This chapter tells us how rubber elasticity was discovered and describes the basic physical principles behind this behavior. We come to understand the role of entropy in rubber elasticity and why rubber, in contrast to all experience with other materials, over a wide temperature range, becomes stiffer when it is hotter and why rubber gives off heat when it is stretched. We discover that nature supplies two diastereomeric polymers from different trees – one an elastomer, and the other a stiff rigid material. And we understand that the diastereomeric difference resides in *cis* versus *trans* configurations around double bonds and how this is related to the biological necessity for *cis* double bonds in fatty acids.

We followed the story of natural rubber and the accidental discovery of vulcanization. We understand how vulcanization acts to crosslink the different polymer chains in natural rubber and turns a curious gummy material into a major industrial commodity that plays an essential role in modern life. In this reaction of sulfur with natural rubber we've seen a parallel to the familiar chemistry involving addition of halogens to double bonds and the nature of carbocations. But there is a difference. 1,2-Addition is not possible because of the difficulty of forming large rings.

Polymers are essential to rubbery behavior and conformation, and the flexibility about single bonds plays a key role in allowing predictions of what kinds of polymers can be elastomers. We study the kinds of polymers that could never form elastomers and discover that a large proportion of aromatic rings along the backbone can be one of the sources of the stiffness that precludes rubbery behavior but which contributes to strength.

We followed the history of rubber and its increasing importance as an underpinning material for modern technology and discovered how the two world wars of the twentieth century played a large role in the development of synthetic rubbers. This was possible because of understanding of the two essential characteristics for elastomeric behavior, chain flexibility and limited crosslinking. The first allowed the polymer chains to stretch and the second restricted the chains from slipping past each other causing undesired permanent deformations.

These insights and the increasing power of chemistry gave industry the power to make purely synthetic rubbers. These take many forms and allow the creation of new kinds of crosslinking mechanisms that do not rely on the use of sulfur or the presence of double bonds. We studied many of these synthetic rubbers and came to understand their structural basis and their uses from tires to gaskets and all things in between. We came to understand that the principles behind rubbery behavior could be expressed in the nature of polymers containing silicone and the roles such materials play.

Physical interactions among polymers, leading to crystallization or glassy regions, may also act to restrict motion to such an extent as to overwhelm the inherent flexibility of the polymer chains. When it was understood that such physical interactions amount to effective crosslinking, polymers were designed in which these physical interactions are controlled and therefore act as the necessary but limited crosslinks to make commercially important elastomers called thermoplastic elastomers. We studied the nature of the glassy states formed by polymers and how glassy behavior was used by Shell to create Kraton™.

We learned how crystallinity may be used to form crosslinks and was used by DuPont to create spandex. And finally, we discovered how the extreme repulsion between ionic moieties and hydrocarbons can be used to form an entirely different kind of reversible crosslink leading to a tough material called Surlyn®.

Some of the subjects treated in this chapter are listed below. These are key words and terms that act as reminders of the chapter's contents and should become a valuable part of your chemical vocabulary.

- Conformational analysis
- *Anti* and *gauche* conformations
- Polymer chain shape
- Rubber elasticity
- Thermodynamic considerations in rubber elasticity: the role of entropy
- Temperature and elasticity
- *Havea brasiliensis*
- *Gutta percha*
- Polymer crystallinity
- Polymer glassy state
- Unsaturated fatty acids
- Polar addition to double bonds
- Closing large rings
- Vulcanization
- Peroxides
- Hydride transfer to carbocations
- Polar chain mechanisms
- Ozone, double bonds and degradation of natural rubber
- Hypalon™, spandex, Kraton™, Surlyn®
- Free radical chain mechanism
- Covalent crosslinking
- Physical crosslinking
- Stiff polymers – anti-elastomers
- Poly(siloxane) rubbers
- Major elastomers
- Glass transition and conformational flexibility
- Block copolymers
- Urethanes and urea linkages in polymers
- Acid-catalyzed ring opening of tetrahydrofuran
- Ionomers

Study Guide Problems for Chapter 7

1. Review the conformational states of ethane and butane and assign the energy changes to the various kinds of strain involved.
2. Why does the temperature of rubber rise on stretching?
3. Why does rubber become stiffer as temperature increases?

4. Demonstrate that randomly incorporating *anti* and *gauche* conformational states along the backbone of polyethylene would lead to a chain whose direction changed in a random manner.

5. Polyethylene and especially high-density polyethylene (HDPE) discussed in Chapter 2 (Section 2.5) is highly crystalline. In polyethylene crystals, the *anti* conformation is predominant, although in the absence of crystallization, in the amorphous regions of the polymer material, both *anti* and *gauche* conformations are commonly encountered along the chain. What is the reason for the conformational difference for the chains in these compared states?

6. Can you find support for the statement that the temperature of the environment in which an animal functions determines the degree of *cis*-unsaturation in the fatty acids in that species and that plant-derived fatty acids are the most unsaturated?

7. Polar addition of bromine to double bonds occurs with *trans* stereochemistry. Show how this can be demonstrated by the addition of bromine to the stereoisomers of 2-butene and also by the addition of bromine to cyclohexene.

8. Addition of bromine to the stereoisomers of 2-phenyl-2-butene occurs with less stereochemical specificity than addition of bromine to 2-butene. Offer an explanation for this fact. What would be the products from addition of ClOH to this alkene?

9. In the hydride transfer step in Fig. 7.7 (step 2), the transferred hydride can arise from either methylene group flanking the double bond of the contributing chain of natural rubber. Why is one methylene group preferred over the other in this step?

10. The mechanistic steps shown in Fig. 7.8 describe a chain mechanism. Identify the initiation and propagation steps.

11. Propose as many termination steps as possible for the chain mechanism described in Fig. 7.8.

12. In the formation of low-density polyethylene (LDPE) substituted with sulfonyl chloride via reaction with chlorine and sulfur dioxide (Fig. 7.10), there are different sites along the chain for the substitution. How do these sites differ and what would be their relative reactivity in the propagation step that determines the substitution pattern along the chain?

13. Propose possible termination steps in the free radical chain mechanism leading to sulfonyl chloride substituted polyethylene in Fig. 7.10.

14. Why is it necessary to keep the ratio of sulfur dioxide to chlorine high in the chain mechanism shown in Fig. 7.10?

15. Use the information in Fig. 7.12 to draw the structures of each of the elastomeric chains to show a significant portion of the chain including, when necessary, the different diastereomeric and constitutional arrangements for the diene monomers incorporated into these chains such as *cis*, *trans*, 1,4-addition and 1,2 addition.

16. How does the fact that there is random incorporation of the different monomers in the chain affect your answer to Question 15? Can a unique structure for the chain be given?

17. Propose mechanisms, along the lines shown in Figs 7.7 and 7.8, for each of the polymers in Fig. 7.12, which form elastomers on vulcanization. In what way does each of these polymers differ in detail from the mechanism for vulcanization of natural rubber?

18. Use of 1,3-butadiene would not work as well as 1,4-hexadiene to form EPDM rubber (Fig. 7.12). Why not?

19. Propose a mechanism for the reaction of water with dimethyldichlorosilane (Fig. 7.14) to yield poly(dimethylsiloxane). In the original research in which this polymer was discovered at General Electric, it was found that limiting the rate of access of water to the growing polymer chain led to a higher molecular weight polymer. What might be the reason for this observation?

20. What might be the reason that bonds between silicone and oxygen allow greater conformational motion than carbon–carbon bonds, and why does this difference lead to poly(dimethylsiloxane) forming a glassy state at a lower temperature than for polyethylene?

21. Can you write a parallel mechanism for peroxide crosslinking of the copolymer of ethylene and propylene (Fig. 7.13), and for poly(dimethylsiloxane) (Fig. 7.14)? Is the mechanism a free radical chain mechanism?

22. Alkyd resins are formed via a free radical crosslinking process (see Chapter 4, Section 4.10). What is the fundamental difference between the mechanism for crosslinking of the polymers in Question 21 and the formation of the alkyd resins?

23. In each polymer shown in Fig. 7.15, only the repeat unit of the chain is shown. Draw out each polymer structure with at least two repeat units to show how the chains are connected.

24. From the repeat units of the chain shown in Fig. 7.15 and your answer to Question 23, imagine what monomers could be used to synthesize such polymers and what reactions could be involved.

25. Some limited number of small molecules form glassy states rather than crystallize. Can you think of a way to form a derivative of benzoic acid to block crystallization and therefore force formation of a glass as temperature is lowered?

26. Can you suggest a reason why saccharides are difficult to crystallize and therefore on cooling from the melt state often form glasses?

27. Poly(siloxanes) (Fig. 7.14) have extremely low glass transition temperatures, far below −100 °C (Fig. 7.16). Propose a polymer structure that would take advantage of the low glass transition temperature of poly(siloxanes) and that would be a thermoplastic elastomer. What might be the commercial advantages of such an elastomer?

28. Do the ideas of conformational analysis support the hypothesis that there is a correlation between T_g and flexibility seen in Fig. 7.16, that is, that more flexible polymer chains have a lower T_g?

29. Small molecules of similar structure such as hexane and pentane are miscible and when mixed together form homogeneous liquids in all proportions. Yet polymers of similar structure are generally immiscible when mixed together and tend to phase separate. Can you think of a reason for this difference in behavior of small molecules versus polymers given the information that the higher the molecular weight of the polymers the more likely the immiscibility?

30. The synthesis of Kraton^TM (Fig. 7.17) involves a technique called "living polymerization," which is described in most textbooks of polymer science. What is "living polymerization?"

31. An exception to the general rule that polymers are immiscible arises when the chains allow a favorable interaction such as hydrogen bonding. Propose two polymer chain structures that could interact in this manner. Why would hydrogen bonding tend to overcome the factors favoring immiscibility of polymers as in Question 29?

32. There are some polymers that are curiously miscible in that their mutual attraction is not readily understandable. One of these pairs is poly(vinyl methyl ether) and polystyrene. One curious aspect of the miscibility of this polymer pair is that substituting deuterium for hydrogen in the aliphatic backbone of the polystyrene makes no difference to the miscibility, while substituting deuterium for the hydrogens on the benzene rings of the polystyrene causes the miscibility to be improved – that is, it occurs over a wider temperature range. Do you have any explanation for this effect, or any ideas for experiments that might yield further insight?

33. In what way does the structure of spandex have parallel features to the structure of nylons?

34. Formation of the "soft" segments of spandex via ring opening of tetrahydrofuran (Fig. 7.20) requires acid catalysis, while ring opening of ethylene or propylene oxide can be accomplished with water in the absence of catalysts. Why is the opening of the ether ring in ethylene or propylene oxide easier than that for tetrahydrofuran?

35. Review the material in Chapter 4 on polyurethanes and propose a mechanism for the formation of the urea and urethane linkages in the spandex structure exhibited in Fig. 7.19 for polymer formation via the reaction of the prepolymer exhibited in Fig. 7.21 with ethylene diamine.

36. Reaction of the prepolymer in Fig. 7.21 with ethylene glycol instead of ethylene diamine yields an alternate version of spandex. Show the structure of this version.

8

Ethylene and Propylene: Two Very Different Kinds of Chemistry

8.1
Ethylene and Propylene

Ethylene and propylene obviously have similar structures, and as we have learned in Chapter 1 (Section 1.11) are made available for industrial use from an identical source – steam cracking of petroleum fractions. But in this chapter we will discover the fundamentally different industrial chemistry of these two alkenes. We have already seen an important part of the chemistry when we described the ability of ethylene, but not propylene, to polymerize via a free radical mechanism (see Chapter 2, Sections 2.2 and 2.3).

Addition chemistry of alkenes is driven by the fact that conversion of a π to a σ-bond releases considerable energy. With ethylene, addition chemistry is, so-to-speak, the "only game in town." In certain examples this addition chemistry is followed by other chemical steps, which can reform the double bond. But the first step is always addition to the double bond. However, for propylene there is another kind of reactivity. The presence of the double bond weakens the carbon–hydrogen bonds in the adjacent methyl group. Breaking of a carbon–hydrogen bond adjacent to a double bond necessarily leads to a resonance-stabilized intermediate, which can be seen in comparing bond dissociation energies.

For example, while breaking a carbon–hydrogen bond in ethane costs 98 kcal/mole, breaking a carbon–hydrogen bond from the methyl group in propylene costs only 87 kcal/mole. And breaking a carbon–hydrogen bond in ethylene is even more difficult at 108 kcal/mole. These differing fundamental aspects of the characteristics of a double bond – addition to the double bond versus enhanced reactivity of bonds adjacent, that is, allylic to the double bond – strongly influence the industrial chemistry of ethylene and propylene.

8.2
The Industrial Importance of Ethylene and Propylene

Industrial importance hardly describes the roles of ethylene and propylene in the chemical industry. Ethylene is the largest volume product of cracking of petroleum fractions, produced in 2003 at the level of 25 *billion* kilograms per year in the USA, and this ethyl-

Tab. 8.1 Percentage use of ethylene and propylene to produce major derived chemicals

Ethylene-derived products	Percentage of ethylene production	Propylene-derived products	Percentage of propylene production
Polyethylene	58	Polypropylene	46
1,2-Dichloroethane, vinyl chloride/poly (vinyl chloride)	15	Acrylonitrile	12
Ethylene oxide and ethylene glycol	12	Propylene oxide and propylene glycol	10
Ethylbenzene/Styrene	6	Cumene for phenol and acetone	9
Vinyl acetate	2	Acrylic acid	5
Ethanol	0.7	n-Butyl alcohol	4
		Isopropanol	3
		2-Ethylhexanol	2

ene is converted to a 115 *billion* kilograms of a wide variety of chemical products. Propylene, the consumption of which is somewhat less, is also produced and used in huge volumes, thereby demonstrating the great importance of this petroleum-derived product. Thus these are the two most important basic chemicals. Although there are seven basic chemicals in all (ethylene, propylene, butadiene, benzene, toluene, *p*-xylene, and methane) which contribute the largest volume to the chemical industry, and each is important in its own right, ethylene and propylene are the most important both from the point of view of volume and correspondingly from the importance of the chemicals derived from them.

The major importance of both ethylene and propylene derives from the production of polyolefins. Approximately half of the production of these basic alkenes is used to synthesize polyethylene and polypropylene. But ethylene and propylene give us much more than the addition polymers described in Chapter 2. Poly(vinyl chloride), PVC, produced via the value chain (the increase in cost associated with production of one chemical from another) of ethylene to 1,2-dichloroethane to vinyl chloride to PVC, consumes about 15% of all ethylene produced. The chemistry associated with the production of vinyl chloride will be discussed in Chapter 10 (Section 10.10).

Including the basic addition polymers of ethylene and propylene, Tab. 8.1 shows a list constituting a *Who's Who* of the chemical industry. The foundations of the chemical industry rest on these basic products of the steam cracking of petroleum. In fact, it is the critical use of ethylene and propylene that makes petroleum fractions and steam cracking, our focus in Chapter 1, so important.

8.3

Ethylene Oxide and Propylene Oxide are Very Large Volume Industrial Intermediates derived from Ethylene and Propylene but must be Industrially Synthesized in Entirely Different Ways

Before the modern method of synthesis of ethylene oxide was invented, this epoxide of ethylene was made by classical chemistry. Chlorine dissolved in water is in equilibrium with hypochlorous acid. In a reaction described in mechanistic detail in Chapter 4 (Section 4.6, Fig. 4.5), hypochlorous acid adds to double bonds to form chlorohydrins, which can be converted to epoxides by treatment with base. Fig. 8.1 shows the conversion of both ethylene and propylene to their epoxides following addition of hypochlorous acid.

But while the industrial route to propylene oxide continued to follow this old path, the route to ethylene oxide made great advances. Catalyst development based on the properties of silver allows the direct reaction of oxygen with ethylene to produce ethylene oxide in a single step in high yield (Fig. 8.2). Union Carbide in 1937, twelve years after they started commercial production of ethylene oxide by the hypochlorous acid route, developed a process for the direct oxidation of ethylene-to-ethylene oxide. By 1940, some 10% of United States capacity for ethylene oxide was using the direct oxidation, and by 1975 all capacity used the new route.

There is a great advantage for production of ethylene oxide by the new method. Chlorine is not wasted. To make 1 ton of ethylene oxide using the hypochlorous acid route (Fig. 8.1) required 2 tons of chlorine and 2 tons of lime. An additional problem is that this method produced huge amounts of the byproduct $CaCl_2$, which had to be disposed of. This can be described differently, as seen in Fig. 8.1. For the hypochlorous acid route, an entire molecule of chlorine, Cl_2, is consumed to produce one molecule of ethylene oxide, which contains no chlorine at all.

We have seen before in the production of epoxy and polycarbonate resins intermediates that there is an economic penalty for wasting chlorine (see Chapter 4, Section 4.22). This economic problem, in addition to the dangerous characteristics of chlorine (a subject to be discussed in Chapter 10) consistently drives innovation to replace chlorine in industrial syntheses.

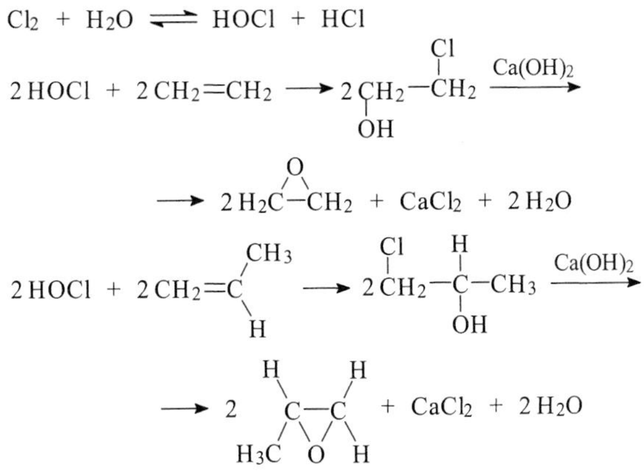

$$Cl_2 + H_2O \rightleftharpoons HOCl + HCl$$

$$2\,HOCl + 2\,CH_2{=}CH_2 \longrightarrow 2\,\underset{\underset{OH}{|}}{\overset{\overset{Cl}{|}}{CH_2}}{-}CH_2 \xrightarrow{Ca(OH)_2}$$

$$\longrightarrow 2\,H_2C\overset{O}{\overset{\diagup\diagdown}{-}}CH_2 + CaCl_2 + 2\,H_2O$$

$$2\,HOCl + 2\,CH_2{=}C\underset{H}{\overset{CH_3}{\diagup}} \longrightarrow 2\,\underset{\underset{OH}{|}}{\overset{\overset{Cl}{|}}{CH_2}}{-}\underset{}{\overset{\overset{H}{|}}{C}}{-}CH_3 \xrightarrow{Ca(OH)_2}$$

$$\longrightarrow 2\,\underset{H_3C}{\overset{H}{\diagup}}\underset{}{\overset{}{C}}{-}\underset{O}{\overset{}{C}}\underset{H}{\overset{H}{\diagdown}} + CaCl_2 + 2\,H_2O$$

Fig. 8.1 Ethylene and propylene are converted to their epoxides via the hypochlorous acid approach.

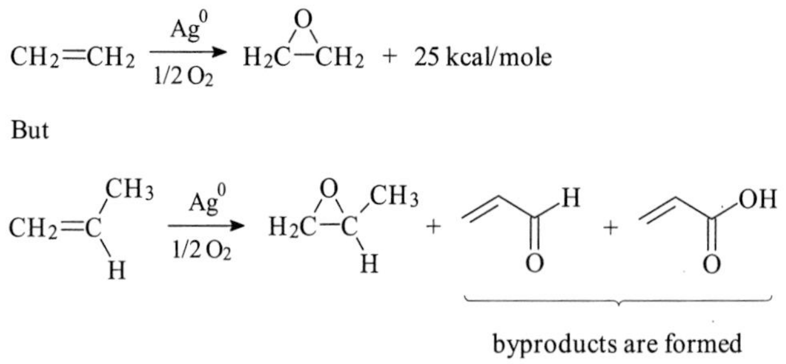

$$CH_2{=}CH_2 \xrightarrow[1/2\,O_2]{Ag^0} H_2C\overset{O}{\overbrace{}}CH_2 + 25\,\text{kcal/mole}$$

But

$$CH_2{=}C\overset{CH_3}{\underset{H}{\diagup}} \xrightarrow[1/2\,O_2]{Ag^0} H_2C{-}C\overset{CH_3}{\underset{H}{\diagup}} + \diagdown\!\!\diagup\!\overset{H}{\underset{O}{}} + \diagdown\!\!\diagup\!\overset{OH}{\underset{O}{}}$$

byproducts are formed

Fig. 8.2 Silver-catalyzed synthesis of ethylene oxide from ethylene and oxygen, and attempted use of this method for propylene.

The reactivity of the resonance-activated methyl group in propylene, however, blocks the silver-catalyzed oxidation path to propylene oxide from propylene since oxidation also takes place on the methyl group in competition with the double bond. Large amounts of unwanted byproducts are produced, especially acrolein (Fig. 8.2). The lower bond energy of the methyl group-bound hydrogens allows the oxygen catalyst system to break these bonds leading to oxidation at the methyl group and conversion to oxidized products.

The plants dedicated to the hypochlorous acid addition approach to ethylene oxide were kept on stream by substituting propylene for ethylene to produce propylene oxide (Fig. 8.1).

8.4
The Production of Propylene Oxide without using Chlorine

As described above, the resonance-activated reactivity of the methyl group in propylene made it necessary to continue to produce propylene oxide via the expensive process that wastes chlorine (see Fig. 8.1). It was natural that the chemical industry would seek a route to propylene oxide that avoided use of chlorine. This route was found in an indirect oxidation based on a reaction in which alkenes are converted to epoxides via reaction with hydroperoxides. The hydroperoxide oxidizes the propylene and is itself reduced to an alcohol.

The key to the reactive characteristic of hydroperoxides is in the weak and polarized oxygen–oxygen bond, RO–OH, which allows this peroxide bond to be broken, under certain conditions, by the π-electrons of a double bond (Fig. 8.3). Peracids, with the same peroxide bond, R–(C=O)–O–OH, behave similarly in their ability to convert alkenes to epoxides. *m*-Chloroperbenzoic acid, although too expensive for large-scale industrial processes, is a classical reagent used in small-scale laboratory syntheses of epoxides (Fig. 8.3).

The reaction path in Fig. 8.3 leads to an opportunity. One could use a hydroperoxide yielding a commercially valuable byproduct alcohol, ROH, in Fig. 8.3, and therefore add further value to this approach to propylene oxide. This is a 2-for-1-reaction idea that was conceived by Ralph Landau who headed a company called Scientific Design. We have seen the value of a 2-for-1 process before in Chapter 3 (Section 3.11) in the linked pro-

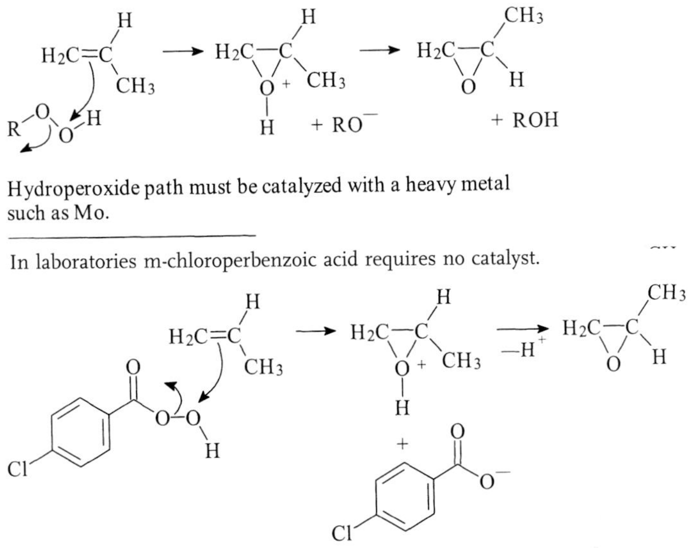

Hydroperoxide path must be catalyzed with a heavy metal such as Mo.

In laboratories m-chloroperbenzoic acid requires no catalyst.

Fig. 8.3 Mechanism for the epoxidation of double bonds using alkyl hydroperoxides and also *m*-chloroperbenzoic acid.

duction of phenol and acetone from cumene. The latter, interestingly enough, also involves a hydroperoxide.

The processes Landau designed were based on old chemistry explored in the 1930s by Shell. ARCO was also working on this technology, and the two companies collaborated.

There are multiple possibilities for Landau's conception of a 2-for-1 hydroperoxide method since any alkyl hydroperoxide may be used in principle. But only two were commercialized and both are used today (Fig. 8.4). The choice resides in the value of the alcohol produced and in the economics of the formation of the hydroperoxide. The latter factor depends on the availability of a hydrocarbon precursor that can be transformed to the hydroperoxide, which means the hydrocarbon should be removed by as few steps as possible from a cracking product of petroleum fractions (see Chapter 1). Reactivity of that hydrocarbon to produce the hydroperoxide means that there should be a weak carbon–hydrogen bond to allow insertion of oxygen from the air.

Oxygen is inexpensive, and therefore it is no surprise that it is a prized reagent for industrial processes – as long as air rather than pure oxygen can be used. This was precisely the situation in the conversion of cumene to its hydroperoxide on the route to phenol and acetone, which also provided the first industrial example of a 2-for-1 process as noted just above.

ARCO chemists studied two hydrocarbons that fit the bill outlined above. One was isobutane with a tertiary carbon–hydrogen bond. The other was ethylbenzene with a benzylic carbon–hydrogen bond. The advantage of using the hydroperoxide derived from isobutane or ethylbenzene to convert propylene to propylene oxide is that it creates two industrial products from one reaction, propylene oxide and the alcohol derived from the hydroperoxide (Fig. 8.4). And the alcohols derived from the hydroperoxides of isobutane

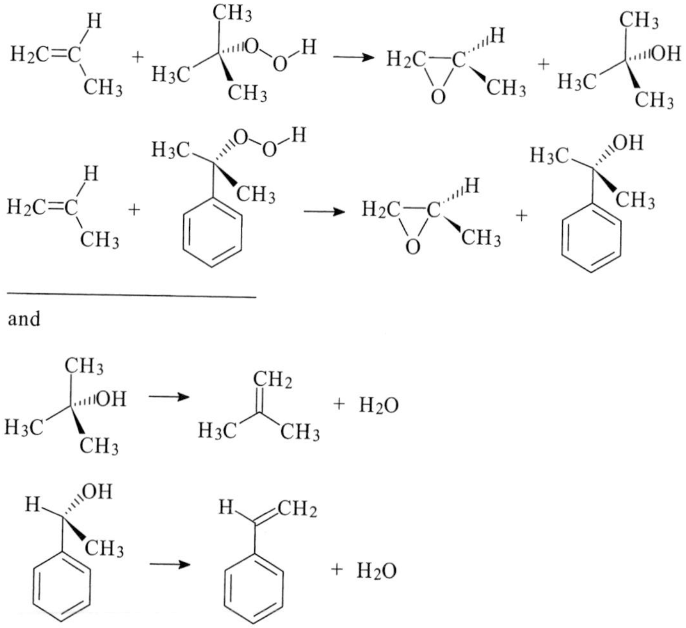

Fig. 8.4 The hydroperoxides of isobutane and cumene react with propylene to form the respective alcohols and propylene oxide.

and from ethylbenzene, *tertiary* butanol and phenylmethylcarbinol respectively, are useful industrial intermediates. By 1975, 40% of all propylene oxide was made in the United States by reacting *tertiary* butyl hydroperoxide with propylene, producing therefore *tertiary* butyl alcohol.

The value of *tertiary* butanol to the chemical industry arises in two ways. The first is its dehydration to isobutene and subsequent oxidation to methacrylic acid. This approach was discussed in Chapter 6 (Section 6.9). The second value of this alcohol is also dehydration to isobutene followed by reaction with methanol under acid conditions to form methyl *tertiary* butyl ether, MTBE, the additive to gasoline to replace tetraethyl lead as an octane improver (see Chapter 6, Section 6.10).

The other path to propylene oxide, based on the hydroperoxide of ethylbenzene, produces 1-phenyl-1-hydroxyethane as the coproduct (Fig. 8.4). The value here is derived from easy dehydration of this alcohol to styrene. In this manner, ethylbenzene produced by the electrophilic aromatic substitution of benzene with ethylene discussed in Chapter 3 (Sections 3.5–3.7) yields styrene in two ways, directly via loss of hydrogen (Chapter 3, Fig. 3.4) or indirectly via loss of water from the alcohol produced in the formation of propylene oxide (Fig. 8.4). It is therefore the value of the ultimate products, isobutene and styrene that stimulate industry to follow the hydroperoxide routes to propylene oxide as well as the elimination of the use and costs associated with chlorine.

8.5
Why did Dow maintain the Hypochlorous Route to Propylene Oxide?

Why did Dow continue with the chlorohydrin route to propylene oxide? The other companies went out of business. Dow produces chlorine in large amounts from the salt beds located near its plant in Michigan. Using the chlorohydrin route to propylene oxide allowed Dow to run its chlorine production facilities at a higher capacity, therefore reducing costs by economy of scale. The extra profit from the chlorine production and sales may have compensated, therefore, for the extra cost of the chlorohydrin route to propylene oxide. Another innovation greatly contributed to continued use of the chlorine-based method. Dow found an effective process allowing replacement of $Ca(OH)_2$ with $NaOH$, therefore producing the more easily disposed of byproduct $NaCl$ instead of the troublesome $CaCl_2$. In the early 2000s Dow, using the chlorine-based method, produces approximately 40% of the USA's propylene oxide.

There is another interesting advantage for Dow maintaining the old approach to propylene oxide. In avoiding the hydroperoxide route Dow avoids a 2-for-1 process. Although 2-for-1 processes may have great economic advantages and are prized for this reason, there is a tyranny associated with them. A market must exist for both products in the ratio in which they are produced. There have been times when excess production of styrene has been an economic burden on the 2-for-1 processes using the ethylbenzene hydroperoxide route (Fig. 8.4).

However, because of the potential profit from styrene production, Dow has nevertheless purchased propylene oxide/styrene technology from a Russian research institute. Economics is a very complex subject that is interwoven in fascinating ways with the process chemistry used by industry.

8.6
Before We continue to investigate the Difference in the Industrial Chemistry of Ethylene and Propylene, let's take a Diversion from the Main Theme of the Chapter.
Why are Ethylene Oxide and Propylene Oxide so Important to the Chemical Industry?
We find out by adding Water

What can be expected from an epoxide on addition of water? We don't have to look further than Chapter 4. In the formation of epoxy resin, the nucleophilic phenolic hydroxyl group of bisphenol A opens the epoxide ring of the epichlorohydrin (Chapter 4, Section 4.4). In the curing of an epoxy resin the nucleophilic curing agent, for example, a diamine, opens the terminal epoxide rings (Chapter 4, Section 4.5). In the formation of the polyether polyols used to crosslink polyurethanes, ring opening of epoxides is the critical reaction (Chapter 4, Section 4.17). Epoxides, which are subject to bond angle as well as torsional strain, are easily opened by nucleophiles. A similar reactivity can be expected with the nucleophile water. If water is added to ethylene oxide at high temperature, as shown in Fig. 8.5, the epoxide ring opens to produce ethylene glycol.

Reasoning by analogy is standard fare in chemistry, and offers a powerful means to predict the course of a chemical reaction. Unfortunately, in the reaction of water with the

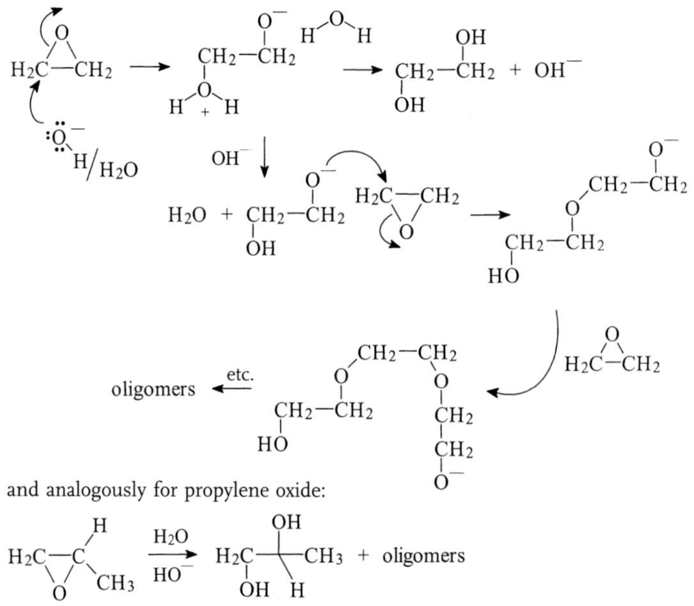

Fig. 8.5 The mechanism for the reaction of water with ethylene oxide. An intermediate leads to byproducts.

epoxides of ethylene and propylene, the predictive power arising from our knowledge of the differential reactivity of nucleophiles shows us that byproducts should be produced together with the desired ethylene and propylene glycol, respectively. The reason is straightforward. The initial product of the reaction of water with either ethylene or propylene oxide is, unfortunately, an alkoxide anion (Fig. 8.5), which is a stronger nucleophile for attacking unreacted starting material than is the water. This predicts that oligomers should be formed and in fact this is exactly what is found in the industrial process for the reaction of water with either ethylene oxide or propylene oxide (Fig. 8.5).

The formation of oligomers from the reaction of ethylene oxide and water requires an extra purification step, especially because an important use of ethylene glycol is to form polyesters, primarily poly(ethylene terephthalate) via transesterification (for parallel chemistry see Chapter 4, Section 4.10) with terephthalic acid, as shown in Fig. 8.6. The polyesters require high uniformity of structure to attain critical levels of crystallinity for their commercial uses as stiff polymers.

This need for uniformity of structure in the polyesters means that the ethylene glycol used in the synthesis must be separated from the oligomers (Fig. 8.5) by repeated fractional distillations. Ethylene glycol can be obtained in this manner because of its significantly lower boiling point compared to the oligomers (Fig. 8.5).

Someone familiar with a laboratory distillation apparatus would not easily recognize the industrial distillation process. The water–glycol mixture from the reactor is fed to multiple evaporators where the pressure is lowered in stages to remove the water for recycling. Vacuum distillation is then used to remove the ethylene glycol. The higher oligo-

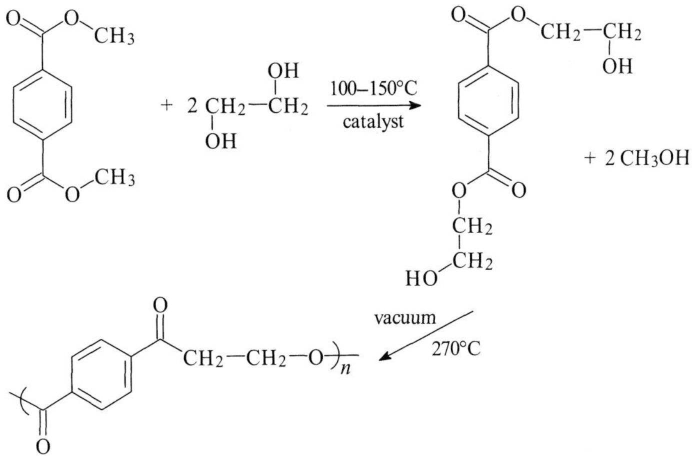

Fig. 8.6 The reaction of terephthalic acid and ethylene glycol to form the polyester.

mers remain in the "column bottoms." Then another distillation under lower pressure and higher temperature is necessary to distill these higher-boiling oligomers.

There is something revelatory about the chemical industry in this distillation. All distillations require the expenditure of considerable energy to gain the necessary temperature. Why pay for the energy if you can get it for free? Why not use the large exotherm from the silver-catalyzed oxidation of ethylene to ethylene oxide (see Fig. 8.2) for the energy source for the distillation? And so be it. Waste nothing!

In recent years, several billion kilograms of ethylene glycol were produced each year, while di- and triethylene glycols production amounted to about one-tenth of this amount. Ethylene glycol is always under price pressure because of excess capacity and profit depends on saving all costs possible. Lowering the cost of the energy for the distillation is therefore very important in this industrial process.

There are many uses of ethylene glycol not requiring this very high level of purity. Antifreeze consumes about 25% of ethylene glycol production and is the second largest use after polyesters (Fig. 8.6). Ethylene glycol finds use in pumps and industrial heating and cooling units. Ethylene glycol is also effective in defrosting aircraft wings and in de-icing runways at airports, although propylene glycol is preferred.

Ethylene glycol is certainly a useful molecule but it is also toxic, whereas propylene glycol is not. The liver oxidizes ethylene glycol to oxalic acid, which is the reason for its toxicity, whereas the same biochemical oxidation mechanism converts propylene glycol to benign lactic acid. Propylene glycol, although useful for de-icing aircraft wings, can also be used as a sweetener for foods – two seemingly incongruent applications.

8.7

Any Process that could produce Ethylene Glycol without Oligomeric Products would be Highly Desirable

How can the formation of oligomers be repressed? Have we seen a parallel problem before, and can we use its solution as an analogy for our current problem? In the formation of ethylbenzene and isopropylbenzene described in Chapter 3 (Sections 3.4 to 3.8), the major problem in the electrophilic aromatic substitution is formation of diethyl and even triethylbenzene and similarly, di- and triisopropylbenzene. The once-alkylated benzene, the desired product, is more reactive for further electrophilic aromatic substitution than benzene. The situation is precisely parallel. Ethylene glycol, the desired product, is more reactive with ethylene oxide than is water.

Raising the concentration of the benzene was the approach taken to solve the ethylbenzene and cumene problem of multialkylation. More benzene compensates for its lower rate constant for electrophilic aromatic substitution compared to ethylbenzene or isopropylbenzene. Similarly, increased proportions of water can compensate for its lower rate constant for reaction with ethylene or propylene oxide compared to the product glycol.

But even with a 20:1 molar excess of water, higher glycols, oligomers (Fig. 8.5) are obtained. Although ethylene glycol can be produced in excess of 90%, this is not good enough. Expensive separation steps to remove the oligomers from the ethylene glycol are still necessary as discussed above. Although alternative ideas exist for the exclusion of oligomers in forming ethylene glycol and propylene glycol from their respective epoxides, these have not been commercialized and so the water-based processes continue. There has not arisen a parallel to the zeolite magic bullet, which solved the multiakylation problems in production of ethylbenzene and cumene.

8.8

We've seen the problems arising from the allylic hydrogens in propylene. Does the Reactivity of these Hydrogens bestow any Advantages?

The short answer to this question is certainly yes, with two of the most important industrial intermediates, acrylic acid and acrylonitrile, produced directly via oxidation chemistry that depends on the reactivity of the methyl group of propylene. Let's look at the basis of the importance of these intermediates before discovering how they are synthesized. We have already seen this consideration to some extent for acrylonitrile in the synthesis of certain nylon intermediates (Chapter 5, Section 5.7). Let's look further at acrylic acid.

8.9

The Importance of Poly(acrylic acid) and its Esters

In the United States in the early 2000s, over 22 billion kilograms of industrial chemicals were derived from propylene. This remarkable number constitutes approximately 10% of all organic chemicals produced in the United States. We have seen these chemicals in Tab. 8.1 (Section 8.2). Of this 22 billion kilograms, approximately 3 billion kilograms are

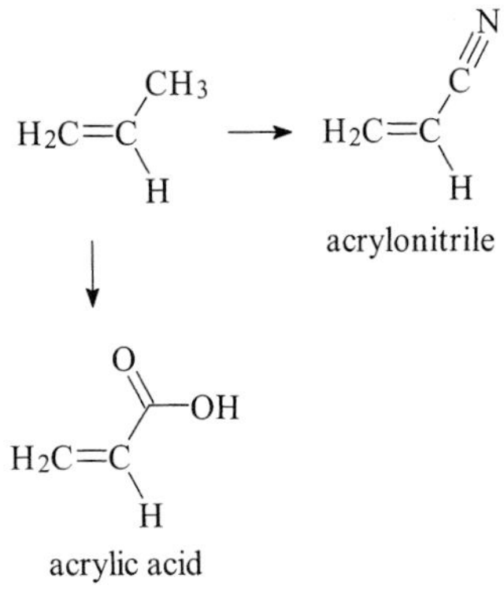

Fig. 8.7 Propylene leads to both acrylic acid and acrylonitrile. Reactivity of the methyl group of propylene yields two important industrial intermediates.

acrylonitrile

acrylic acid

chemicals that derive directly from the reactivity of the methyl group of propylene, which leads to acrylic acid and acrylonitrile (Fig. 8.7).

Many things around us result from acrylic acid and its esters. We have what is thought to be an essential for modern parenthood, that is, poly(acrylic acid) (Fig. 8.8), which is the basis for superabsorbent polymers for diapers. If you wish to convince yourself of the water-loving properties of the polyacrylic acid in diapers, simply add water to a bit of the powder, which can be obtained by tearing a diaper. Stir and observe the formation of a gel strong enough to allow you to turn the glass of water upside down. But a more remarkable use for this water-loving property of poly(acrylic acid) is a potential use in deserts to absorb water so that plants could grow. At this time, strips are available for home gardening containing polyacrylic acid and seeds. One critical factor here is that the plant has to attract the water more strongly than the poly(acrylic acid).

When detergents were first developed, phosphates were added to tie up the calcium and magnesium ions in the hard water that is commonly found in many parts of the world. Without the phosphates the detergents did not work well. Many "old-timers" can remember collecting tubs of rainwater to allow soap to be used for washing. This would avoid the insoluble calcium and magnesium fatty acid salts that would form from the soap if used in the hard water obtained from wells.

But the phosphates in detergents cause eutrophication – the formation of a scum of green algae on the surface of lakes that received the drainage from detergent use. The phosphates served as nutrients for the algae. This problem has been addressed in several interesting ways, but one important solution has been to add poly(acrylic acid) to detergents instead of phosphates to capture the calcium and magnesium ions in hard water. And there are many other uses of poly(acrylic acid) such as serving as viscosity control agents for clay slurries so that pipelines from where the clay is mined to where it is used

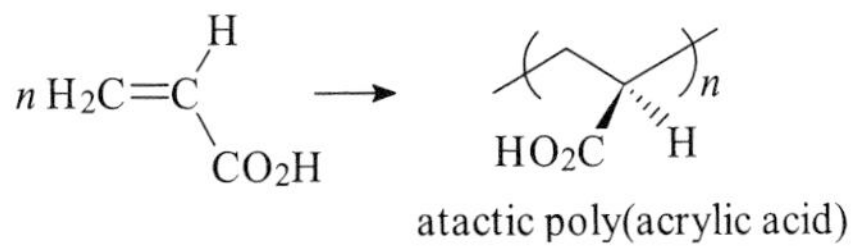

atactic poly(acrylic acid)

Fig. 8.8 The polymerization of acrylic acid.

can ship the slurry. Poly(acrylic acid) is also useful in boilers to control scale. And there are a variety of miscellaneous uses.

Poly(acrylic acid) is, however, hardly the major use for acrylic acid. Various esters of poly(acrylic acid) including the methyl, ethyl, *n*-butyl and 2-ethylhexyl esters are randomly copolymerized with vinyl acetate or methyl methacrylate and used as the binding agents for latex paints. Copolymerization was a subject we looked at in Chapter 7 for its importance in making elastomers (Sections 7.9, 7.11, and Fig. 7.12). And we learned of differing kinds of copolymers. In random copolymers, the overall properties of the macromolecule are the average of the properties of the copolymerized units – a full mixing one might say. Mixing of different monomers is used to give the macromolecule, the copolymer, properties that arise from desirable characteristics of both monomers. The esters of acrylic acid are prized as comonomers because they confer on the paint film the ability to withstand the deteriorating effect of UV light. In fact, more esters of acrylic acid are used in the chemical industry than is acrylic acid itself.

8.10
The Importance of Acrylonitrile and Polyacrylonitrile

The first interesting fact about acrylonitrile is that more than half of the United States production is exported. There is an interesting reason for this. The synthesis of acrylonitrile from propylene produces byproduct HCN, as we shall see below (Section 8.14), and this dangerous byproduct has blocked production of acrylonitrile in many countries of the world. Acrylonitrile however is quite important for industry world-wide. It is polymerized together with other vinyl monomers to copolymers with fiber-forming properties. These are wool-like fabrics discovered by DuPont with the well-known brand names Orlon[TM], Acrilan[TM], and Courtelle[TM].

At the time the fiber-forming discovery was made, there was no good way to make acrylonitrile on a large scale. DuPont set up three independent teams to find a good process, but Standard Oil of Ohio (SOHIO), a small regional oil company, beat DuPont to the draw. SOHIO discovered that acrylonitrile could be produced by catalytic oxidation of propylene in the presence of ammonia, by what is now called ammoxidation, a process closely related to the oxidation of propylene to acrylic acid. These processes to be discussed in detail in Section 8.14 displaced all earlier methods for producing acrylonitrile and acrylic acid (Sections 8.11 and 8.12).

Ammoxidation was also discovered in England by Distillers, a whiskey company with a talented chemical staff. This same group also engineered the large-scale production of phenol from cumene (see Chapter 3, Section 3.11) and the oxidation of naphtha to acetic

polyacrylonitrile **Fig. 8.9** The polymerization of acrylonitrile.

acid. Not bad for a small player in the chemical industry, but clearly a player specializing in oxidation! Naturally the competitive discovery of ammoxidation by two companies gave rise to a patent suit, which was settled in a way no other patent suit was ever settled. British Petroleum, BP, bought both companies!

The second problem concerned with polyacrylonitrile was that DuPont could not make good fibers on a large scale. Fibers are made by pushing the polymer through a plate with lots of little holes. The plate is called a spinneret. This technique was also used to make rayon and nylon fibers. The rayon required that cellulose be converted to a soluble xanthate, which was pushed through the holes as a solution into an acid bath to decompose the xanthate and regenerate the cellulose. The nylon, on the other hand, was melted and manipulated as discussed in Chapter 5, Section 5.4. However, polyacrylonitrile decomposes at its melting point and no solvent could be found for it. What could be done?

Carl Marvel, who we learned earlier was a longstanding DuPont consultant (Chapter 5, Section 5.8), addressed this problem of forming a fiber from polyacrylonitrile. He suggested creating new kinds of solvents. These would be highly polar but incapable of hydrogen bonding. We call these aprotic solvents. One of these, dimethylformamide, $(CH_3)_2NCHO$ – the amide formed by reacting dimethylamine with carbon monoxide (the anhydride of carbonic acid) – was successfully used to dissolve polyacrylonitrile for formation of fibers. A host of other aprotic solvents are used today in the synthesis of all kinds of molecules including polymers. Engineers also use them to separate, by distillation, compounds with similar vapor pressures. As an example, butadiene is separated from 1- and 2-butenes and isobutene by adding an aprotic solvent like acetonitrile, CH_3CN, to the mixture. The acetonitrile is attracted most strongly via polar interactions with the butadiene and allows the other C_4 olefins to distill off.

The third problem is that the fiber from polyacrylonitrile could not be dyed. A fiber without the ability to accept a color is worthless. The structure of polyacrylonitrile does not have an acidic, basic, or hydrogen bonding functional group to attract and hold the dye molecules. Incorporating in the chain of acrylonitrile, co-monomer units with dye-attracting properties, such as vinylpyridine, could in principle solve this problem. But incorporation of different groups in the chain causes loss of the fiber stiffness, a property highly prized. An excellent fiber must have what is called a *high modulus*, which means that very little elongation occurs when force is applied to stretch the fiber (with the notable exception of spandex (Chapter 7, Sections 7.12 and 7.13). This problem was solved when DuPont discovered how to create extremely high molecular weight polymers. Polymer molecular weight generally correlates with stiffness, so that some stiffness could be sacrificed allowing incorporation of the dye attracting units along the chain. Another benefit allowed incorporation of additives to dissipate any build up of electric charge. One no longer gets a shock when touching a metal doorknob after walking on an Orlon carpet.

These fibers, as well known as they are to the general public, are not the most important use of acrylonitrile. A large volume of acrylonitrile is also terpolymerized with 1,3-butadiene and styrene to form an important plastic called ABS resin (Fig. 8.10). ABS resin is widely used for electronic equipment including telephones, which have casings made from it. In addition to its plastic characteristics that can be beautifully tuned by adjustments in its three components, lipstick resistance gives it a big advantage for telephone casings over a cheaper competitor, polypropylene. ABS is also used for products as

ABS resin is a complex material produced by copolymerization of styrene with acrylonitrile

Fig. 8.10 The production of ABS resin.

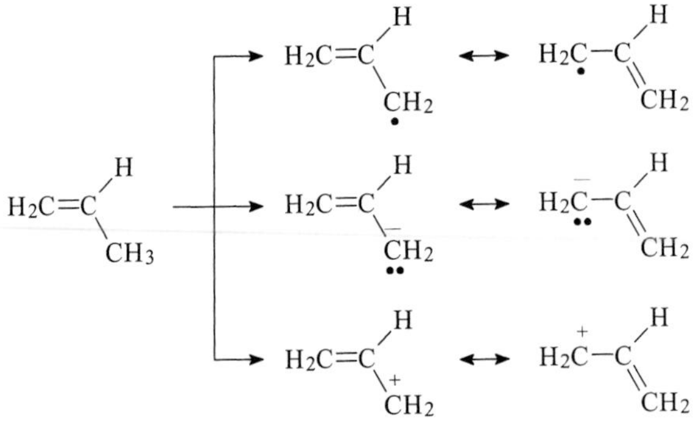

$\longrightarrow$ SAN (commercial name)

But the copolymerization is carried out in the presence of an elastomer such as SBR (Figure 7.12 (b)) or Butadiene Rubber (Figure 7.12 (c))

So that the ABS resin is a physical mixture of SAN with the elastomer, but also the SAN grafted (covalently bound) to the elastomer.

Fig. 8.11 Resonance stabilization accounts for the reactivity of the methyl group of propylene.

disparate as toilet seats and canoes, the latter application depending on its "tear resistance." This is a property that most plastics do not have, and it is used to describe the fact that on impact the ABS, like metal, will dent before it tears. White-water canoeists prize this property, which is also found in the polymer alloy of ABS and polycarbonate (Chapter 4, Section 4.19) used for side panels and doors of automobiles. ABS is also used for refrigerator doors. There are thousands of other uses with numerous household items being made all or in part from ABS resin.

Acrylonitrile is also copolymerized with butadiene to give an oil-resistant elastomer (see Chapter 7, Fig. 12), while copolymerization with styrene gives a clear polymer that is useful for decorative bottles, for example, cosmetic bottles. Polyacrylonitrile fibers, on heating to very high temperatures, yield carbon fibers used to reinforce epoxy resins for purposes as disparate as golf club handles, parts for stealth bombers, spaceship shuttle doors and many other uses where high strength and light weight are essential.

We set out to make a convincing case to you the reader of the importance of both acrylic acid and acrylonitrile to the chemical industry and to consumers of the world. As we shall see below, both chemicals are derived from the controlled oxidation of the methyl group of propylene – a property that arises directly from the resonance stabilization of the intermediates formed on breaking the carbon–hydrogen bonds of this methyl group (Fig. 8.11).

Fig. 8.12 The addition of HCN to acetylene yields acrylonitrile.

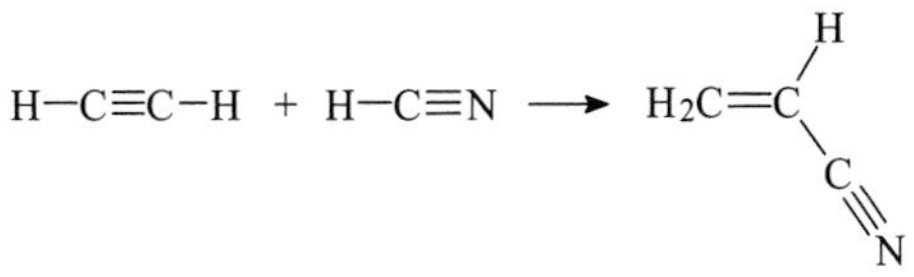

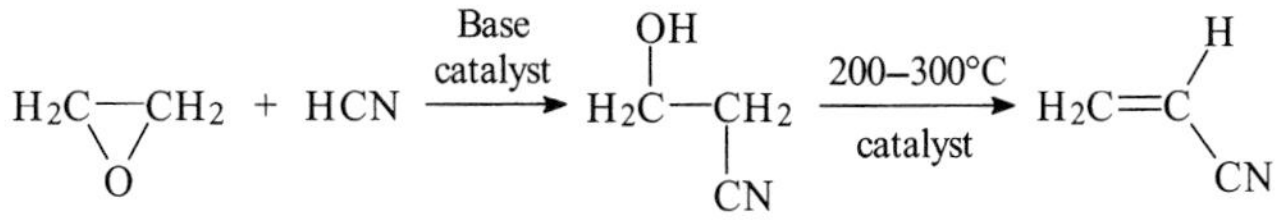

$$H-C\equiv C-H \; + \; H-C\equiv N \; \longrightarrow \; H_2C{=}C\underset{C\!\!\!\equiv\!\!\! N}{\overset{H}{\big\langle}}$$

$$H_2C\underset{O}{\overset{\diagup\!\!\!\diagdown}{-}}CH_2 \; + \; HCN \xrightarrow[\text{catalyst}]{\text{Base}} H_2C\underset{CN}{\overset{OH}{-}}CH_2 \xrightarrow[\text{catalyst}]{200-300°C} H_2C{=}C\underset{CN}{\overset{H}{\big\langle}}$$

Fig. 8.13 Reaction of ethylene oxide with HCN to yield the cyanohydrin, followed by elimination to acrylonitrile.

8.11
How was Arylonitrile produced in the "Old Days" before the "Propylene Approach" took over?

Although known since 1893, acrylonitrile's value was not appreciated until the 1940s when it was used as a component of nitrile rubbers produced for the war effort by both the United States and Germany (see Chapter 7, Section 7.9). It was not much later that acrylonitrile was discovered to form polymers with excellent fiber applications.

By 1950 acrylonitrile had become an important industrial intermediate, but its production was mired in "old-fashioned chemistry." One of the earliest of these routes to acrylonitrile was simply addition of HCN to acetylene (Fig. 8.12). There will never be a simpler and more direct route to acrylonitrile, but it has long been abandoned. This will be discussed in Chapter 10, where we focus on the evolution of the chemical industry, and discover all that is wrong with any process based on acetylene. Another early route without the disadvantages of acetylene but requiring the use of another dangerous chemical and also an extra step involves the reaction of ethylene oxide with HCN (Fig. 8.13).

Could one invent a single-step process to produce acrylonitrile directly from a product of steam cracking of petroleum? As we have noted above (Fig. 8.7, Section 8.10), and as we shalll discuss below, the answer is a resounding yes (Section 8.14).

8.12
How was Acrylic Acid produced in the "Old Days" before the "Propylene Approach" took over?

The first method of preparation of acrylic acid (structure shown in Fig. 8.7) was by air oxidation of the corresponding aldehyde, acrolein. This was first performed in the 1840s. It is interesting that the modern industrial method of producing acrylic acid by direct oxidation of propylene (Section 8.14) takes place in two steps producing first acrolein, which then must undergo air oxidation. However, in the 1840s acrolein was produced in an entirely different manner than it is today.

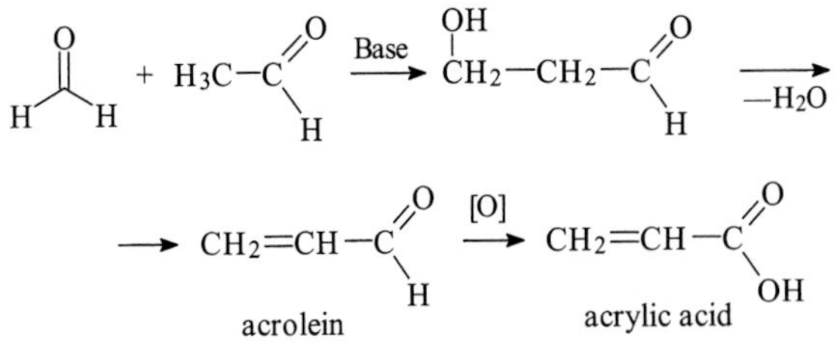

Fig. 8.14 The acetaldehyde-formaldehyde route to acrolein and to acrylic acid.

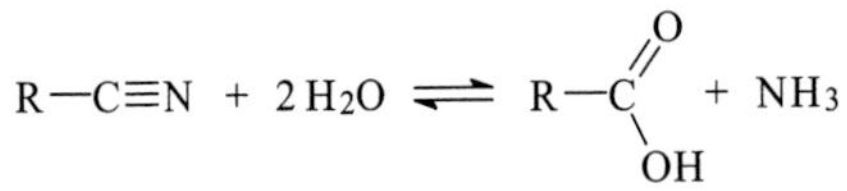

Fig. 8.15 The interconversion of nitriles and carboxylic acids.

Aldehydes are one step below carboxylic acids on the oxidation ladder, and are easily oxidized by air to carboxylic acids. An aldehyde that is stored in the presence of air will soon be contaminated with the carboxylic acid formed by its oxidation. There is an interesting aside to this fact in the wine industry. Wine has many aldehydes contributing to its flavor, and their easy conversion to the less-desirable carboxylic acids, which have a sour taste, is one of the reasons why wine once opened will go "bad" in a relatively short time. On the other hand, it is the carboxylic acids, particularly acetic acid, which converts wine to vinegar, adding value to "poor" wine.

Therefore, given acrolein, one could easily obtain acrylic acid. But where did the acrolein come from 160 years ago – this first and structurally simplest example of a conjugated carbonyl compound? It was derived from 3-hydroxypropanal, which is the product of one of the earliest examples of the aldol condensation (ol for alcohol and al for aldehyde), specifically a mixed aldol condensation, by reaction of formaldehyde and acetaldehyde (Fig. 8.14). The aldol condensation is a reaction that has played important roles in modern industrial chemistry and will be discussed in detail in Chapter 9 (Sections 9.3 and 9.4). As a reminder, we have come across another example of a mixed aldol condensation using formaldehyde in a proposed synthesis of methyl methacrylate (see Chapter 6, Fig. 6.12).

Because of the importance of acrylic acid as an industrial intermediate (Section 8.9) and the simplicity of its structure, there have been many industrial routes to its production. Among these was the obvious approach of hydrolysis of acrylonitrile to acrylic acid. Nitriles and carboxylic acid groups are in the same oxidation state, as we observed in the transformation of the cyanohydrin of acetone to methacrylic acid in Chapter 6 (Section 6.3) and in the conversion of adipic acid to hexamethylene diamine in Chapter 5 (Section 5.5, Fig. 5.7).

Water will convert a nitrile to a carboxylic acid (Fig. 8.15). Acrylonitrile can therefore be easily hydrolyzed to acrylic acid, so that adding a water-addition step to the syntheses in Figs 8.12 and 8.13 produces acrylic acid. Industry did take this path for a while.

8.13
An Early Example of Transition Metal Catalysis led to a Direct Route
from Acetylene to Acrylic Acid

The historically most important industrial method, now obsolete, produced acrylic acid from acetylene, but not via the addition of HCN to acetylene followed by hydrolysis (Figs 8.12 and 8.15). Rather, it was one of the earliest examples of the power of transition metal chemistry. But it was an exceptionally dangerous process involving acetylene, which was known to explode at times without apparent cause (see Chapter 10) and a nickel carbonyl (Reppe) catalyst, an exceptionally toxic chemical.

Chapter 10, which is concerned with the evolution of the chemical industry driven by a desire to eliminate danger to the industry's workers and to the environment, will focus on many obsolete processes based on acetylene, but that leading to acrylic acid was one of the most dangerous. Walter Reppe created the synthesis of acrylic acid from acetylene in the 1940s in Germany. The advantage of a direct route from acetylene made overwhelming sense in those days and the process was widely praised.

The key mechanistic step in the Reppe process (Fig. 8.16) is the formation of the nickel–hydrogen bond, which then adds across the triple bond of acetylene. The nickel, complexed with carbon monoxide molecules adds to one end of the triple bond and the hydrogen to the other end (Fig. 8.17).

In the Reppe process the weak bond between carbon and nickel formed after the initial addition step is replaced by a bond between carbon and one of the carbon monoxide molecules complexed to the nickel (Fig. 8.17). The rearrangement is driven by the far larger bond strength between a carbon–carbon bond versus a Ni–carbon bond. Replacement of metal–carbon with carbon–carbon bonds is a common theme in organometallic chemistry. In fact, we have seen such bond strength differences driving the rearrange-

$$Ni(CO)_4 + HCl \longrightarrow HNi(CO)_2Cl + 2\,CO$$

Fig. 8.16 The Reppe process for the synthesis of acrylic acid from acetylene and nickel carbonyl.

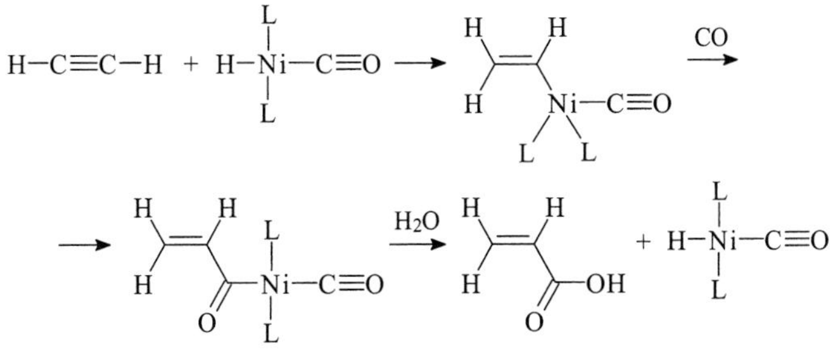

"L" is a ligand to the Ni

Fig. 8.17 Mechanistic steps for nickel-catalyzed conversion of acetylene to acrylic acid.

ment from a metal–carbon to a carbon–carbon bond previously in the Ziegler–Natta poly-merization of ethylene or propylene (see Chapter 2, Fig. 2.10, Section 2.10).

The path to acrylic acid is then finally completed by hydrolysis of the nickel-to-carbonyl bond, which regenerates the catalyst (Fig. 8.17).

For many years the Reppe process (Figs 8.16 and 8.17) was the only competitive route to acrylic acid. But a single example shows how eager the chemical industry was to re-place acetylene. Rohm and Haas, a company we have heard a great deal about for their development of methyl methacrylate chemistry (see Chapter 6, Sections 6.2 and 6.3), in-stalled the Reppe process at a plant in Houston, Texas in 1948, which eventually pro-duced almost 10 million kilograms per year of acrylic acid. In 1976, the direct oxidation of propylene (Section 8.14) pioneered by the SOHIO group and developed by the Japa-nese company, Nippon Shokubai, was introduced at the same site in Houston, and this led to closure of the Reppe process within one year. This is an example of shutdown eco-nomics – a subject focused on in Chapter 9, where we learn how economic forces driven by efficiency and environmental concerns, which are inexorably linked, push the evolu-tion of the chemical industry.

Now we are finally ready to come to the last act of our story – an act that has been pre-saged by our introduction to transition metals and an act that we have told you to expect – the direct oxidation of propylene. With all three carbons and the double bond already given to us in its structure, the only step necessary is the controlled oxidation of the methyl group. And the magic of transition metal catalysis is just right to do the job!

8.14
The Oxidation of Propylene to Acrylic Acid and to Acrylonitrile shuts down all Previous Processes. The Catalyst is the Key, but the Allylic Hydrogens are Essential

To create acrylic acid or acrylonitrile from propylene requires maintaining the double bond in propylene while replacing the three carbon–hydrogen bonds of the methyl group in a way to produce a carboxylic or a nitrile group. In such a process, the methyl group – which is the lowest oxidation state of carbon – is replaced with functional groups that are the highest oxidation state of carbon. The breaking of the three carbon–hydrogen bonds, the necessary oxidation, must be accomplished, however, in a highly controlled manner. Heating propylene with oxygen to high temperatures, for example, in the range of sev-eral hundreds of degrees would certainly oxidize the methyl group, but would also cause extensive decomposition of the entire molecule and could not be used as an industrial procedure.

It was not until 1959 that a catalyst was developed which had the required specificity and could be used in a commercial process with high output of product. The catalyst had the chemical characteristic of repressing the oxidative decomposition of the propylene and allowing control over the reactivity of the allylic hydrogens. There are several formu-lations of successful catalysts that have evolved over the more than forty years since the first successful catalyst was introduced, including one based on uranium, but they all ac-complish the same goal. And this goal is only possible because of the special characteris-tic of the carbon–hydrogen bonds on the methyl group of propylene, their allylic relation-

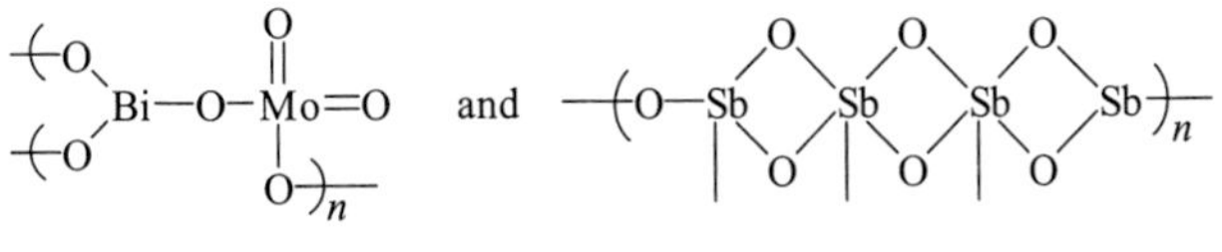

Fig. 8.18 Catalyst systems producing acrylonitrile and acrylic acid.

ship to the double bond (Section 8.1). Let's look at the structures of two of these catalysts as shown in Fig. 8.18.

Although each catalyst system will produce both acrylonitrile and acrylic acid, the conditions of use differ greatly. For acrylonitrile, propylene is mixed with air and ammonia and is passed over a fluid bed reactor containing the catalyst at between 400 °C and 500 °C. The role of the ammonia is to replace some of the oxygen atoms in the catalyst with nitrogen, which is necessary to produce the nitrogen-containing nitrile group. The catalysts containing only oxygen and the metal produce acrolein (Fig. 8.14), which, as we will see below, is the primary product of the oxidation of propylene. It must be oxidized further to acrylic acid in a second step.

For acrylonitrile, a single pass with an effective industrial catalyst will convert over 98% of the starting material to a mixture of the desired product and the byproducts, HCN and acetonitrile, CH_3CN. This industrial process is called ammoxidation. There is an interesting aside to this. An important source of HCN for the classical industrial synthesis of methyl methacrylate (see Chapter 6, Section 6.3) comes from ammoxidation. This inexpensive source of HCN – inexpensive because it is a byproduct of the oxidation of propylene leading primarily to acrylonitrile – helps to sustain the old approach to methyl methacrylate in the face of the greatly improved methods developed to compensate for its deficiencies as discussed in detail in Chapter 6.

The other byproduct of the conversion of propylene to acrylonitrile, acetonitrile, CH_3CN, also finds some value as an industrial solvent, although its production as a byproduct has been markedly reduced by improvement of catalysts for ammoxidation. Even the off-gases from acrylonitrile production are useful. The unreacted ammonia is reacted with sulfuric acid to produce ammonium sulfate, which is a fertilizer useful for soils that are too basic, a problem that is especially common in the Far East. Waste nothing!

The production of acrylic acid is a more complex operation than that for acrylonitrile for two reasons. First, two steps are required. Propylene is first oxidized to acrolein, which is subsequently oxidized further to acrylic acid. This means more complex engineering. Producing acrylic acid in one step from oxidation of propylene gives reduced yields. This is because the higher temperatures required to convert propylene to acrolein also bring about the more nearly complete oxidation of both the acrolein and the acrylic acid to CO_2 and H_2O. The initial oxidation from the hydrocarbon to the aldehyde is more difficult then the step from the aldehyde to the carboxylic acid.

In the two-step process, the catalysts and the temperature can be fine-tuned in each step to give optimum yields. The catalyst used in the step that produces acrolein is designed, under the process conditions, not to catalyze the further oxidation to acrylic acid. For the oxidation of propylene to acrolein, a catalyst useful at 305 °C is $Mo_{12}BiFe_2CO_3$-$NIP_2K_{0.2}$. An 88% yield of acrolein results, together with a 3% yield of acrylic acid. Another catalyst, $Mo_{12}BiFeW_2CO_4Si_{1.35}P_2K_{0.06}$ requires a temperature of 325 °C and yields

90% of acrolein and 6% of acrylic acid. The second phase conversion of acrolein to acrylic acid requires a lower temperature of 220–280 °C. Typical catalysts, which give yields above 90%, are $Mo_{12}V_{4.6}Cu_{2.2}W_{2.4}Cr_{0.6}$, $Mo_{12}V_2W_2Fe_3$, and $Mo_{12}V_3W_{1.2}Ce_3$.

Comparison of the stoichiometry of these two complex catalysts demonstrates the kind of empirical manipulation carried out by industrial chemists and engineers involved in developing catalysts for the oxidation of propylene. Chemists who develop catalysts are of critical importance to industry. Catalysts are at the heart of the chemical industry just as their biochemical cousins, enzymes, are the essence of life processes.

The oxidation of acrolein to acrylic acid is hardly new, having been known for over 160 years as the earliest path to acrylic acid (see Fig. 8.14). Certainly modern catalysts are used for the conversion, as we have just seen, but it is the first step – the highly selective reactivity of propylene to acrolein – that is the key part of the modern method. We have found out what several of the most important catalysts are made of (Fig. 8.18), but now the question arises as to how do the catalysts do such a specific job? How do they work?

All the very different catalyst structures work essentially identically. How could this be so, considering the different elements and different structures involved? The reason is that the necessary characteristics for catalyzing the selective oxidation of propylene can usually be found in a variety of highly oxidized metallic elements. Two characteristics of metals are important: (1) available empty orbitals to form π-complexes with alkenes; and (2) variable oxidation states.

Both characteristics of metals are associated with the partially filled or empty orbitals that are of low enough energy to accept electrons. The metal acts as a Lewis acid; that is, the π-electrons of a double bond, for example, can interact with the available orbitals of the metal. One could say there is coordination between the metal and the electron donor. The bond is appropriately called a coordinate bond. We have seen this bonding characteristic before in several examples of catalysts based on transition metals. Nickel, in the Reppe process, is already involved in a coordinate bond with carbon monoxide molecules when it forms a coordinate bond with the π-bond of acetylene (Fig. 8.17). We encounter a coordinate bond in the initial interaction of the π-bond of propylene or ethylene with the transition metal of the Ziegler–Natta catalyst (see Chapter 2, Fig. 2.10, Section 2.10). The coordinate bond, in which one of the partners contributes both electrons, is a weaker bonding force than a covalent bond, in which each partner contributes one electron, but is widely found in the interactions of organic molecules containing available electrons such as lone pair electrons or π-electrons with metals with valence electrons at high quantum levels.

The partially filled orbitals of many metals – especially those with high quantum levels – can easily add or lose electrons in moving toward a fully filled or completely empty state of the inert gas structure. Take for example the highest oxidation state of antimony, Sb, that is, +5. Inspection of a periodic table demonstrates that the neutral state of antimony is $4s^2\,4p^3$ $4d^0$ so that the +5 oxidation state brings the oxidized metal to the electron configuration of the noble gas, krypton. The next most stable electron configuration is +3 in which the 4p orbital is empty and the 4s orbital is filled, another commonly observed stable state of metals. As we will see below in an analysis of how antimony oxides act as oxidation catalysts for propylene, these two oxidation states, +5 and +3 play essential roles [factor (2) above], as do the available empty orbitals of the metal [factor (1) above].

Fig. 8.19 Antimony-catalyzed oxidation of propylene to acrolein.

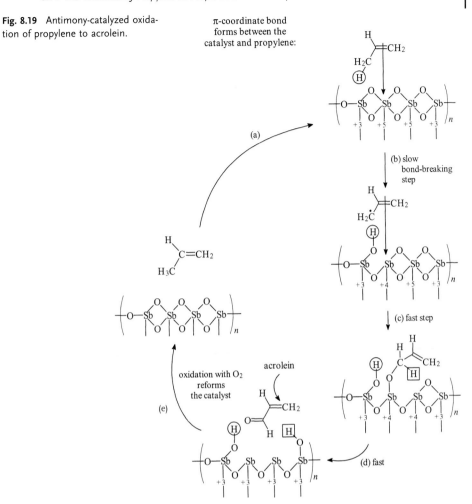

Contrast antimony with elements on either the extreme right or left of the periodic table, as for example an alkali metal such as sodium, where loss of a single electron leads to the configuration of the inert gas neon. Or consider chlorine where addition of a single electron forming the Cl^{-1} state attains the argon electron configuration. Variable oxidation states, which are so essential in the use of antimony oxides as oxidation catalysts, are unknown in those elements that "exist" closer to the edges of the periodic table.

The mechanistic flow chart for the catalyst based on oxides of antimony, Sb (Fig. 8.18) is presented in Fig. 8.19. The information in this chart is generated by several decades of work by chemists and chemical engineers in industry and academia to figure out how the catalytic oxidation of propylene works. This kind of work is exceptionally important because the knowledge gained can yield insight into how to improve these and related catalysts.

Focusing on the transformations of the propylene molecule reveals an essential understanding of how the catalyst works. The first step (Fig. 8.19 step a) is the entry of the propylene molecule into the catalytic cycle. This leads to a complex, an interaction or a coordination of the π-bond of the propylene with low energy available orbitals in the +5 oxidation state of the antimony (Fig. 8.19). This is factor (1) noted above.

What happens to the coordinate bond between the propylene and the metal? The antimony (Fig. 8.19a) is in the highest possible oxidation state, and highly substituted with oxygen. This structural characteristic coupled with the high temperatures used starts the oxidation of the propylene by homolytic breaking of one of the three equivalent allylic carbon–hydrogen bonds (Fig. 8.19b). The circled hydrogen in step (b) is transferred to oxygen. Here, as in all the chemistry of propylene, the key factor is the homolytic cleavage of the allylic carbon–hydrogen bond driven by the lowered bond strength caused by the resonance stabilization of the resulting radical. This step is the slowest in the process because a covalent bond, the C–H bond must be broken. However weakened a C–H covalent bond may be, a great deal of energy is still required to break it. In addition, an unstable oxidation state, Sb +4, is formed as the antimony is reduced following the breaking of one of the bonds between the antimony and oxygen (Fig. 8.19, step b).

While the allylic radical formed in step (b) maintains its coordinate bond to the metal, this coordinate bond is very weak compared to a covalent bond between carbon and oxygen. This weak bonding drives the next step (Fig. 8.19, step c) in which the coordinate bond is broken and replaced with the far stronger antimony-to-oxygen-to-carbon bond (Fig. 8.19, step c). Step (c) occurs at the expense of breaking another antimony-to-oxygen bond, therefore reducing a second antimony from +5 to +4. At this stage in the process, there are two antimony atoms in unstable +4 oxidation states; and, as can be seen, there remain two equivalent C–H bonds (one of these hydrogens is boxed), which are allylic to the double bond and therefore weakened. The next step (Fig. 8.19, step d), which produces the target product acrolein, is therefore a very fast one.

In the highly oxidized and high-temperature environment in which the catalyst is used we have seen that one of the allylic carbon–hydrogen bonds on the methyl group in propylene was broken with the hydrogen (circled) transferred to oxygen in the catalyst structure (Fig. 8.19). The carbon–hydrogen bonds in the methylene group in the oxygen-linked structure (Fig. 8.19, boxed hydrogen) are even more susceptible to homolytic cleavage. Why?

There are four reasons: (1) A secondary carbon–hydrogen bond is weaker than one in a methyl group by about 4 kcal/mole. (2) The allylic effect from the adjacent double bond is still present. (3) A carbon–hydrogen bond adjacent to an oxygen is further weakened for homolytic cleavage by about 12 kcal/mole. (4) On breaking one of the carbon–hydrogen bonds to the methylene group in the intermediate, the boxed hydrogen (Fig. 8.19), the antimony-to-oxygen bond breaks forming acrolein in which the aldehyde group is stabilized by conjugation with the double bond. And we have also taken the oxidation state of the antimony in steps (c) and (d) (Fig. 8.19) from the unstable +4 to the $4s^2$ $4p^0$ +3 state. Everything is pushing in the same direction.

The whole process is conducted in oxygen-enriched air to reoxidize the catalyst back to the original state, Sb +5 (Fig. 8.19, step e). The reoxidized catalyst then allows transformation of another molecule of propylene to acrolein.

Although it is not necessary to discuss the process in detail, the identical principles and driving forces are involved in the use of the identical catalyst for conversion of propylene to acrylonitrile with ammonia acting to replace two of the oxygen atoms bridging the two Sb +5 atoms with NH groups.

8.15
Summary

In focusing on ethylene and propylene and the differences in their industrial chemistry we have taken a complex path through a great deal of chemistry. The essential difference between the two alkenes arises from the presence of the relatively weak allylic carbon–hydrogen bonds in the methyl group of propylene. For the same reason that ethylene but not propylene can be polymerized via a free radical initiation, ethylene can be oxidized directly to ethylene oxide, a reaction path that is not available to propylene. For this reason, propylene was initially constrained to a path to propylene oxide wasting chlorine while ethylene oxide could be converted to its epoxide more efficiently. But this blocked path to propylene oxide led to innovation that allowed the chemical industry to produce propylene oxide using hydroperoxide chemistry, which then allowed other useful co-products to be produced in the same process.

And this focus on production methods for ethylene and propylene oxide motivated us to look in some detail at the industrial use of these chemicals, which led to an understanding of how the three-membered ring is opened with water, and how the glycols produced have value. In following this chemistry of the production of ethylene and propylene glycol we found analogies to problems faced by industrial chemists in the production of both ethylbenzene and isopropylbenzene by electrophilic aromatic substitution.

Then we discovered how the methyl group of propylene and its relatively easily broken allylic C–H bonds, which caused difficulties in polymerization and epoxide formation, could lead to industrial intermediates of great importance, acrylic acid and acrylonitrile. After looking at several of the commercial products arising from these chemicals, our attention was directed to the historical background of these monomers and how they were formerly produced. This examination took us on a path through industrial chemistry that is highly informative but nevertheless obsolete and that finally led us to acetylene and transition metal chemistry. All of these obsolete methods were eventually replaced by a catalytic method that allowed direct conversion of propylene to acrylic acid and acrylonitrile. And although this direct oxidation of propylene is only possible because of the relatively weak C–H bonds of the methyl group we came to appreciate the effort necessary to develop commercially viable catalysts.

In focusing on one of the catalysts used to oxidize propylene to acrolein, which can then be easily oxidized to the valuable acrylic acid, we discover the essential characteristics of all the varied oxidation catalysts that can be used for this purpose. For this understanding we had to review some fundamental inorganic chemistry of the oxidation states of transition metals. Then we came full circle to understand how the basic principles involved in conjugation and resonance, projected onto the allylic bonds of propylene, drove these very efficient industrial processes, the catalytic oxidation of propylene to both acrylic acid via acrolein and to acrylonitrile.

Some of the subjects treated in this chapter are listed below. These are key words and terms that act as reminders of the chapter's contents and should become a valuable part of your chemical vocabulary.

- Epoxides
- Hypochlorous acid addition to double bonds
- Resonance stabilization and allylic bonds
- Hydroperoxides and formation of epoxides
- 2-for-1 reactions
- Ethylene glycol
- Propylene glycol
- Oligomerization
- Relative nucleophilicities
- Poly(ethylene terephthalate), Dacron®, Mylar®, Terylene®
- Acrolein
- Acrylic acid and its polymers and commercial uses
- Acrylonitrile and its polymers and commercial uses
- Reppe process
- Acetylene chemistry
- Transition metals
- HCN
- Aldol condensation
- Interconversion of nitriles and carboxylic acids
- Nickel carbonyl
- Propylene oxidation and ammoxidation

Study Guide Problems for Chapter 8

1. Use a standard table of bond dissociation energies in any organic chemistry textbook to evaluate the difference between reaction of propylene with chlorine to yield: 3-chloro-1-propene and HCl; 1-chloro-1-propene and HCl; and 1,2-dichloropropane.

2. Tab. 8.1 lists the major industrial products synthesized from either ethylene or propylene. Draw chemical structures for each of these products. For the several of these products that have been discussed in Chapters 1–7, outline the industrial synthesis. For those that have not already been presented, attempt to develop a synthetic path.

3. Show the chemical reactions that interfere with the attempted free radical polymerization of propylene. What is the probability that 3-phenyl-1-propene or 1-butene could form addition polymers via free radical initiation?

4. What mechanistic conclusions can be drawn from the mode of addition of the hypochlorous acid to propylene in Fig. 8.1? Show the step leading to this addition and how this intermediate is converted to propylene oxide.

5. While the use of alkyl hydroperoxides to synthesize epoxides requires catalysis (Fig. 8.3), perbenzoic acids such as the commonly used m-chloroperbenzoic acid, converts alkenes to epoxides without a catalyst. Offer an explanation for the enhanced reactivity of perbenzoic acids.

6. Assuming that the energy of activation for conversion of a hydrocarbon to a hydroperoxide by reaction with oxygen is directly related to the bond dissociation energy of the carbon–hydrogen bond involved, judge the relative reactivities of the weakest bond in: methane, ethane, toluene, isobutane, ethylbenzene. Use one of these hydrocarbons to write a free radical chain mechanism for the formation of a hydroperoxide.

7. The two byproducts of the hydroperoxide route to propylene oxide are *tertiary* butyl alcohol and 1-phenylethanol, which are both converted to olefins by loss of water. How might the details of the elimination of water from these alcohols be different? Which might require the higher temperature process? Could acid catalysis be effective in this process? If so, write a mechanism for such a catalyzed process.

8. Propose a process by which Dow converts NaCl to Cl_2. What happens to the sodium?

9. If propylene oxide were produced from a hydroperoxide made from isotopically labeled oxygen and then heated with water to produce propylene glycol, where would the isotopic oxygen reside in the final product?

10. Although the internuclear angles of an epoxide ring are 60°, the interorbital angles are far larger. What is the difference between these angular measures? Epoxides are subject to torsional as well as angle strain. Describe the source of this torsional strain. Can the torsional strain energy be estimated from the conformational properties of ethane and if so, how?

11. In the synthesis of poly(ethylene terephthalate) in Fig. 8.6, the first step is conducted with a temperature high enough to distill methanol but not ethylene glycol, while the second step occurs at a higher temperature and under vacuum so that ethylene glycol can be removed. Write a mechanism for both steps consistent with and explaining the necessity for the varying temperature conditions.

12. The transformations shown in Fig. 8.7 arise via large-scale industrial oxidations that would be difficult on a laboratory scale. Propose a synthesis from propylene to acrylonitrile and also to acrylic acid using reactions suitable for an undergraduate organic chemistry laboratory.

13. What is a reasonable structural basis to explain why poly(acrylic acid) is so much more water absorbent than poly(methacrylic acid)?

14. Acrylonitrile is polymerized by a free radical chain mechanism just as for polyethylene in Chapter 2. Suggest what steps of the chain mechanism should be adjusted to attain the extremely high molecular weights that DuPont needed, as discussed in the text, to allow incorporation of a dye-attracting monomer into the polymer. Do you have any ideas how changing the process conditions might affect the steps that need to be adjusted?

15. If one replaced acetylene with methylacetylene in Fig. 8.12, and propylene oxide for ethylene oxide in Fig. 8.13, predict the products using the identical procedures for reaction with HCN.

16. Looking ahead to discussion of the aldol reaction in Chapter 9, write a mechanism for the formation of the aldol derived from formaldehyde and acetaldehyde in Fig. 8.14. What kind of catalyst is necessary? Are there any other products that could reasonably form in this reaction?

17. Write reasonable mechanisms for the transformations between nitriles and carboxylic acids shown in Fig. 8.15. Suggest a catalyst that would work well for each step, and specify expected intermediates.

18. The addition across the double bond by Ni and H in the Reppe process in Fig. 8.17 could remind you of hydroboration in which B and H also add across a double bond in the first step. Although the end products of hydroboration and the Reppe process are entirely different, what experiment could you carry out to test the similarity or difference for the first addition step? Could acetylene serve for this experiment?

19. Arrange the atomic orbitals in proper sequence to be filled with electrons and determine the stable oxidation states of bismuth and molybdenum. How do these states compare with those of antimony for use as an oxidation catalyst of propylene as shown in Fig. 8.19.

9

The Demise of Acetaldehyde:
A Story of How the Chemical Industry Evolves

9.1
An Interesting Example of Shutdown Economics

Shutdown economics is a term widely used in the chemical industry. It's meaning is simple. A new economical process is devised against which older, less economical processes cannot compete. The old process is shut down. We saw shutdown economics in Chapter 8 when many companies that produced propylene oxide via the chlorohydrin route, which wasted expensive chlorine, had to close down their plants after the ARCO hydroperoxide route went on stream (Chapter 8, Section 8.4). Or we saw it again when Rohm and Haas shut down their acetylene-based Reppe process and replaced it with the direct oxidation of propylene to produce acrylic acid (Chapter 8, Section 8.13).

Shutdown economics also applies to acetaldehyde. We are going to study in this chapter how the earliest methods for production of acetaldehyde were first shut down by newer production methods for this chemical. And then how the major need for acetaldehyde as a high volume chemical disappeared because more economical processes, which did not require acetaldehyde, were discovered for making its two most important derivatives – n-butanol and acetic acid. Hence, the production of acetaldehyde decreased from 675 million kilograms in the United States in 1975 to 135 million kilograms in 2000. This is an important story in the chemical industry. n-Butanol and acetic acid are produced by the chemical industry in large amounts. Both are important chemicals that were formerly made from acetaldehyde but are now produced more economically without acetaldehyde.

The major use for n-butanol is to make esters of acetic acid, acrylic acid (see Chapter 8, Section 8.9) and methacrylic acid (see Chapter 6, Section 6.2), that is, n-butyl acetate, n-butyl acrylate, and n-butyl methacrylate. n-Butyl acetate is an important solvent. The other two are monomers that form copolymers with vinyl acetate, which are then emulsified with water and other additives and pigments to make water-based paints. The role of the polymer is to hold the pigment in the film that coats the surface after the water evaporates. The butyl group is ideal for this job. A longer alkyl group would be so hydrophobic that it could not be emulsified with water in the paint. A shorter alkyl group would not be hydrophobic enough to protect the painted surface against moisture. n-Butanol and its esters, like n-butyl acetate, are also excellent solvents, and n-butanol reacts with ethylene or propylene oxides to form still other types of solvents.

The new replaces the old:

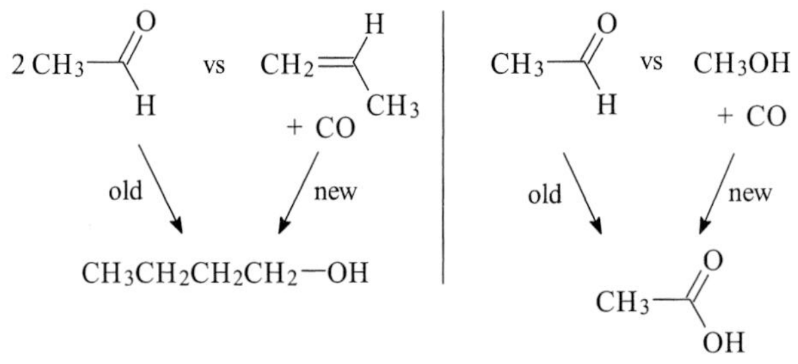

Fig. 9.1 Acetaldehyde versus new processes for the source of the carbon atoms in *n*-butanol and acetic acid.

Acetic acid is important because it reacts with ethylene to form vinyl acetate (see Chapter 10, Section 10.8). This is a variation of the Wacker reaction of ethylene with water to produce acetaldehyde, the process (Section 9.8) we shall focus on in this chapter as one example of shutdown economics. Most vinyl acetate is polymerized, as noted above, for use in water-based paints. Poly(vinyl acetate) is also a component of adhesives. "Elmer's Glue" is one example (Chapter 10, Section 10.9). Acetic acid has other important uses, one of which is as a solvent for the oxidation of *p*-xylene to terephthalic acid. As discussed in Chapter 8 (Fig. 8.6), terephthalic acid reacts with ethylene glycol to form commercially important polyester that finds uses as varied as fibers, films, and plastic bottles. The polyester is by far the most important synthetic fiber. Acetic acid may also be converted to acetic anhydride, a reactant for the syntheses of aspirin made in thousands of kilograms per year, and for cellulose acetate, produced in hundreds of millions of kilograms per year.

Because of the importance of *n*-butanol and acetic acid, acetaldehyde became an important chemical intermediate. The question is, what happened to replace acetaldehyde?

The aldol condensation, which we have already come across in Chapter 6 (Section 6.11), and which will be discussed further below (Section 9.3), had been used to convert the two carbons of acetaldehyde to the four carbons of *n*-butanol. It has now been replaced by a reaction between propylene, carbon monoxide and hydrogen, the so-called oxo or hydroformylation reaction to which we have also been introduced (Chapter 6, Section 6.11). The other major use of acetaldehyde, conversion to acetic acid by direct oxidation, has been replaced by a reaction in which carbon monoxide is inserted between the hydroxyl and methyl groups in methanol (Fig. 9.1). These modern processes for production of *n*-butanol and acetic acid have shut down the old routes from acetaldehyde and therefore the need for acetaldehyde as a large-scale intermediate in the chemical industry has diminished. At the same time the production of acetic acid and *n*-butanol have greatly increased.

9.2

An Aspect of the Evolution of the Chemical Industry that begins with World War I

The story of the rise and subsequent demise of acetaldehyde starts with a need that arose in World War I, not for *n*-butanol or acetic acid, but for acetone. Many years ago, cordite was a new explosive made from nitrocellulose, nitroglycerin (see Fig. 4.14) and petroleum jelly and was a great advance over previous "powders" used as propellants for everything from small arms to artillery. The British used cordite to fire some 258 million shells in World War I. Cordite was more reliable and, most important, gave off very little smoke so that snipers' positions were not revealed and machine guns could be fired without so much smoke that the gunners could not see past the haze.

Cordite was known as "smokeless powder." But cordite had to be formulated with acetone, a solvent that could only be made by a cumbersome process that involved the dry distillation of calcium acetate or the destructive distillation of wood. An additional difficulty was that much of the acetone in England had to be imported from the United States via shipping lanes that were under attack by the growing power of German submarines.

A better process and one that could be centered in England was desperately needed. Almost simultaneously with this rising need for a reliable source of acetone, a young chemist, Chaim Weizmann, who had emigrated to England from the Russian village Motol, was working at the University of Manchester. Weizmann published a paper describing the fermentation of carbohydrate substances with an organism named *Clostridium acetobutylicum*. A mixture of ethanol, acetone and *n*-butanol resulted. Weizmann was approached by the British war office and asked if his process could be scaled up to levels necessary for cordite production for the war effort. The answer was that the fermentation could be conducted on a far larger scale and if the engineering problems could be solved the answer would certainly be yes.

Although the acetone could be easily separated from the other components with their differing boiling points and a solution to the cordite problem was at hand, there were some unexpected bumps in the road. The carbohydrate source initially used was corn that was imported from America. The German submarine offensive that had threatened the importation of acetone now threatened to cut off the supply of corn. This problem led to a huge campaign in England to collect horse chestnuts. School children all over the country participated and horse chestnuts became a major source of carbohydrates for Weizmann's process and therefore played an essential role in the war effort. Inspection of school record books from the time documents the importance of the children's participation.

Lloyd George, the prime minister at the time of the war, was so grateful to Weizmann for the discovery of this route to acetone that he introduced him to the foreign secretary, A. J. Balfour who invited Weizmann to ask for his reward. Balfour was not prepared for Weizmann's request. There was a movement afoot called Zionism. Its objective was to create a homeland for the Jews in what was called Palestine, which had been under British control for many years. From this came the Balfour declaration of 1917, which stated that Britain looked with favor on the founding of a Jewish state in that area. This declaration played a key role in the creation of that state, which was founded 31 years later after another war in which Nazi concentration camps reinforced the desperate need of the Jews for their own state. Weizmann became the first president of Israel.

In the fermentation process producing acetone for the manufacture of cordite, the recovered *n*-butanol had only few uses, and most of the *n*-butanol was therefore burned as a fuel. Did *n*-butanol ever find a use? The answer to this question arises from an unlikely direction. Henry Ford conceived the idea of assembly line production for automobiles but the stumbling block was the last step, the painting of the automobile. Linseed oil-based paints (Chapter 4, Section 4.10) dried slowly, whereas assembly-line production requires that each operation proceed quickly. The Dupont Company came to the rescue with nitrocellulose-based lacquers. Although nitrocellulose was itself an explosive, when properly compounded it made excellent automobile paint. This paint, however, required *n*-butyl acetate as a solvent. This was a major and increasingly important use for *n*-butanol, which was the starting material from which *n*-butyl acetate was made by esterification with acetic acid. And to obtain the *n*-butanol the Weizmann process was used throughout the world.

But as is the nature of the chemical industry, the importance of *n*-butanol stimulated development of improved methods for its production and soon it became apparent that this essential molecule could be made in a better way from acetaldehyde rather than by fermentation. How can acetaldehyde be converted to *n*-butanol? This involves one of the classical reactions of organic chemistry, the aldol condensation, a reaction we have twice fleetingly introduced as we've noted in Section 9.1.

Let's first look at the industrial use of the aldol condensation and in this way come to understand its importance. Then we can step back and find out how to produce acetaldehyde, the molecule that undergoes the aldol condensation for the production of *n*-butanol. And finally we will discover the new chemistry that finally shut down the large-scale production of acetaldehyde.

When the whole story unwinds we will find shutdown economics galore and we shall follow a path through the chemical industry that gives its managers headaches and a yearning for what are called robust processes, that is, processes that are not easily shut down by dislocations in the industry. But let's not get ahead of the story.

9.3
The Aldol Condensation leads to n-Butanol

A necessary structural feature for all aldol condensations is that one of the aldehyde molecules taking part in the reaction has hydrogen bound to the α-carbon, the carbon adjacent to the aldehyde group. Carbon-bound hydrogens of this type are called active hydrogens. In the self-condensation of acetaldehyde both partners have active hydrogen atoms in the methyl groups, while in the acetaldehyde–formaldehyde condensation (Chapter 8, Fig. 8.14) only the acetaldehyde has this structural feature. Such carbon bound hydrogen is acidic enough to have a proton abstracted even by aqueous bases. In fact in the self-condensation of acetaldehyde, sodium hydroxide in water even at near to ice temperatures can cause the reaction by initiating loss of a proton from the methyl group of one of the acetaldehyde molecules (Fig. 9.2).

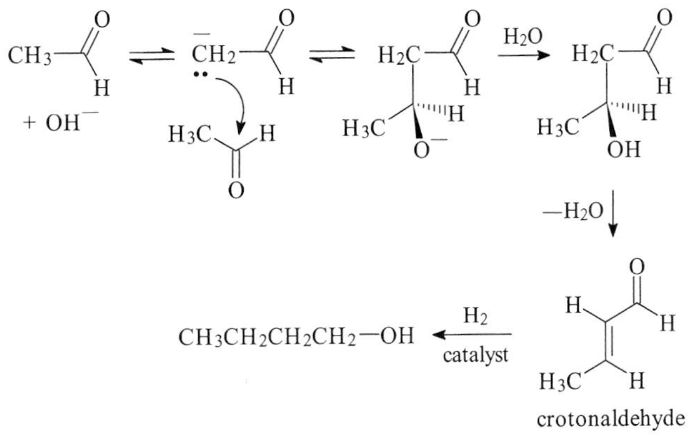

Fig. 9.2 Aldol condensation of acetaldehyde leading to crotonaldehyde and *n*-butanol.

The acidity of carbon-bound hydrogen adjacent to a carbonyl group is remarkable compared to the hydrogen atoms of alkanes. For example, the loss of a proton in an unfunctionalized hydrocarbon is subject to an equilibrium constant more than fifty orders of magnitude smaller than for acetaldehyde. No known base can act to abstract a proton from an unfunctionalized alkane. But resonance is at work in the loss of a proton from the α-carbon of acetaldehyde. The negative charge – the carbanion formed at this α-carbon as a consequence of the proton loss – can be delocalized into the carbonyl group.

Many organic molecules with a C–H group adjacent to a functional group allow for such resonance stabilization of negative charge. Examples are esters, nitriles (see Chapter 5, Section 5.7) and nitro groups, among many others. These C–H bonds act as proton donors with appropriate bases to provide a large number of important chemical reactions. In all of these reactions the resonance-stabilized carbanion acts as a nucleophile, allowing formation of new carbon–carbon bonds. As we have seen in Fig. 9.2, this activity is precisely the role of the carbanion formed from acetaldehyde.

The carbonyl function group is well situated for nucleophilic attack, and we have seen this characteristic in nucleophilic addition to carbonyl functions in ketones, as in the synthesis of methyl methacrylate (Chapter 6, Section 6.3) and in substitution reactions at an acyl carbonyl in the synthetic path of methyl methacrylate (Chapter 6, Fig. 6.5). We have and shall see it again in the synthesis of polycarbonates (Chapter 4, Section 4.20, Chapter 10, Section 10.13). Consistent with this reactivity pattern of carbonyl compounds, the nucleophile formed from one acetaldehyde reacts by attacking the aldehyde carbon of a second acetaldehyde molecule (Fig. 9.2) producing a product that is both an aldehyde and an alcohol, an aldol.

Crotonaldehyde is produced from the aldol derived from acetaldehyde via elimination of water (Fig. 9.2). The driving forces for all elimination reactions are identical, the entropy gain by elimination of a small molecule, and the stability of the eliminated molecule, in this case water. The stability of the conjugated double bond is high, and loss of a molecule of water is therefore inevitable under the mildest stimulation.

Crotonaldehyde is not far removed from *n*-butanol (Fig. 9.2). Both the double bond and the aldehyde group can be catalytically reduced with hydrogen under mild conditions to produce *n*-butanol, and in fact for many years this approach was the standard method for production of *n*-butanol in the chemical industry. However, the self-aldol condensation of acetaldehyde (Fig. 9.2) to form *n*-butanol is no longer used, having been shut down by improved methods to be discussed later in this chapter. Before we look at these new methods for producing *n*-butanol, let's see how the aldol reaction survives in other areas of the chemical industry.

9.4
Reactivity Principles associated with the Aldol Condensation

Following the industrial definition, a polyhydric alcohol is a molecule with multiple hydroxyl groups. As discussed in Section 4.17 of Chapter 4, the hydroxyl groups of the polyhydric alcohols presented in Fig. 4.18 were used as nucleophilic groups to open the epoxide rings of ethylene and propylene oxide to form what are industrially known as polyether polyols. The polyether polyols were then used to synthesize polyurethanes by reaction with isocyanates (Chapter 4, Section 4.16). Three of these multiple hydroxyl-containing molecules that are used to form polyether polyols (Fig. 9.3), pentaerythritol (Chapter 4, Fig. 4.18), trimethylolpropane, and neopentyl glycol are made by the aldol condensation. For this reason, let's look more closely at the synthesis of these industrial chemicals.

The molecules in Fig. 9.3 are all made by aldol condensations between two different aldehydes in a so-called mixed aldol condensation. Mixed aldol condensations give rise to problems if both aldehyde molecules have active hydrogens, that is, if each aldehyde has a C–H bond on the carbon adjacent to the carbonyl group. In this situation, the product mixture will depend on the relative reactivity of each of the aldehydes to act as nucleophile and electrophile. Which aldehyde will form the carbanion that will attack the carbonyl of the other aldehdye? This is a difficult situation to control. However, even in a mixed aldol condensation between two aldehydes in which only one has active hydrogens, multiple products may arise. And the latter variety of aldol condensation is the situation for the production of the polyhydric alcohols shown in Fig. 9.3.

For a single product to be produced in an aldol condensation, certainly a highly desirable situation for any industrial synthesis, certain conditions must be satisfied. If only one of the aldehydes has active hydrogens, the process must be engineered so that the carbanion formed from the aldehyde with the active hydrogen can only react with the aldehyde *without* the active hydrogens. The latter aldehyde is formaldehyde, as we will see for the formation of all the industrial intermediates in Fig. 9.3.

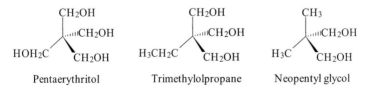

Fig. 9.3   Structures of pentaerythritol, trimethylolpropane, and neopentyl glycol.

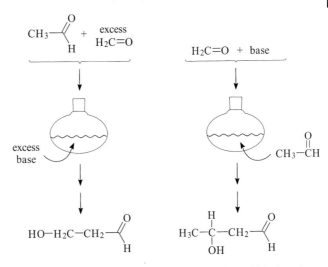

Fig. 9.4 Reaction conditions to favor reaction of acetaldehyde with formaldehyde and avoid self-condensation of acetaldehyde.

This engineering goal can be accomplished, as expressed in the mixed aldol condensation of acetaldehyde and formaldehyde, by using a large excess of the formaldehyde. Alternatively, one could engineer the reaction so that all of the active hydrogen-containing aldehyde, that is, acetaldehyde, is rapidly converted to a carbanion so that there is no free acetaldehyde to react with the carbanion. These ideas are expressed in Fig. 9.4.

Fortunately, there is another factor at work to help achieve the desired reaction path between the competitive reactions (Fig. 9.4). The reaction of the acetaldehyde-derived carbanion with formaldehyde is intrinsically far faster than with unionized acetaldehyde. Even if the acetaldehyde and formaldehyde were in equal concentrations, the reactive preference would be with formaldehyde.

Various reasons can be given for the difference in reactivity of the carbonyl groups of acetaldehyde and formaldehyde. One is that the groups surrounding the central carbon in the four-coordinate intermediate formed on nucleophilic addition to formaldehyde are smaller than those from any other carbonyl compound including acetaldehyde (Fig. 9.5). This circumstance makes formaldehyde faster-reacting because there is less steric hindrance, less crowding, in forming the four coordinate intermediate.

Alternatively, one could suggest that the electron-deficient carbonyl carbon is stabilized in acetaldehyde compared to formaldehyde by the higher level of substitution of the former. Just as carbocations are stabilized by higher substitution (see Chapter 1, Section 1.11), an electron-deficient carbonyl group is subject to the same relief. Because it is the electron deficiency that gives the carbonyl group its reactivity to nucleophiles, this relief of the electron deficiency would reduce the reactivity.

From either of the two points of view – take your choice or take both – formaldehyde should be more susceptible to nucleophilic addition, as can be well documented by the data in Fig. 9.6. In the equilibrium for the hydration, that is, the addition of water to formaldehyde and acetaldehyde, formaldehyde hydration is favored by a factor of 1000.

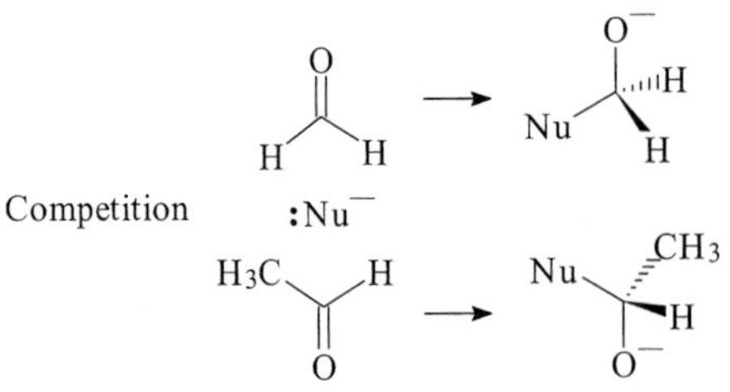

Fig. 9.5 Four-coordinate intermediate formed on nucleophilic addition to aldehydes.

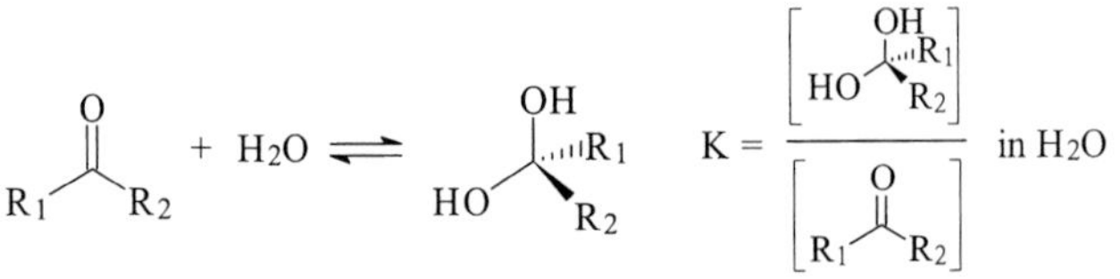

Competition

Fig. 9.6 Hydration equilibrium constants for addition of water to carbonyl compounds.

Structure	K
$CH_2=O$	2.3×10^3
$CH_3CH=O$	1.1
$(CH_3)C-CH=O$	0.23
$H_3C-\overset{O}{\overset{\|}{C}}-CH_3$	1.4×10^{-3}
$FCH_2-\overset{O}{\overset{\|}{C}}-CH_3$	1.1
$F_3C-\overset{O}{\overset{\|}{C}}-CF_3$	1.2×10^6

And ketones undergo hydration to a far smaller extent than aldehydes (except in special circumstances) – a fact connected to the lower rate of addition of nucleophiles to ketones compared to aldehydes, which can be ascribed to the same two reasons noted above.

The kinds of reactivity considerations discussed above are very important for the industrial processes that produce the three most important polyhydric alcohols, pentaerythritol, trimethylolpropane, and neopentyl glycol (Fig. 9.3). In all three, one of the components is formaldehyde while the other aldehydes, those with the active hydrogens, are respectively, acetaldehyde, n-butyraldehyde, and isobutyraldehyde. The higher reactivity of formaldehyde means that a very large excess of formaldehyde does not have to be used to gain the desired reaction path; that is, to form the aldol product from reaction of the higher aldehyde with formaldehyde rather than reaction of the higher aldehyde with itself (Fig. 9.4). Formaldehyde cost is lowered this way, and byproducts arising from the reaction of formaldehyde (which would take place if the formaldehyde were in very large excess) with the final polyhydric products are suppressed. Everything works in the right direction!

The discussion above also pertains to the first aldol condensation, carried out in the nineteenth century, which was the mixed aldol condensation of acetaldehyde and formaldehyde, leading to the first synthesis of acrolein (see Chapter 8, Fig. 8.14). Here, 1 mole of acetaldehyde was reacted with 1 mole of formaldehyde. A high yield of the mixed aldol product, from the reaction of the enolate of acetaldehyde with formaldehyde, was driven by the same principle, the higher reactivity of formaldehyde.

9.5
Pentaerythritol and other Polyhydric Alcohols synthesized via Aldol Condensations

Although it may not be obvious on looking first at the structure of the highly symmetrical polyhydric alcohol, pentaerythritol (Fig. 9.3), the central carbon is derived from the methyl group in acetaldehyde. Three of the –CH$_2$OH groups arise from three molecules of formaldehyde, while the remaining –CH$_2$OH group is derived from the aldehyde group in acetaldehyde. How did this happen? It becomes clearer on inspection of the intermediate state shown in Fig. 9.7. In a series of three steps, at moderate temperature in the aqueous Ca(OH)$_2$ used in the industrial process, each one of the hydrogens on the methyl group of acetaldehyde is removed sequentially, followed in turn by attack of the resulting nucleophilic carbon on a formaldehyde molecule. In this way the three formaldehyde molecules become the three –CH$_2$OH groups in the aldehyde intermediate, **1**, shown in Fig. 9.7.

The industrial process then turns to a reaction discovered by Stanislao Cannizzaro, an Italian whose life spanned the last three-quarters of the nineteenth century. Cannizzaro was responsible for the not inconsequential advance that led to the ability to determine the molecular weight of molecules. Without knowing the molecular weight, the proportion of the elements in a molecule, which was available information before his discovery, would never allow determination of the structure of a molecule. In other words, without Cannizzaro's discovery, the field of chemistry would have stopped dead in its tracks, at

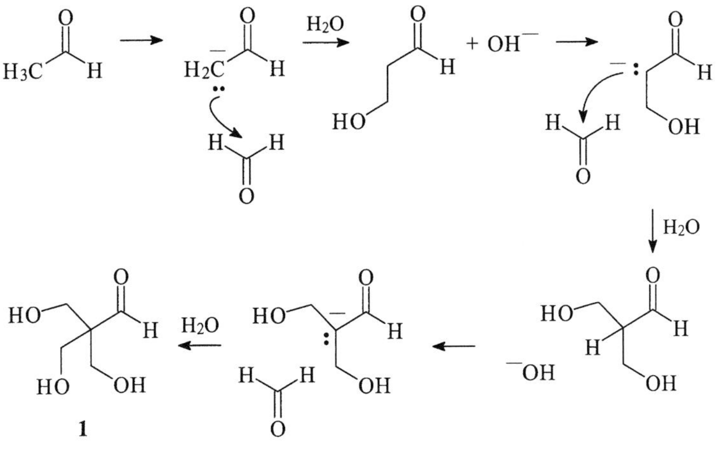

Fig. 9.7 Synthesis of the penultimate intermediate for the production of pentaerythritol.

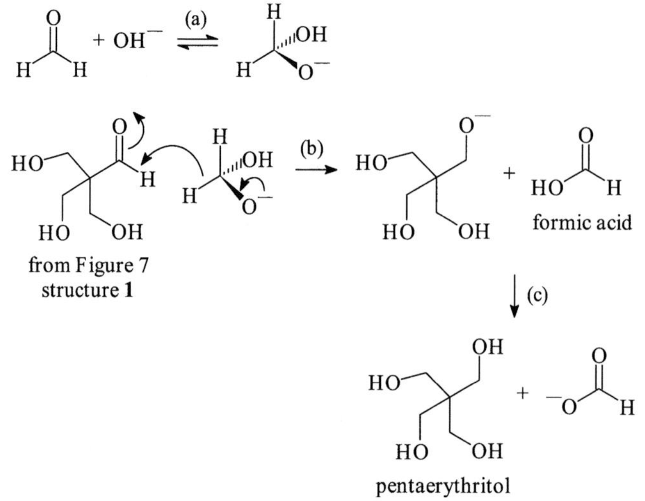

Fig. 9.8 The mechanism of the Cannizzaro reaction to form pentaery-thritol.

least until someone else came along who made that discovery. Cannizzaro also did other things, and one of these was the invention of a reaction which has been named after him and which is used for the last step in the production of pentaerythritol.

In the basic aqueous solution in which pentaerythritol is produced, hydroxide anions, OH⁻, are nucleophilic enough to add to the carbon of formaldehyde as shown in step (a) of Fig. 9.8. This attack creates the conditions for the Cannizzaro reaction. The intermediate alkoxide formed from formaldehyde (Fig. 9.8, step (a)) reforms the carbonyl group and in doing so transfers hydrogen with two electrons, a hydride ion (Fig. 9.8, step (b)), to the aldehyde group of the aldol intermediate (1, in Figs 9.7 and 9.8). Formic acid is formed in this step (Fig. 9.8, step (b)), which immediately transfers a proton to the alkoxide and forms pentaerythritol (Fig. 9.8, step (c)).

The Cannizzaro reaction is exceptional in two regards. First, addition of OH⁻ to an aldehyde, formaldehyde, leads not to the usual addition but rather to substitution. Nucleophilic attack at aldehydes and ketones usually leads to addition, with substitution occurring only for acyl carbonyl groups (Chapter 6, Section 6.3, Fig. 6.3). Second, the group that leaves arises from breaking a carbon–hydrogen bond with the bonding electrons leaving with the hydrogen.

The aldehyde group of the aldol, which is the aldehyde group of the starting acetaldehyde, is reduced to the fourth –CH₂OH group, while the formaldehyde, which donated one of the hydrogens, is oxidized to formic acid. This not only forms pentaerythritol but the formic acid produced also neutralizes the Ca(OH)₂ solution (Fig. 9.8).

A prominent use of pentaerythritol is reaction with concentrated nitric acid to form pentaerythritol tetranitrate (PETN). This molecule is a highly explosive compound belonging to the same class as nitroglycerin, which we have learned about in the story of dynamite and the Nobel Prize (see Chapter 4, Section 4.13). PETN is not as old as nitro-

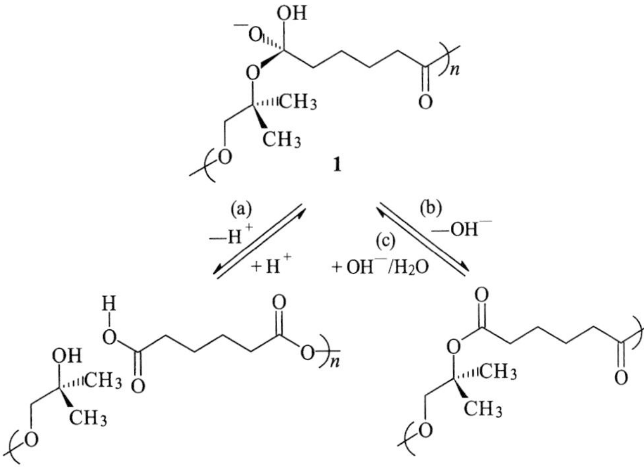

Fig. 9.9 The structure of polyester made from neopentyl glycol.

glycerin, and only became available after World War I. Perhaps it was because of PETN's well-known extreme shattering force when it explodes, coupled with a low sensitivity to explosion until initiated, that PETN was chosen by a terrorist, known as the "shoe bomber" who packed his sneakers with the explosive and tried to set them off on a flight from Paris to Miami on December 22, 2001.

As with the related nitroglycerin, PETN can also be used to relieve muscle pain in certain heart conditions (see Chapter 4, Section 4.13), thereby giving another face to this otherwise dangerous material. It also demonstrates the point that chemicals are neither "bad" nor "good." The moral aspect comes not from the chemical, but from the value judgments of the humans who use the chemical: relieve your pain or blow you up!

The other two polyhydric alcohols in Fig. 9.3, neopentyl glycol and trimethylolpropane, are produced from mixed aldol condensations of formaldehyde with the two isomeric butyraldehydes, n-butyraldehyde and isobutyraldehyde by closely parallel processes to that used to make pentaerythritol.

The special properties of neopentyl glycol (Fig. 9.3) have earned for it a commercial niche, which arises – as surprising as it may sound – from the advantage of steric hindrance. The two hydroxyl groups in neopentyl glycol allow reaction with dicarboxylic acids to form polyesters, a class of condensation polymers we have discussed earlier in the section on alkyd resins (Chapter 4, Section 4.10) and also in Carother's earliest polymerization experiments (see Chapter 5, Section 5.2). Forming the four-coordinate intermediate (1 in Fig. 9.9) in this particular situation (Fig. 9.9, step (a)) is, however, difficult and requires special conditions. The presence of the two methyl groups interfere sterically with the esterification reaction by making the four-coordinate intermediate (1, Fig. 9.9) too crowded.

However, from this steric difficulty in forming the four-coordinate intermediate (1, Fig. 9.9), which then forms the ester in step (b) (Fig. 9.9), comes value. Once the polyester is formed it is difficult to hydrolyze to the ester, which also occurs through the same four-coordinate crowded intermediate (1, Fig. 9.9). Adding to the difficulty of hydrolysis is the hydrophobic character of the methyl groups adjacent to the ester bond, which repels the water molecules necessary for hydrolysis of the ester (step (c), Fig. 9.9). The polyesters therefore produced from neopentyl glycol are exceptionally stable and can be used under conditions of exposure to moisture and heat where other polyesters would fail. Hydrolysis of the polyester leads to failure because the polymer chain is broken into smaller fragments with subsequent loss of polymer properties. This is precisely the reason why water had to be removed so rigorously in the first synthesis of polyesters at DuPont by the Carothers' group (see Chapter 5, Section 5.2). Those polyesters, however, were not protected by the steric and hydrophobic effects encountered in the polyesters synthesized from neopentyl glycol (Fig. 9.9).

9.6
A Prominent "Plasticizer" is synthesized via an Aldol Condensation

Among the many industrial products produced using the aldol condensation, even in addition to the polyhydric alcohols, no single one is more important than 2-ethylhexanol, which comes from the self-aldol reaction of *n*-butyraldehyde. This follows a path parallel to the self-condensation of acetaldehyde in Fig. 9.2. Enolate formation from *n*-butyraldehyde is followed by attack of the enolate at the aldehyde carbon of a second *n*-butyraldehyde forming the aldol, which then loses water to form the conjugated double bond, which is subsequently catalytically saturated with hydrogen to form 2-ethylhexanol (Fig. 9.10).

Most 2-ethylhexanol is used to form esters to be used as plasticizers, which are used primarily with poly(vinyl chloride). Plasticizers have been called "internal lubricants," and their function is to disrupt the forces that attract the chains to each other, the kinds of forces that were discussed in Chapter 7 (Sections 7.11, 7.12) as interfering with chain motion. In the presence of the plasticizers, the polymer chains can more easily slip by

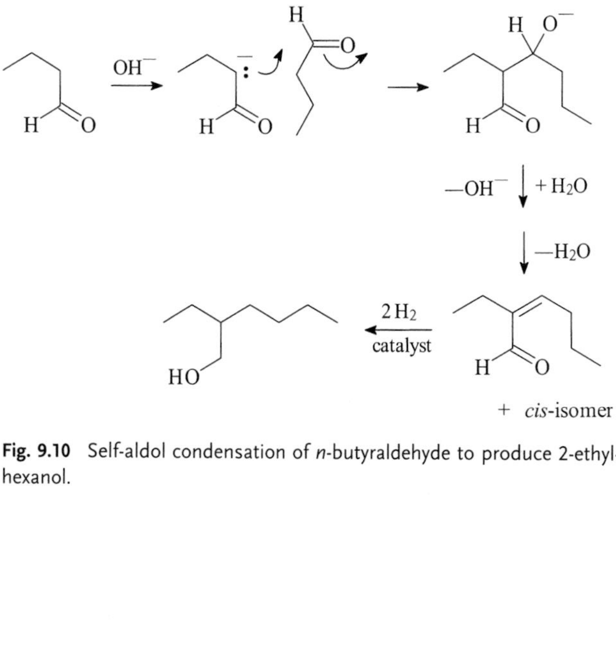

Fig. 9.10   Self-aldol condensation of *n*-butyraldehyde to produce 2-ethyl-hexanol.

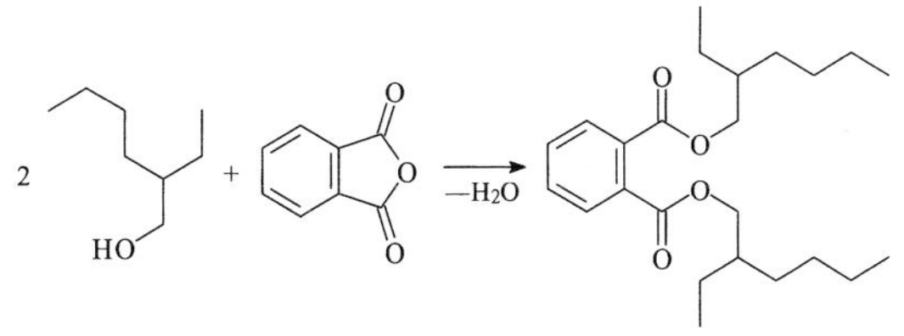

Fig. 9.11 Structure of the phthalic
ester of 2-ethylhexanol.

Di-2-ethylhexyl phthalate
a "plasticizer"

each other. The successful plasticizer is therefore a molecule designed to produce disorder, and for such a role a disorderly molecule works best. What is a disorderly molecule? A general answer is a molecule with an irregular structure, a molecule that has difficulty crystallizing and/or melts at a low temperature. Such molecules generally have high solubility – a principle that was discussed relative to the crystallization properties of the various isomers of bisphenol A (see Chapter 3, Section 3.10).

Some polymers can be changed greatly by introduction of plasticizers. For example, poly(vinyl chloride) (Chapter 10, Section 10.10) is a hard solid, but once plasticized it becomes leather-like. Some 80% of all plasticizers produced find use in combination with poly(vinyl chloride).

The earliest use of a plasticizer appears to be the use of camphor added to cellulose nitrate to allow easier fabrication. This process was patented in 1870, thereby demonstrating the long history of plasticizer use deriving from the realization that adding small molecules could inexpensively alter the properties of polymers. But the successful plasticization of cellulose nitrate had mixed results. It allowed the use of this polymer to make motion picture film (celluloid) in the early part of the twentieth century. However, as for most multiply nitrated compounds such as nitroglycerin (see Chapter 4, Section 4.13), trinitrotoluene (TNT), and pentaerythritol tetranitrate (PETN) (Section 9.5), nitrated cellulose was subject to flammability and violent decomposition, which led to a number of serious fires in motion picture theaters. In fact, at the same time nitrocellulose was being used for motion pictures, it was used in a different formulation for production of the smokeless powder, cordite, of importance to World War I (see Section 9.2). Switching to cellulose acetate eventually solved the motion picture problem.

2-Ethylhexanol becomes an excellent plasticizer when it is esterified with phthalic anhydride, as shown in Fig. 9.11. Two "disorderly" molecular moieties combine in a single molecule and in addition a stereoisomeric mixture results, which depresses the melting point even further. Moreover, it has a low volatility and therefore does not easily escape from the formulation, which would make it far less useful and also potentially an environmental hazard.

9.7
The Grandfather Molecule of the Aldol Condensation is Acetaldehyde.
How was and is Acetaldehyde produced?

Acetaldehyde's long history starts in the 1700s when it was first produced by oxidation of ethanol. However, it took until 1835 to determine the structure that led to the name, (al)cohol (dehyd)rogenated. The structural knowledge arose from the work of Justus von Liebig, an influential German professor of chemistry who created combustion analysis, a technique still used today to determine the relative amounts of the elements in organic molecules. In Liebig's method the sample is burned in a stream of oxygen, thus producing water from the hydrogen as well as carbon dioxide from the carbon present. These combustion products are collected separately and weighed, and from this information the molecular formula is calculated.

It was in this manner that Liebig determined that ethanol and acetaldehyde differed by a molecule of hydrogen – in other words, that the ethanol was dehydrogenated. This first method for production of acetaldehyde from ethanol is in use to a limited extent even today, although the reaction hardly resembles those early procedures. In modern industrial processes, ethanol is passed over a silver catalyst at nearly 500 °C in a stream of air, releasing water in the form of superheated steam and producing acetaldehyde.

In 1916, a new method for production of acetaldehyde was introduced involving mercuric ion-catalyzed addition of water to acetylene. The acetylene-based process is an early example of a well-known reaction called oxymercuration, by which water or alcohols can be added to both alkenes and alkynes under mild conditions. As shown in Fig. 9.12, when

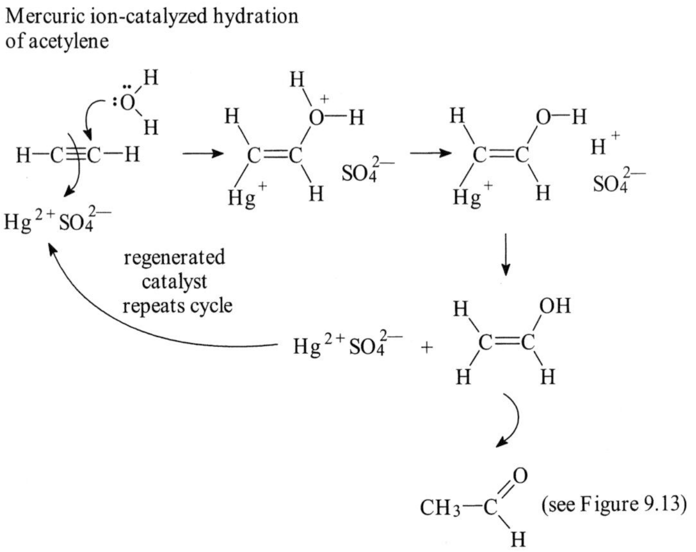

Mercuric ion-catalyzed hydration
of acetylene

Fig. 9.12 Hydration of acetylene to form vinyl alcohol, which then tautomerizes to acetaldehyde.

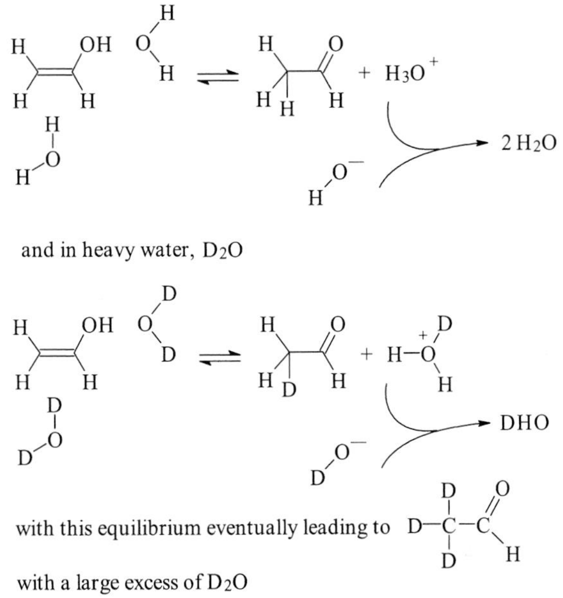

Fig. 9.13 Uncatalyzed tautomeric equilibrium between vinyl alcohol and acetaldehyde.

and in heavy water, D_2O

with this equilibrium eventually leading to $D-\overset{D}{\underset{D}{C}}-C\overset{O}{\underset{H}{}}$

with a large excess of D_2O

water is added to acetylene, vinyl alcohol is initially formed, and undergoes instantaneous rearrangement to acetaldehyde.

Conversion of vinyl alcohol to acetaldehyde (Fig. 9.12), its constitutional isomer, is an example of tautomerism (see Chapter 10, Section 10.9). The hydrogen on the oxygen of the vinyl alcohol is released to the aqueous medium, forming the aldehyde carbonyl group. The aqueous medium releases a different proton to the double-bonded CH_2 group, forming the methyl group of the acetaldehyde (Fig. 9.13). These proton exchanges occur if one dissolves acetaldehyde in water. It is a dynamic equilibrium in which the equilibrium constant strongly favors the aldehyde, K=[vinyl alcohol]/[acetaldehyde] is 10^{-5}, which is typical of the equilibria between enols and their carbonyl tautomers. For example, the analogous equilibrium constant for acetone is in the range of 10^{-6}. Enols, except with special structural features, are very unstable and interconvert rapidly to the carbonyl tautomers – a process that is speeded by either mild acid or base catalysts.

Because the tautomers, the enol and the carbonyl form, interconvert while exchanging protons with the surrounding water molecules (Fig. 9.13), placing a carbonyl compound in heavy water, D_2O, will cause exchange of deuterium for the active hydrogens (Section 9.3) in the carbonyl compound. For acetaldehyde, this means exchanging deuterium for the hydrogens of the methyl group. If the enol of acetaldehyde is created in D_2O, as in the hydration of acetylene (Fig. 9.12), deuterium will appear in the isolated acetaldehyde immediately because a proton from the surrounding water must be incorporated in the enol structure to form acetaldehyde. As we shall see in the Section to follow, the discussion above about isotope exchange in heavy water has a great deal to do with an important detail of the process that shut down earlier acetaldehyde processes, the Wacker reaction.

9.8
**A Palladium-based Process, the Wacker Reaction, shuts down
all Older Industrial Methods to Acetaldehyde**

The birth of the process that shut down production of acetaldehyde via either ethanol oxidation or hydration of acetylene (Section 9.7) took place in 1894, when it was observed that palladium chloride in aqueous solution formed acetaldehyde directly from ethylene. However, it took until 1960 for this chemistry to find its way into an industrial process, and ironically the development took place in a German laboratory dedicated to advancing the use of acetylene in the chemical industry.

There were two problems that delayed the palladium-based process. First, ethylene did not become widely available to the chemical industry until after steam cracking was introduced in the 1930s (Chapter 1, Section 1.4). In fact, the availability of large amounts of ethylene via steam cracking of petroleum fractions spelled the end to the chemical industry's dependence on dangerous acetylene for many processes (Chapter 1, Section 1.6 and Chapter 10, Sections 10.2 and 10.3). What we see in acetaldehyde production is one example of that trend. The widely known danger of acetylene as an industrial intermediate has a long history.

Second, palladium is an expensive metal, and in the reaction conducted in 1894 one mole of palladium was used for each mole of acetaldehyde produced. The oxidation of ethylene to acetaldehyde caused the reduction of Pd^{++} to Pd^0. The process was economically unacceptable. Palladium is a precious metal, and part of the platinum group; in fact, it is even more rare than platinum, existing at approximately 10^{-7} percent of the earth's crust. In 1894 and until the advance described below, this reaction was an experimental curiosity with no practical potential.

But German chemists in the late 1950s, working at Wacker-Chemie and Farbwerke Hoechst, conceived a reversible way to reoxidize the Pd^0 back to Pd^{++}. It turned out not to be difficult. Addition of a cupric salt, Cu^{++}, causes any Pd^0 produced, before it precipitates out of the system, to be oxidized back to the active Pd^{++}. The Cu^+ produced in this redox reaction is then reoxidized back to its active oxidation state,

$$2CuCl + O_2 + HCl \Rightarrow 2CuCl_2 + H_2O$$
$$Pd^0 + 2CuCl_2 \Rightarrow 2CuCl + Pd^{++}$$

Cu^{++}, by atmospheric oxygen. A stoichiometric reactant, palladium, was converted to a catalyst. George W. Parshall in his highly esteemed book on homogeneous catalysis called this development "a triumph of common sense."

And so the older processes based on either ethanol or acetylene were shutdown and industry turned to this palladium-based method, designated the Wacker process (Fig. 9.14), for producing acetaldehyde.

The first step in the Wacker process is coordination of ethylene with palladium in its +2 oxidation state to form a coordinate bond in which the π-electrons of ethylene are donated to the partially filled d-orbitals of the palladium. This process is the same kind of coordinate covalent bond we have seen in different reactions of transition metals and have discussed in Chapter 8 (Section 8.13). This complex accomplishes a transformation of the ethylene com-

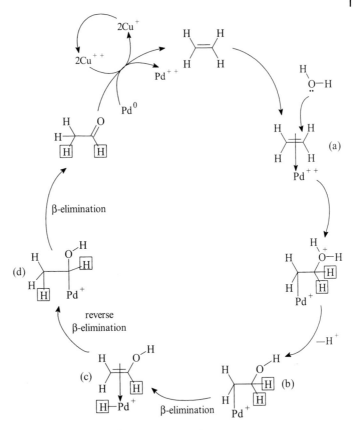

Fig. 9.14 Mechanism of the production of acetaldehyde via the Wacker reaction.

parable to the transformation of acetylene arising from complexation with mercuric ion (see Fig. 9.12). Ethylene, instead of being a source of electrons – that is, a nucleophile, as for example in reaction with HBr and for that matter in forming the palladium complex (Fig. 9.14 (a)) – now becomes an electrophile. With the positively charged palladium draining electron density from the double bond, ethylene's character is reversed. This role reversal lowers the barrier for reaction with the electron-rich water molecules of the aqueous medium. In this way, a palladium-to-carbon bond is produced (Fig. 9.14 (b)).

In the chemistry of compounds in which palladium is bonded to carbon, the so-called α-carbon, there is a common reaction path known as β-elimination, which "simultaneously" breaks the palladium-to-α-carbon bond and the C–H bond on the β-carbon. An alkene is formed. β-Elimination, which is the path from (b) to (c) in Fig. 9.14 is hardly unique to palladium, and in fact is a common reaction path found in the chemistry of several transition metals bound to carbon. Moreover, we have seen β-elimination as the most important reaction in the steam cracking of petroleum, the reaction responsible for breaking larger to smaller molecules from the radicals formed (see Chapter 1, Section 1.11).

In either situation, β-elimination from free radicals or from metal alkyls, a double bond is formed. However, the alkene formed by β-elimination in the Wacker process (Fig. 9.14 (b) to (c)) is vinyl alcohol complexed to palladium. If vinyl alcohol were released from this complex it would immediately undergo transformation to its tautomer acetaldehyde (Figs 9.12 and 9.13). If this path were followed, we would apparently be at the end of our reaction.

And indeed, this was thought, for many years, to be the end of the story. Vinyl alcohol would be released into the aqueous medium and would rapidly tautomerize to acetaldehyde. However, when the Wacker reaction is carried out in heavy water (D_2O), no deuterium appears in the acetaldehyde isolated immediately from the reaction medium, as should have happened if, as we have discussed above (Fig. 9.13), a molecule of vinyl alcohol were released. Following the path of the boxed hydrogens in Fig. 9.14 shows that incorporation of hydrogen from the aqueous medium (or deuterium from D_2O) can only appear in the acetaldehyde product via the tautomerism path from vinyl alcohol. No deuterium from the heavy water therefore means no vinyl alcohol and no tautomerism.

It took considerable effort to understand that the palladium hydride formed in the β-elimination (Fig. 9.14 (c)) remained coordinated to the newly formed double bond. In other words, vinyl alcohol is not released to the aqueous medium. This coordinated state (Fig. 9.14 (c)) then leads to acetaldehyde via re-addition of the palladium hydride to the double bond of the coordinated vinyl alcohol (Fig. 9.14 (c) to (d)) followed by expulsion of palladium metal, Pd^0, and a proton. Did you notice that this last step to form the acetaldehyde is also β-elimination, while the step from (c) to (d) is the reverse of a β-elimination?

The oxidation of ethanol and the hydration of acetylene were both shut down by the palladium-based route to acetaldehyde. But the palladium-based route to acetaldehyde, however beautiful and economical, is less important today, not because there is a more effective path to acetaldehyde but because there is no longer a need for large amounts of acetaldehyde. This outcome was discussed in the very beginning of this chapter (Section 9.1, Fig. 9.1). The two large-volume industrial products, acetic acid and *n*-butanol, which are formed from acetaldehyde can now be made more effectively by methods that do not require acetaldehyde. Let's focus here on one of the most important – the production of *n*-butanol and its relative *n*-butyraldehyde. We learned earlier in this chapter that these chemicals were made via aldol condensation from acetaldehyde (Section 9.3). But they can now be produced directly from propylene. How is this done?

9.9
Hydroformylation – Another Triumph for Transition Metals

As we have noted in Chapter 6 (Section 6.11), where we discussed one of the routes to methyl methacrylate, a German chemist, Otto Roelen at Ruhrchemie in 1938, was investigating the reactions between carbon monoxide, hydrogen, and ethylene with solid cobalt salts. He discovered that propionaldehyde was formed. Roelen had discovered a reaction now known as hydroformylation, one of the most important reactions in industrial chemistry.

$$CH_2CH_2 + CO + H_2 \Rightarrow CH_3CH_2CHO \text{ (catalyzed by cobalt salts).}$$

On the chance that it may have something to do with his ability as a chemical inventor, one of us, and our family, took Otto Roelen out to a trout dinner many years ago. Dr. Roelen started working on the trout with surgical intensity. When we expressed wonderment at his seeming commitment, he told us that his efforts were aimed at isolating the cheek meat, which is the most delectable food in the world.

Although at first the production of propionaldehyde from ethylene was seen as another example of heterogeneous catalysis involving an uncertain ionized state of cobalt, it was eventually realized that the active catalyst was in fact a well-defined cobalt complex $HCo(CO)_4$. This soluble complex was formed by reaction of hydrogen and carbon monoxide with the solid state Co^{+2} salt as shown in the following equations. An additional point is that the $HCo(CO)_4$ produced is in equilibrium with $HCo(CO)_3$ + CO.

$$Co^{+2} + H_2 + CO \Rightarrow Co^{\circ}$$

$$Co^{\circ} + CO \Rightarrow Co_2(CO)_8$$

$$Co_2(CO)_8 + H_2 \Rightarrow HCo(CO)_4$$

Carbon monoxide is an effective ligand for a wide variety of metals with available empty orbitals and we have seen this activity before in the Reppe reaction (Chapter 8, Section 8.13) as well as in a reactive path to methyl methacrylate (Chapter 6, Section 6.12). In both resonance forms of carbon monoxide there are unbonded electrons on the carbon available for donation to the open orbitals of the metal. In the Reppe reaction and in hydroformylation, as we shall see, the coordination complexes with carbon monoxide are easily reversible. This aspect is essential to the chemistry involved.

Knowledge of the chemical structure of the catalyst was a big breakthrough in uncovering the mechanism of the catalytic process in hydroformylation (Fig. 9.15). In $HCo(CO)_3$, the resulting open coordination site on the cobalt that had been occupied by the CO lost from $HCo(CO)_4$ is then available for forming a π-complex with a wide variety of alkenes (Fig. 9.15 (a)). Again, we find in hydroformylation, as in all the metal-catalyzed industrial processes we studied in Chapter 2 (Section 2.10) and Chapter 8 (Section 8.13), that the donation of π-electrons from an unsaturated organic molecule to available empty orbitals on the metal is the first step. In hydroformylation the several steps to follow lead to addition of hydrogen to one side of the double bond and CHO to the other, forming therefore aldehydes, for example, propionaldehyde from ethylene and butyraldehyde isomers from propylene.

The first step in the overall reaction path following formation of the π-complex with the alkene (Fig. 9.15 (a)) parallels the first step in the addition of nickel hydrides to acetylene in the Reppe process for acrylic acid (Chapter 8, Section 8.13). The cobalt bearing its attending ligands adds to one end of the double bond and the hydrogen to the other (Fig. 9.15 (b)). It is in the formation of this intermediate that a point of mechanism becomes critical to shutting down of the production of *n*-butyraldehyde and the derived *n*-butanol from acetaldehyde.

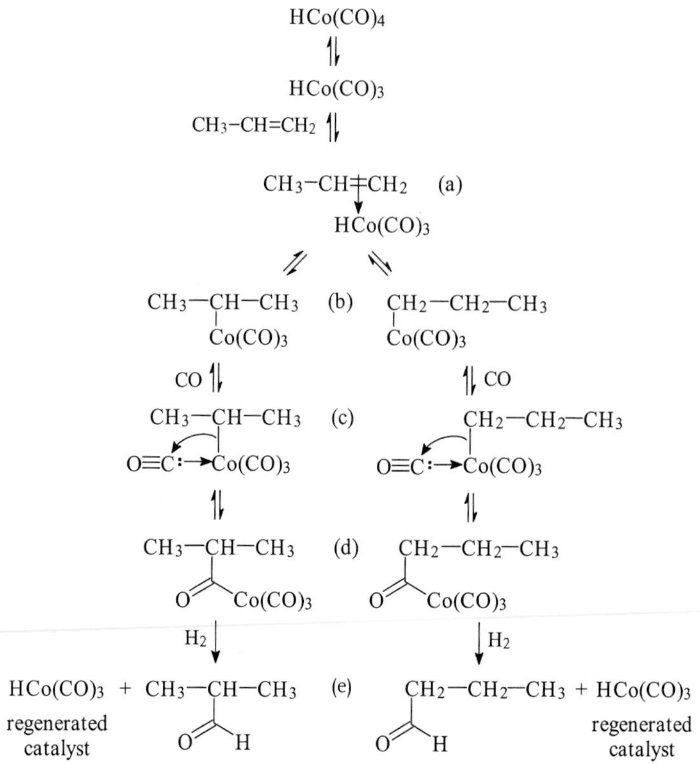

$$HCo(CO)_4$$

$$\updownarrow$$

$$HCo(CO)_3$$

$$CH_3-CH=CH_2 \updownarrow$$

$$CH_3-CH\!=\!CH_2 \quad \text{(a)}$$

$$HCo(CO)_3$$

$$CH_3-CH-CH_3 \quad \text{(b)} \qquad CH_2-CH_2-CH_3$$
$$\quad\;\; | \qquad\qquad\qquad\qquad\quad\; |$$
$$\quad\; Co(CO)_3 \qquad\qquad\qquad Co(CO)_3$$

$$CO \updownarrow \qquad\qquad\qquad\qquad \updownarrow CO$$

$$CH_3-CH-CH_3 \quad \text{(c)} \qquad CH_2-CH_2-CH_3$$
$$O\!\equiv\!C\!:\!\rightarrow\!Co(CO)_3 \qquad\quad O\!\equiv\!C\!:\!\rightarrow\!Co(CO)_3$$

$$\updownarrow \qquad\qquad\qquad\qquad\qquad \updownarrow$$

$$CH_3-C-CH_3 \quad \text{(d)} \qquad CH_2-CH_2-CH_3$$
$$\qquad O \diagup \diagdown Co(CO)_3 \qquad\qquad O \diagup \diagdown Co(CO)_3$$

$$H_2 \downarrow \qquad\qquad\qquad\qquad\qquad \downarrow H_2$$

$$HCo(CO)_3 + CH_3-CH-CH_3 \quad \text{(e)} \qquad CH_2-CH_2-CH_3 + HCo(CO)_3$$
regenerated $\qquad\qquad O \diagup \diagdown H \qquad\qquad\qquad O \diagup \diagdown H \qquad\qquad$ regenerated
catalyst $\qquad\qquad\qquad\qquad\qquad\qquad\qquad\qquad\qquad\qquad\qquad\qquad\qquad\qquad$ catalyst

Fig. 9.15 Mechanism of the hydroformylation of propylene leading to isobutyraldehyde and *n*-butyraldehyde.

Only if the addition is primarily anti-Markovnikov, with hydrogen adding to the most substituted carbon in propylene, will *n*-butyraldehyde, the precursor of *n*-butanol, be produced. We already know the answer because hydroformylation *does* shut down the need for acetaldehyde. But let's discover in the details of this chemistry the driving forces that control which way the addition to the double bond occurs. And let's also find out how the metal-to-carbon bond in the isomeric intermediates (Fig. 9.15 (b)) is converted to the aldehyde. These are questions about what is called "regiochemistry," that is, the result of chemical reactions in which constitutional isomers can be formed.

The original catalyst discovered by Roelen does in fact produce an excess of the desired *n*-butyraldehyde but only in a ratio of about 4:1 to the branched regioisomer, isobutyraldehyde (Fig. 9.15). What are the factors controlling the regioselectivity, and can they be manipulated to increase the proportion of the desired isomer?

A hint at the factors at work in the addition of the cobalt–hydrogen bond across the double bond of propylene (Fig. 9.15 (a) to (b)) can be found in hydroboration in which the addition of a boron–hydrogen bond across a double bond also takes place in an anti-Markovnikov manner, although with a far higher regiospecificity. Following this first ad-

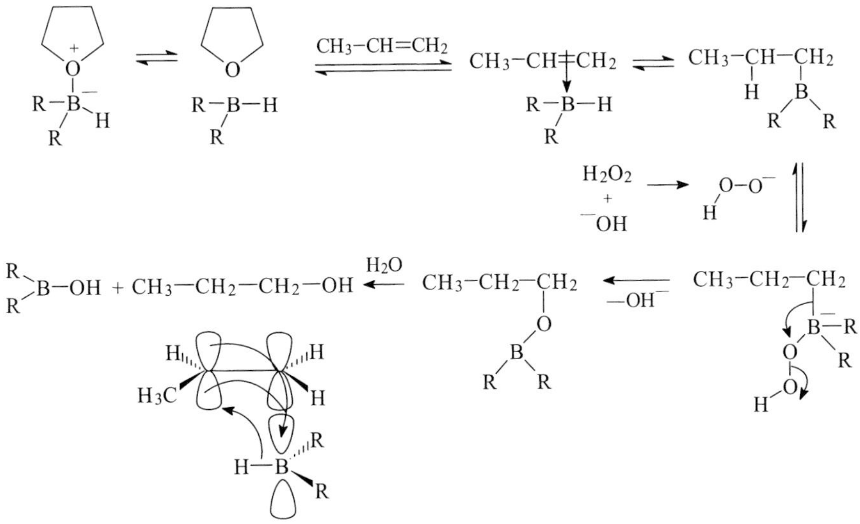

Fig. 9.16 Mechanism of hydroboration.

dition step in hydroboration, shown for propylene above, reaction with base and hydrogen peroxide yields *n*-propanol (Fig. 9.16).

In hydroboration, two factors appear to be at work in controlling the anti-Markovnikov regiochemistry. First, the size of the boron group compared to hydrogen tends to place the boron at the less hindered site of the double bond. Second, the addition of the trivalent boron with its empty p-orbital tends to drain π-electrons from the double bond in the initial orbital overlap that stimulates the addition. Because the more substituted carbon of the double bond can bear the resulting electron deficiency to a greater extent, the carbon-to-hydrogen bond is made at this site and the boron adds to the less substituted carbon as shown in the orbital formulation (Fig. 9.16). A similar orbital formulation could be made for the comparable step (a) to (b) in Fig. 9.15 except that the orbital accepting electrons would be from a far higher quantum level than the p-orbital involved in boron.

Support for the analogy between hydroboration and hydroformylation arises from observations made by industrial chemists as quoted by Weissermel and Arpe. In the discussion of the effect of replacing the carbon monoxide ligands on the hydroformylation catalyst (Fig. 9.17), these authors point out that reduction in electron density, combined with sterically larger ligands on the metal (cobalt or rhodium), both act to increase the ratio of *n*-butyraldehyde to isobutyraldehyde. In other words, as in hydroboration, steric effects and electron deficiency work together to control the regiochemistry.

However, why is the production of *n*-propanol over isopropanol overwhelming in hydroboration of propylene, while in hydroformylation, *n*-butyraldehyde is produced by a ratio of only about 4 to 1 over isobutyraldehyde? Organic chemists love to speculate so why shouldn't we? Perhaps because the bond length between carbon and the metal is far longer than that between carbon and boron, the steric effect favoring addition to the less-hindered position is less important in hydroformylation than in hydroboration. This aspect could be one factor in the difference and if true suggests using even larger ligands around the metal to try to compensate for the longer bond.

In the many years since the discovery of hydroformylation, the industrial process has been modified to produce the far more valuable linear aldehyde at the expense of the branched aldehyde. This was accomplished by modifying the metal complex catalyst with phosphorus-based ligands that replace some of the carbon monoxide. Trivalent phosphorus, as is carbon monoxide, is an excellent nucleophile for formation of coordinate bonds with metals. Because of a lone pair of electrons that is readily polarizeable and because a higher quantum level is involved on phosphorus compared to the carbon of carbon monoxide, the coordinate bonds between phosphorus and metals are far more stable than those between metals and carbon monoxide. Phosphorus-based ligands are therefore widely used to complex with transition metals and can replace carbon monoxide in coordinating to the metal.

The selectivity became better when industrial chemists discovered in the 1970s that hydroformylation would also work by stepping down one quantum level in the periodic table just below cobalt to rhodium. They used triphenylphosphine as the ligand to replace some of the carbon monoxide molecules, $HRh(CO)(P(C_6H_5)_3)_2$. Even better catalysts have been developed, and these are in wide use for producing large ratios of n-butyraldehyde to isobutyraldehyde. Rhodium, however, is very expensive and therefore catalyst recovery becomes important. This area is still under active development, producing even greater regiospecific phosphorus-based ligands than triphenylphosphine.

It's interesting to compare the structure of some of these proprietary ligands and how effective they are in enhancing formation of linear over branched aldehydes (Fig. 9.17).

But we have more of the story to tell in following the mechanism from our point of departure (Fig. 9.15 (b)). How does the cobalt-to-carbon bond or alternatively the rhodium-to-carbon bond lead to the aldehyde group? The basic mechanism, shown in Fig. 9.15, for the original cobalt catalyst discovered by Roelen in 1938 is essentially the same for all the hydroformylation catalysts independent of the metal or the ligands. Let's follow the mechanism for cobalt. The cobalt atom in Fig. 9.15 (b) in both regioisomers is coordinatively unsaturated and therefore, in the carbon monoxide-rich industrial process, adds another carbon monoxide ligand to produce the regioisomers (c) in Fig. 9.15.

Again, we encounter a parallel to what we have seen in the Ziegler–Natta polymerization of alkenes (Chapter 2, Fig. 2.10) and in the Reppe process for production of acrylic acid (Chapter 8, Fig. 8.17). In both cases a weak carbon-to-metal bond drives a rearrangement in which this weak bond is replaced by a stronger carbon-to-carbon bond. In both the Reppe process and in hydroformylation the rearrangement brings about a bond between carbon and carbonyl, with the carbonyl group arising from a complexed carbon monoxide molecule. In this manner, the regioisomers (d) in Fig. 9.15 are formed, and these are then easily reduced by the abundant hydrogen gas. The two aldehydes are formed as is the original cobalt catalyst, $HCo(CO)_3$ (Fig. 9.15 (e)), which is then available to initiate another catalytic round.

The n-butyraldehyde produced in the hydroformylation can be easily reduced by H_2 to form n-butanol. With hydroformylation we certainly no longer need aldol condensations of acetaldehyde to make C-4 aldehydes and their derived alcohols, given the chemistry in Fig. 9.15.

Catalyst structures	Ratio of 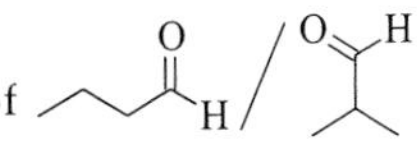 $\underset{H}{\overset{O}{\wedge}}$ / $\overset{O}{\underset{H}{\wedge}}$
$Co_2(CO)_6$	3.4 / 1
$Co_2(CO)_8(C_5H_9)_3P$	5.7 / 1
$Rh(H)(CO)(P(C_6H_5)_3)_3$	10 / 1
$Rh_4(CO)_{12}$	1.3 / 1
and the best:	30–50 / 1

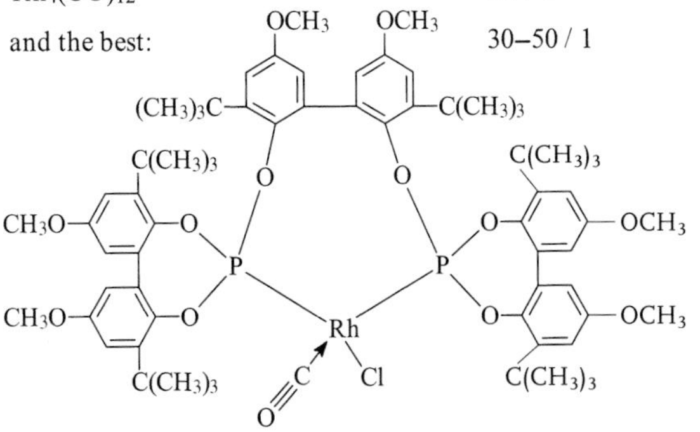

Fig. 9.17 Ligands used for hydroformylation and their effect on regioselectivity.

9.10
How is the Other Product, Acetic Acid, which formerly was made from Acetaldehyde, now produced?

The first process for making acetic acid was by fermentation of carbohydrates. This route is still practiced and the product, a 3% solution of acetic acid in water, is what we call vinegar. Today it is ethanol rather than carbohydrate that is fermented. This was the first aerobic fermentation – aerobic meaning that it requires oxygen. It was carried out in a column filled with wood chips. The alcohol, with the proper organism, was dribbled over the chips as a film thin enough to allow absorption of oxygen from the air. (As an aside, the second important aerobic fermentation was developed during World War II for penicillin production). Subsequently, liquid phase hydrocarbon oxidation became a commercial process for acetic acid production. If naphtha (see Chapter 1, Section 1.3) is used, substantial amounts of byproducts are produced including formic, propionic, and succinic acids as well as acetone. If *n*-butane is used as the staring material for oxidative production of acetic acid, even more byproducts result including acetone, methyl ethyl ketone, methyl and ethyl acetates and formic and propionic acids. Although the starting materials for these processes are inexpensive and the separation of the byproducts is expensive, hydrocarbon oxidation is still used for the production of acetic acid. But the importance of the process relates to the value of the byproducts more than to the value of the acetic acid.

We have discussed earlier the ease of oxidation of aldehydes to carboxylic acids in the transformation of acrolein to acrylic acid (Chapter 8, Section 8.12) and the route from isobutene to methyl methacrylate (Chapter 6, Section 6.9). It is therefore hardly a surprise that acetaldehyde was early on seen as a source of acetic acid, a process that became commercially important in the 1950s. This contributed much to acetaldehyde's fortunes and therefore to the success of the Wacker process (Fig. 9.14). The oxidation of acetaldehyde to acetic acid occurs easily, usually in a solvent such as ethyl acetate, at room temperature with air rather than pure oxygen as the oxidant, and at somewhat elevated pressure.

$$CH_3CHO + {}^1\!/_2\,O_2 \Rightarrow CH_3COOH$$

This process, relatively simple though it is, was shutdown by the Monsanto Chemical Company who used homogeneous catalysis to cause methanol and carbon monoxide to react to provide acetic acid.

$$CH_3OH + CO \Rightarrow CH_3CO_2H \text{ via catalysis.}$$

This certainly is one of the "great" reactions in industrial organic chemistry, and it shows the nature of the intense competition in this industry. Actually, the conversion of methanol to acetic acid was pioneered by BASF using a cobalt catalyst but requiring rather high temperature and pressure, for example, 250 °C and 680 atmospheres. Monsanto's contribution was their discovery that rhodium with iodine catalyzed the reaction at lower temperatures, 150–200 °C, and at close to ambient pressure – conditions that are more economical. Yields were very high and the byproducts were CO_2 and H_2. The Monsanto process shut down the acetaldehyde-based processes as well as the plants that carbonylated methanol with a cobalt catalyst. Nevertheless, the route to acetic acid via oxidation of hydrocarbons, as indicated above, continued largely because of the value of the byproducts.

And what is happening now? Saudi Arabia, with its very cheap ethane, is planning to institute a newly developed ethane-based process. BP has discovered that iridium gives better results than rhodium. Celanese has found that by reducing water content in the reaction, throughput is increased and economics are improved. And, to go full circle – remember that the first process involved fermentation – Celanese has patents on a vastly improved fermentation process with engineered organisms. It's apparent that constant innovation to improve the production of acetic acid is ongoing, giving industrial managers headaches because their process may be subject to shutdown economics in the future.

We have come to the end of our classical story of shutdown economics, which has a happy or sad ending depending on how much investment you had in a plant producing acetaldehyde.

9.11
Summary

In a story with many twists and turns we have discovered how acetaldehyde became important to the chemical industry for the production of the linear C-4 compounds, particularly *n*-butyraldehyde and *n*-butanol, via aldol condensations. We discovered the mechanistic foundations of this important reaction of aldehydes and found that the essential chemical characteristics of the carbonyl functional group are involved. We focused our attention on how the aldol condensation is used to produce other important industrial intermediates including the polyhydric alcohols and 2-ethylhexanol, and how one deals with aldol condensations in which multiple products are possible. We learned something about plasticizers for polymers along the way.

All this development caused acetaldehyde to become a central player in the chemical industry, and focused attention on how to produce it most economically. A method based on acetylene was developed that led us to an understanding of the nature of tautomers. But this acetylene-based chemistry was to be displaced by the development of a wonderful transition metal-based process, the Wacker reaction based on palladium. This outcome led us to understand tautomerism and how conducting a reaction in heavy water can lead to mechanistic insight.

The future seemed bright for acetaldehyde until another transition metal reaction based on cobalt, hydroformylation, turned out to be an even better way to produce the C-4 compounds that had been made from acetaldehyde. And if that were not enough, the chemical industry evolved improved methods to synthesize acetic acid – another industrial intermediate for which acetaldehyde had been important.

In studying both the Wacker reaction and hydroformylation we find important common elements in these reactions with other organic reactions of alkenes catalyzed by transition metals such as Ziegler–Natta polymerization and Reppe chemistry and we even find parallels to the chemistry of boron in hydroboration.

And in the end there is the theme of shutdown economics where we learn that the chemical industry is driven by constant innovation so that one process replaces another in a constant drive for economy and efficiency. The better shuts down the good, and so the industry goes on. And in the chapter to follow, Chapter 10, Doing Well by Doing Good, we shall discover how environmental and safety considerations play an important role in shutdown economics and how the industry profits by paying attention to what is best for the environment.

Some of the subjects treated in this chapter are listed below. These are key words and terms that act as reminders of the chapter's contents and should become a valuable part of your chemical vocabulary.

- Shutdown economics
- Carbanions
- Enolates
- Carbonyl chemistry
- Aldol condensation
- Mixed aldol condensation
- Active hydrogens alpha to carbonyl
- Polyhydric alcohols
- Nucleophilic addition to carbonyl
- Carbonyl hydration
- Plasticizers

- Cannizzaro reaction
- Sterically hindered ester hydrolysis
- Mercuric ion-catalyzed addition to triple bonds
- Tautomers
- Deuterium incorporation
- Beta-elimination
- Palladium catalysis
- Oxo chemistry
- Hydroformylation
- Rearrangements of metal carbon bonds
- Hydroboration
- Catalyst ligands
- Fermentation
- Oxidation production of acetic acid

Study Guide Problems for Chapter 9

1. What products would arise from heating n-butanol with ethylene oxide under similar conditions used for the reaction of water with ethylene oxide to form ethylene glycol? Would the ratio of the n-butanol to ethylene oxide cause the product ratio to change? If the products of this reaction were to act as solvents in which both hydrophobic and hydrophilic qualities were important, how would the ratio of n-butanol to ethylene oxide make a difference in the solvent quality?

2. How does resonance theory account for enhanced acidity for C–H groups adjacent to esters, nitriles and nitro compounds compared to hydrocarbons?

3. Replace acetaldehyde with acetone in the reaction sequence shown in Fig. 9.2 and suggest which step might be slower for the ketone than the aldehdye and why.

4. Show all steps and predict all possible products in the aldol condensation of acetaldehyde with propionaldehyde.

5. Show all steps and predict all possible products in the aldol condensation of acetaldehyde with formaldehyde. Why might the mode of addition and the amount of base used affect the reaction product mixture?

6. Fig. 9.6 shows that the hydration of α-fluorinated carbonyl compounds is attended by a larger equilibrium constant than the analogous α-hydrogen-substituted carbonyl compounds. Could this observation be related to an argument in the text about the relative reactivity of nucleophilic addition to formaldehyde versus acetaldehyde?

7. The Cannizzaro reaction shown in Fig. 9.8 occurs for other aldehydes than formaldehyde, but all aldehydes subject to this reaction have a structural similarity. What is the structural feature necessary to undergo this reaction?

8. Propose synthetic approaches, including the mechanism of each step, for the polyhydric alcohols trimethylolpropane and neopentyl glycol, shown in Fig. 9.3.

9. Considering the detailed mechanism of the hydrolysis, explain why the ester shown in Figure 9.9 should be more resistant to hydrolysis than, for example, the ester formed from adipic acid and ethylene glycol. In a question with a related answer, why

is it exceptionally difficult to hydrolyze poly(methyl methacrylate) (Chapter 6, Section 6.2) to poly(methacrylic acid).

10. Judge the relative advantage of forming a plasticizer by esterifying 2-ethylhexanol with: phthalic anhydride; terephthalic acid; isophthalic acid (1,2-benzene dicarboxylic acid); or benzoic acid. What is meant in the discussion in the text about Fig. 9.11 that states that the ester formed from 2-ethylhexanol with phthalic anhydride is a stereo-isomeric mixture?

11. Draw structures for cellulose nitrate and for the material that replaced it for motion picture film, cellulose acetate. Offer an explanation for why the nitrate is flammable while the acetate is far less so.

12. Using boron-based and also mercury-based chemical reactions, outline the syntheses of both *n*-propanol and isopropanol from propylene, showing all mechanistic steps in both methods.

13. In the oxidative conversion of ethanol to acetaldehyde, two hydrogen atoms are lost from the ethanol. What possibilities exist for the source of these two hydrogen atoms.

14. Both acids and bases catalyze the interconversion between enols and their carbonyl tautomers. Show how the mechanism presented in Fig. 9.13 justifies this catalytic activity in the transformation.

15. In a cyclohexane ring with carbonyl groups located on carbons 1 and 3, the equilibrium constant for tautomeric equilibrium is very large, $K = 20$, and favors the enol form. What could be the reason for this large difference compared to acetone with $K = 10^{-6}$?

16. If the mercuric ion-catalyzed hydration of acetylene shown in Fig. 9.12 occurs in D_2O, and the produced acetaldehyde is immediately removed from the reaction medium, it is found that two of the three hydrogens on the methyl group are deuterium. How is this fact consistent with the mechanism of the reaction?

17. If one dissolves acetaldehyde in acidic heavy water, D_2O, the hydrogens on the methyl group are slowly exchanged for deuterium but not the single hydrogen bound to the carbonyl carbon. Offer a mechanistic explanation for this experimental fact.

18. Recognizing the Lewis acid character of borane, BH_3, trace the movement of electrons in the addition of borane to propylene and account for the observation that addition is anti-Markovnikov without using any steric argument. Apply the same reasoning to addition of the Co–H bond across the double bond in propylene. In what way are the first steps of hydroboration and hydroformylation analogous?

19. Why does cobalt in Fig. 9.15 tend to return to the five-coordinate state, in which there is one hydrogen and four carbon monoxides or four carbon monoxides and one cobalt-to-carbon bond?

20. In the industrial oxidation of acetaldehyde to acetic acid with oxygen, peracetic acid is a proposed intermediate. Is it possible to write a chain mechanism for the reaction of peracetic acid, CH_3CO_3H, with acetaldehyde to form acetic acid?

10
Doing Well by Doing Good

10.1
Many Companies in the Chemical Industry have been Amazed to Learn that Replacement of Dangerous and/or Toxic Chemicals Leads not only to Safety but also to Greater Profit

Many processes in the chemical industry, particularly older ones, make use of chemicals that are dangerous and environmentally hazardous. Accordingly, an important objective is to invent processes that circumvent the use of such chemicals. The point has been made by many who study these kinds of problems that solutions will be much more palatable and therefore more effective if accompanied by increased profit. And in fact an often-unexpected byproduct of solving these problems has been improved economics. Thus, the new process does "good" ecologically, but also it does "well" for the profit and loss sheet.

In this chapter we attempt to show how the chemical industry has handled four of its worst actors: acetylene, hydrocyanic acid, chlorine, and phosgene.

The first, acetylene, is explosive and expensive. The second, HCN, is the deadly infamous Zyklon B used by the Germans in the holocaust in Europe to kill millions of innocent people during World War II. The third and fourth chemicals, chlorine and phosgene, were hardly angelic substances, having both been used as poison gases in World War I.

Because of their inherent danger, all four chemicals add excessive expense to chemical processes that use them, and for this reason the industry has tried to find ways around their use. But, because of their useful reactive properties in the production of important industrial materials, they have continued to play essential roles in the chemical industry.

While acetylene has been replaced to the extent of about 80%, the other three are still used, though their uses have been shortened by innovative chemistry. Let's discover how the industry has tried to replace these noxious chemicals, beginning with acetylene.

10.2
What's the Problem with Acetylene? First, it is Explosive

It's not necessary to have an advanced degree to know about the explosive power of acetylene – it's a famous characteristic of this dangerous chemical. As an example, some bored teenagers in Lovell, Wyoming, recently peppered the town with balloons filled with acetylene and oxygen, using acetylene-welding torches as their source of the dangerous gas. They then set off explosions with a crude fuse using a simple trail of gasoline on the ground lit with a match. The local newspaper reported that the police were befuddled by complaints from local residents of loud explosions, which rattled eardrums and windows. Although these teenagers may have thought it all was very funny, acetylene is no laughing matter. At nearly the same time as the Lovell "town joke" several workers across the world in Luhansk, Russia were killed in an explosion at a plant while handling an acetylene container. What is the source of the explosive power of acetylene and how might this property be related to its useful characteristics?

Acetylene is an excellent example of a molecule with a broad range of important reactive characteristics allowing it to play a central role in the production of many important chemical intermediates. For many years it was therefore almost inconceivable that new methods could be found to replace acetylene. But acetylene is not only dangerous – it also must be synthesized by difficult and energy intensive methods.

The industrial literature on acetylene is filled with warnings of its danger, which is the other side of the coin of acetylene's value. Both acetylene's usefulness and its danger arise fundamentally from the "loosely held" electrons in the triple bond allowing for a variety of interesting, easily performed chemical reactions as well as for explosive decompositions. The bond energy of a π-bond is far less than that of a σ-bond. In fact, the π-bond is only in the range of about two-thirds of the energy of a σ-bond. The fact that acetylene has a relatively large ratio of π to σ-bonds in such a small molecule, C_2H_2, is responsible for the fact that acetylene has a large positive energy of formation over a wide temperature range (Fig. 10.1). This circumstance means that acetylene, in contrast to most organic compounds, in the temperature ranges of its use and storage is highly unstable with respect to its constituent elements. This thermodynamic fact, in combination with the velocity with which acetylene reacts, means that under certain conditions, acetylene is a time bomb waiting to explode at the slightest provocation.

The most comprehensive work describing the chemical industry is the *Kirk-Othmer Encyclopedia of Chemical Technology*, first created in Brooklyn at the Polytechnic University. In the chapter on acetylene many pages are devoted to its explosive dangers. Because of this danger, cylinders of acetylene – used only for the transport of small amounts of the chemical – are filled not with the pure gas but with a solution of acetylene in acetone. The acetone is dispersed in a matrix, a honeycomb-like material that fills the cylinder. This matrix is so important that it is very carefully tested to make sure it is of sufficient strength and durability so that it will not crumble even if the cylinder is subjected to shock. If the matrix did crumble, pools of acetone could accumulate leaving pockets of the potentially explosive acetylene gas, a situation to be avoided at all cost.

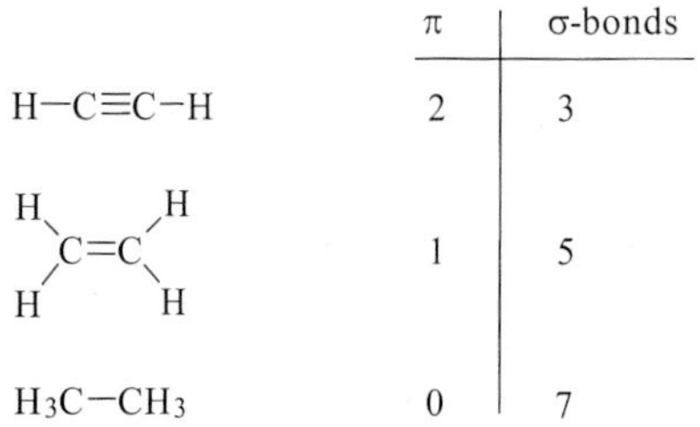

	π	σ-bonds
H−C≡C−H	2	3
H₂C=CH₂	1	5
H₃C−CH₃	0	7

Fig. 10.1 The relatively large ratio of π to σ-bonds in acetylene is the source of the instability.

Another consequence of the explosive danger is that acetylene is never produced in large amounts that require shipping to other sites for use. It is made and then piped to the chemical reactor where it is needed, which is as close as possible. Acetylene is a chemical that is mostly used "over-the-fence" in industrial parlance.

The conditions necessary for detonation and the means for protection against it have been widely studied and all acetylene-based industrial processes take this information into account. One wonders why such a chemical would ever be used in the first place in the tonnage quantities necessary for large-scale chemical production. But in fact acetylene was the fundamental building block of the chemical industry prior to the 1960s, even though there were many instances of damage and death arising from its capricious nature.

10.3
What Else is Wrong with Acetylene?

Adding to the explosive nature of acetylene is the fact that it is expensive to produce. Both acetylene's production cost and its ease of detonation depend on a high positive energy of formation. It could not be stated better than did Parshall: "Its major advantage chemically, the large amount of energy stored in the triple bond function, became a disadvantage economically." Parshall goes on to state: "The 1973 "energy crisis" made high-energy materials such as acetylene ($\Delta F^0 = +50.84$ kcal/mole) extravagantly expensive as feedstocks."

It was in 1973 that production of oil was severely restricted by producers around the world as part of the action of the OPEC cartel. The large increase in the price of oil that followed this action forced the chemical industry to try to replace processes that were energetically expensive. The fact that this increase in the price of oil and therefore the cost of acetylene coincided with an increasing availability of alkenes from steam cracking, which could replace acetylene in many chemical processes, sealed the doom of acetylene.

All methods to produce acetylene, in fact, do demonstrate the large amount of energy necessary to create the triple bond. This property could first be seen in the process by which acetylene was discovered by an Irishman, Edmund Davy, a younger cousin of the far more famous Sir Humphrey Davy, one of the giants of the chemical sciences of the nineteenth century. Edmund Davy reported in 1836 that heating potassium carbonate and carbon to very high temperatures yielded a residue (we would call it potassium carbide) with the formula K_2C_2.

Davy had two surprises when he added water to this residue. He obtained a gas; and this gas was capable of exploding violently. In this way acetylene made its explosive entrance to our earthly stage. The more famous carbide of calcium was discovered by another notable chemist of the nineteenth century, Friedrich Woehler. Woehler reported in 1862, again by reaching very high temperatures, that an alloy of calcium and zinc mixed with carbon produced a compound with the formula CaC_2. Analogous to the related potassium salt prepared by Davy many years before, addition of water produced the same gas, acetylene.

In both potassium carbide and calcium carbide the triple bond of acetylene, as we will see in detail below, is present between the two carbons of the carbide portion of the salt, that is, Ca^{++} $C_2^{=}$. The high energy of this triple bond in $C_2^{=}$ must be supplied in the process of forming it, which is the reason why such high temperatures must be used to form these carbide salts.

The most inexpensive way to produce the carbide group and therefore the method adapted industrially is to use two readily available materials, calcium oxide (lime, CaO) and coke. Heating these materials to 2500 °C, which pays the thermodynamic price of producing the triple bond, yields calcium carbide – the same compound produced originally by Woehler. The cost of acetylene must reflect the cost of the high energy input necessary to reach such an extreme temperature.

10.4
What is the Precise Chemical Nature of these Carbide Salts?

The reaction of calcium carbide with water to release acetylene is an exothermic process in which the heat released can become so extreme as to cause detonation of acetylene. Why does water form acetylene from calcium carbide? A characteristic of acetylene and indeed of any terminal alkyne, that is, a compound with hydrogen bound to one end of the triple bond, is that for a hydrocarbon it is a relatively strong acid. With a pK_a of about 25, hydrogen bound to a triple-bonded carbon is distinguished from all other kinds of hydrogen bound to carbon in hydrocarbons, which are far weaker acids. The acidity of acetylene comes into the range we have discussed in Chapter 9 for the active hydrogens in carbonyl compounds (Section 9.3) or for the acidity of carbon–hydrogen bonds a to nitrile groups (see Chapter 5, Section 5.7). But the acetylene carbanion does not offer the resonance stabilization for carbanions adjacent to carbonyl groups or nitriles. What is the source of the acidity of hydrogen bound to a triple bond?

In acetylene, there are two hydrogens bound to a triple bond, and both are acidic allowing production of a dianion. In calcium carbide, one doubly charged calcium ion is associated with the negative charges at the ends of two acetylene dianions, forming a highly crystalline structure (Fig. 10.2). The acetylide dianion is the conjugate base of a weak acid, acetylene, which means that adding any acid that is stronger than acetylene would necessarily reform the acetylene and produce the conjugate base of the added stronger acid. One does not have to search very far for a stronger acid then acetylene and certainly the most abundant candidate among the multitudes of stronger acids is simply water.

Addition of water will therefore reform the terminal acetylenic carbon–hydrogen bond and in turn produce OH^-, which is the conjugate base of water. In replacing the conjugate base of the weaker acid, acetylene, with the conjugate base of a stronger acid, water, a great deal of heat is released. This effect is an important source of the large amount of heat evolved on treating calcium carbide with water, which is 134 kJ/mole, or to put it in engineering terms, 900 Btu/lb. of calcium carbide. A great deal of energy indeed!

While no accessible base can form a carbanion from ethylene or methane or other hydrocarbons that do not have a terminal triple bond, the acetylide anion can be formed by bases which, although powerful, are readily available. The carbanion formed by the reaction of terminal alkynes with strong bases can act as a nucleophile allowing formation of carbon–carbon bonds by reaction at primary carbon-bearing leaving groups (Fig. 10.3).

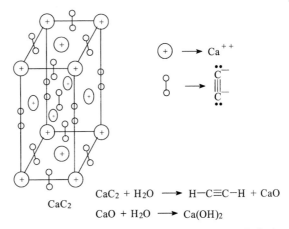

$$CaC_2 + H_2O \longrightarrow H-C{\equiv}C-H + CaO$$

$$CaO + H_2O \longrightarrow Ca(OH)_2$$

Fig. 10.2 Representation of the unit cell of the crystal of calcium carbide and addition of water to form acetylene and calcium hydroxide.

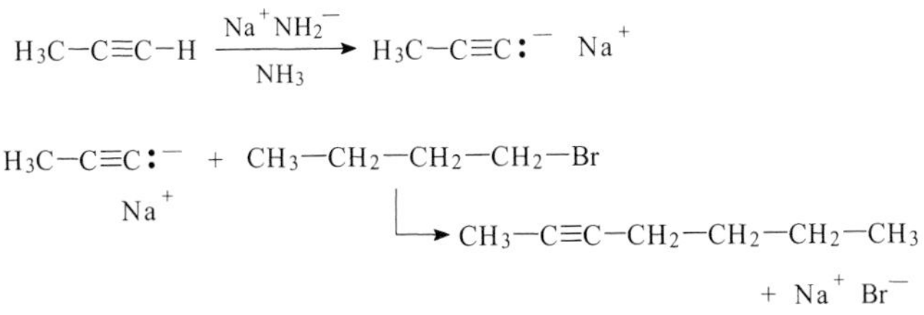

$$H_3C-C{\equiv}C-H \xrightarrow[\text{NH}_3]{\text{Na}^+ \text{NH}_2^-} H_3C-C{\equiv}C\!:^- \ \text{Na}^+$$

$$H_3C-C{\equiv}C\!:^- \ + \ CH_3-CH_2-CH_2-CH_2-Br$$
$$\underset{\text{Na}^+}{}$$

$$\longrightarrow CH_3-C{\equiv}C-CH_2-CH_2-CH_2-CH_3$$
$$+ \ \text{Na}^+ \ \text{Br}^-$$

Fig. 10.3 Example of synthetic use of terminal alkynes: the reaction of the anion of methylacetylene with butyl bromide.

Later in this chapter we will discover how this reactive characteristic of acetylide anions with formaldehyde leads to an important industrial intermediate, 1,4-butanediol (Section 10.11).

Although hydridization of orbitals appears far afield from the 2500 °C processes that produce metallic carbides, both formation of the carbides (Fig. 10.2) and the synthetically powerful acetylide anions of organic molecules (Fig. 10.3) exist because of the nature of the hybridization of the triple-bonded carbon. The electron density of the s-orbital is higher at the nucleus of the carbon atom then the p-orbital, bringing the negative charge in this orbital close to the positive charge of the nucleus. With 50% s-character in the sp-hybrid of the triple-bonded carbon (Fig. 10.1), this type of carbon has the highest electronegativity of any carbon found in hydrocarbon molecules. Put in another way, a negative charge formed on an sp-hybridized carbon will be more strongly stabilized than a negative charge on carbon with less s-character. The correlation between pKa and percentage s-character is demonstrated in Tab. 10.1.

Tab. 10.1 Correlation between pK$_a$ and percentage s-character in C–H bond

Hydrocarbon	pK$_a$	% s-Character in C–H bond
Acetylene	25	50
Ethylene	44	33
Methane	60	25

10.5
Is Acetylene derived from Calcium Carbide of Commercial Importance?

The temperature rise associated with the large exotherm resulting from the mixing of calcium carbide and water, the problems of the labor-intensive mixing of a solid, and the production of an explosive gas, provide a combination fraught with engineering difficulties that argue for operation on a small scale. But even on a small scale, many details such as the size of the calcium carbide particles and the mixing speed and temperature are critical to the control of this dangerous procedure.

For these reasons, calcium carbide is avoided as a source of acetylene on a large scale and this traditional method's only advantage over competitive methods is the convenience for small-scale production of acetylene. In fact it is still used to prepare acetylene for torches. At one time the calcium carbide production of acetylene was widely used for home, street and industrial lighting. But the widespread advent of electrical power about 120 years ago entirely displaced these applications. There is one fuel use of acetylene, however, that still persists and that is fusion welding – heating the surfaces of metals until they flow together. This is the role of the oxyacetylene torch which, while more expensive than cheaper fuels such as propane, is capable of attaining otherwise astonishing temperatures in the range of 3200 °C. Oxyacetylene flames can therefore be used in working most metals for welding and cutting and other more specialized uses.

There is an interesting boutique use for acetylene. One early small-scale use of the calcium carbide method that has persisted to the present day is for lighting purposes. Starting in the earliest years of the twentieth century, miners worked with small acetylene lamps on their hard hats as long as the mines were "non-gassy." The non-gassy mine was essential to avoid explosions of gases such as methane often encountered in coal mines. The intense temperatures of the acetylene flame associated with the desirable brilliant white light could trigger such explosions. In the photograph from an English mine in Fig. 10.4, three miners out of four were wearing acetylene lamps. A close-up of an acetylene lamp is also shown in Fig. 10.4, which clearly reveals the water container above and the calcium carbide particle container below. Even today, one can buy a hat with an acetylene lamp to be used for cave exploration, spelunking. These, however, hardly resemble the lamps the miners wore.

Fig. 10.4 Caving lamps based on acetylene. (Reprinted with permission of the East Lothian Council Museums Service.)

10.6
Large-scale Production of Acetylene

The large-scale production of acetylene is based on an observation in the nineteenth century, as is also true for the calcium carbide process (Section 10.5). The experiment involved either high temperature or an electric arc, and was carried out by the great French chemist Pierre Eugene Marcellin Berthelot around 1860 as part of his intense effort to produce organic compounds of great variety by methods involving both inorganic and organic starting materials.

Berthelot was born in 1827 and lived until his 80th year. He was quite a remarkable person, for not only did he make essential contributions to chemistry – being called by some the father of organic chemistry – but he contributed heavily to all known fields of chemistry at that time, in some cases laying foundations that we build upon now. In addition he was a man of government, playing an important role in the defense of Paris in the Franco-Prussian war of 1870 and becoming Minister of Public Instruction in 1886, and even Foreign Minister of France in 1895. His name is widely known in France today, even by those who may not know the details of his accomplishments. Streets in Paris bear his name. One is hard-put to think of a scientist who could rival Berthelot's breadth of accomplishment.

Berthelot, who gave acetylene its name, played a large role in nineteenth century science. He is credited with putting the final nails in the coffin of vitalism, a theory that life processes were necessary for the synthesis of organic molecules. It is interesting that Woehler, who synthesized acetylene by adding water to calcium carbide (Section 10.3), had put the first nail in that same coffin with his synthesis of urea from ammonium cyanate, demonstrating for the first time that a chemical substance closely associated with life could be obtained without the intervention of life processes.

In the 1920s, Badische Anilin und Soda Fabrik, known better as the giant German chemical company BASF, started to investigate the possibility of producing acetylene on a large scale, and initiated research based on Berthelot's experiments of the 1860s. The presumably important industrial products that could be derived from the many chemical reactions possible with this molecule stimulated this research. Here began the effort that triggered the ascent of acetylene as an important industrial intermediate. But the ascent become a descent as shutdown economics, later in the twentieth century, played its role in using ethylene and propylene to replace acetylene in the production of important industrial intermediates. But this is getting ahead of our story.

At approximately the beginning of World War II, in 1940, the research effort initiated in the 1920s at BASF had led to the opening of a plant at Huels to produce acetylene. Such a plant, as for all large-scale production of acetylene, is based on two essential considerations: (1) Only at very high temperatures does acetylene become more stable than the chemicals formed in competition with it; and (2) acetylene is unstable with respect to the elements that make it up at all temperatures.

The reaction must be "quenched," that is, a sudden lowering of the temperature which takes the acetylene formed at the high temperature and brings it rapidly to a temperature where it does not have the energy to be transformed to other molecules, even if these other molecules are more stable. The quenching blocks the decomposition of the acetylene to carbon and hydrogen. These factors could hardly have been considered in the first experiments conducted by Berthelot, therefore accounting for his very small yields. But we must appreciate the skills in the laboratory of this great French chemist in identifying acetylene in such small amounts with the experimental equipment available at the time.

The data in Fig. 10.5 are critical to the manufacture of acetylene. Among the small hydrocarbon molecules produced at high temperatures from hydrocarbons, acetylene is the

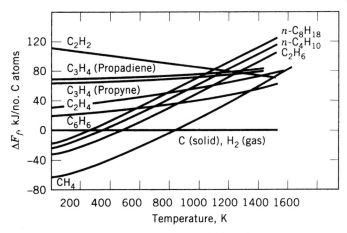

Fig. 10.5 Free energy of formation of several hydrocarbons.

most stable only above 1400°K (Fig. 10.5). Therefore, only above this temperature could significant amounts of acetylene be produced at equilibrium. Because the formation of acetylene from its elements is highly endothermic, its equilibrium concentration increases with increasing temperature. This is not the situation for the other hydrocarbons shown in Fig. 10.5.

However, the data in Fig. 10.5 also show that at 1400°C acetylene is unstable compared to solid carbon and gaseous H_2. Many experiments have shown that acetylene, at the temperatures necessary to produce it, rapidly decomposes to its elemental components. Taking these thermodynamic considerations into account in the industrial processes, the necessary temperature – above 1400°K – is reached very quickly so that acetylene is produced, but this temperature is maintained for only a fraction of a second so that the acetylene is not decomposed.

This approach was first accomplished successfully in the Huels' process in Germany in 1940 by using an electric discharge in a hydrocarbon atmosphere. Natural gas, which is mostly methane, is fed into a reactor to intersect with an electric arc, which can reach, in its vicinity, the astonishing temperature of 20,000°K. The average temperature, however, is far lower, in the range of 1500°K – precisely in the range that Fig. 10.5 showed us is necessary to produce significant quantities of acetylene.

Engineers developing the electric arc process for acetylene production wanted to speed the quench step. Thus, they decided to add inert gases such as argon, helium, or nitrogen. They reasoned that these small molecules with their high translational energies would carry away the excess heat and lower the temperature quickly. In fact, to their great surprise, such gases did the opposite by stimulating the decomposition of acetylene. Hydrogen, on the other hand, was very effective in maximizing the yield.

The mystery of the specific effect of hydrogen was solved by an ingenious series of experiments. The process was carried out with D_2 in place of H_2, which gave rise to large amounts of deuterated acetylene, C_2D_2. In a related experiment, the hydrocarbon precursors for acetylene production were replaced in the electric arc process with a mixture of acetylene made only from carbon-13 mixed with acetylene made only from carbon-12. The recovered acetylene from this process was then studied in a mass spectrometer, which showed that the two isotopes of carbon became mixed in each molecule.

These labeling experiments with carbon-13 and deuterium demonstrated that the extreme temperature conditions broke down the hydrocarbon molecules, as well as the acetylene produced, into fragments of C_2H, C_2, CH and H species, which then collided with each other thousands of times forming acetylene, breaking the formed acetylene into these molecular fragments, reforming acetylene again, and so on. In 0.1 milliseconds it was estimated at these high temperatures that each molecule of acetylene underwent tens of thousands of collisions, even while moving only centimeters downstream. Thousands of these collisions result in acetylene molecules breaking into molecular fragments shown above and reforming acetylene again and again.

If an inert gas such as nitrogen is present instead of hydrogen, the molecular fragments can react together not only to produce acetylene, but also the more stable elements, solid carbon, graphite, and H_2 and as well benzene. As seen in Fig. 10.5, benzene is the only relatively small molecule that is more stable than acetylene at the production temperatures. But in an atmosphere of hydrogen the carbon fragments will collide most frequently with H and H_2 to produce acetylene rather than with other carbon fragments, which would lead to the growth of larger entities such as benzene and higher unsaturated hydrocarbons, the constituents of sticky carbon deposits and soot.

There are many variations for producing acetylene pioneered by Berthelot in 1860 and leading to the first industrial process at Huels in 1940. Each has its own unique variations according to how the high temperatures and quenching are achieved and what kinds of hydrocarbon sources, from as simple as methane to as complex as coal, may be used. Methane, however is the most important. While acetylene production in Japan and in Eastern Europe is still based mostly on calcium carbide, all producers of acetylene on a large scale in the United States and Western Europe rely on methane via high temperature processes. There is some recovery of acetylene produced in steam cracking (Chapter 1) of petroleum, and although this route is the cheapest source only small quantities are produced.

10.7
How was Acetylene Used to Produce Industrial Intermediates?

Chemicals formerly produced from acetylene but made in a different way had a value of about 12 billion dollars in the United States in the early 2000s. Studying how these chemicals were derived from acetylene (Fig. 10.6) will lead to appreciation of the chemical versatility of this former building block of the chemical industry. And observing how much more complex is the chemistry and engineering involved in producing these chemicals from sources other than acetylene allows one to appreciate the danger and expense that caused the chemical industry to turn away from acetylene.

Let's begin with four venerable industrial intermediates that were made from acetylene via a common mechanism. Acetaldehyde, vinyl chloride, vinyl acetate, and acrylonitrile were all formed by addition of H_2O, HCl, HO_2CCH_3, and HCN to acetylene respectively, catalyzed by a transition metal.

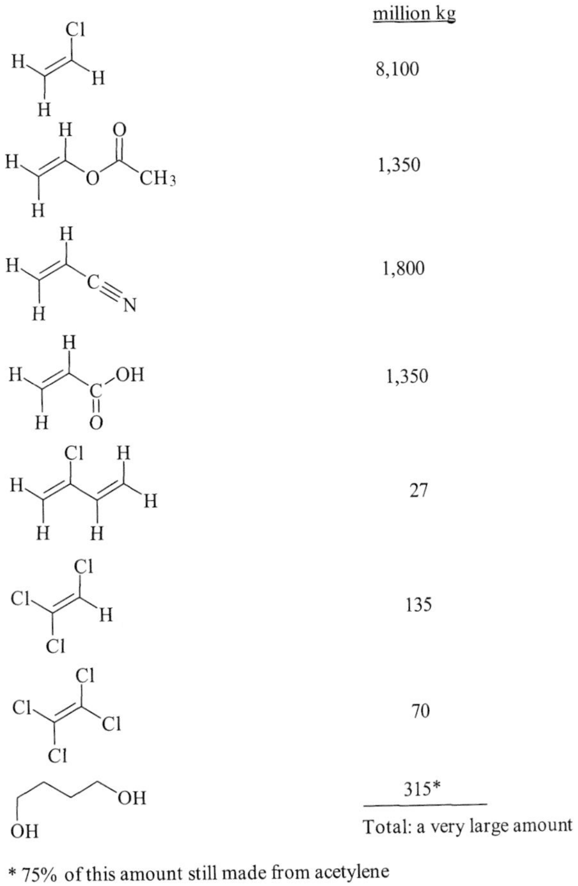

Fig. 10.6 Approximate amounts of chemicals produced today that were formerly produced from acetylene.

	million kg
	8,100
	1,350
	1,800
	1,350
	27
	135
	70
	315*
	Total: a very large amount

* 75% of this amount still made from acetylene

The unifying element in all this chemistry is the instability of acetylene, which has been the focus of much discussion in this chapter. The π-bonds in acetylene are easily perturbed by electrophiles, which means that acetylene forms coordinate bonds (Chapter 8, Section 8.13) with positively charged transition metals, particularly from Hg^{+2}, Zn^{+2} and Cu^{+1} (Chapter 9, Section 9.7).

Coordination of Hg^{+2}, Cu^{+1}, and Zn^{+2} are all used industrially to activate acetylene for nucleophilic attack (Fig. 10.7). If the nucleophile is water, then acetaldehyde results. Acetic acid provides vinyl acetate; HCl gives vinyl chloride; and HCN provides acrylonitrile. A proton then replaces the metal-to-carbon bond with a carbon-to-hydrogen bond, which completes the process yielding the target molecule and releasing the metal ion, which can activate another acetylene molecule giving rise to a catalytic cycle.

Because we have focused on both acrylonitrile and acetaldehyde in Chapters 8 (Section 8.13) and 9 (Section 9.7) we shall limit our discussion here to how vinyl acetate and vinyl chloride are produced using ethylene in place of acetylene (Fig. 10.7). In studying how

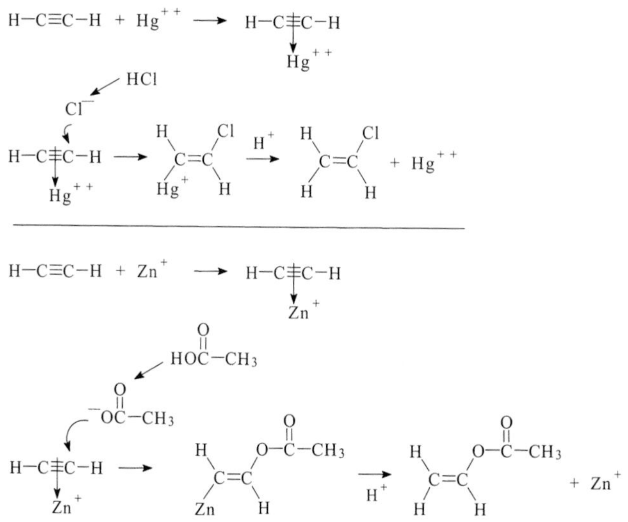

Fig. 10.7 Transition metal-catalyzed addition to acetylene to form vinyl acetate and vinyl chloride.

these industrial intermediates are produced we will see that the same general principles apply to all these reactions.

10.8
Replacing Acetylene with Ethylene and Zinc with Palladium for the Production of Vinyl Acetate

In principle, the Wacker process could be used to produce vinyl acetate instead of acetaldehyde by replacing water (Chapter 9, Section 9.8, Fig. 9.14) with acetic acid. The overall chemistry to produce vinyl acetate (Fig. 10.8) parallels the Wacker process including the reduction of cupric ion, Cu^{+2} to cuprous ion, Cu^{+1} to reoxidize the produced Pd^0 to Pd^{+2} as can be seen in comparing the catalytic cycles in Fig. 10.8 and Fig. 9.14.

A difference between production of acetaldehyde and vinyl acetate by the palladium-catalyzed method is that in the reoxidation of the cuprous salt by air (compare Fig. 10.8 to Fig. 9.14), water is produced. This difference does not affect acetaldehyde production (Chapter 9, Fig. 9.14) because water is the reaction medium. But the water produced in the vinyl acetate process (Fig. 10.8) competes with acetic acid in the addition to palladium-complexed ethylene so that some acetaldehyde is also produced together with vinyl acetate.

There is an interesting advantage here because of the ease of oxidation of acetaldehyde to acetic acid. This side production of acetic acid reduces cost by reducing the need to supply acetic acid for the process. By varying the conditions, industrial engineers could vary the ratio of vinyl acetate and acetaldehyde, which is then oxidized to acetic acid, and benefit from this favorable raw material factor. And, in fact, the whole process seemed to

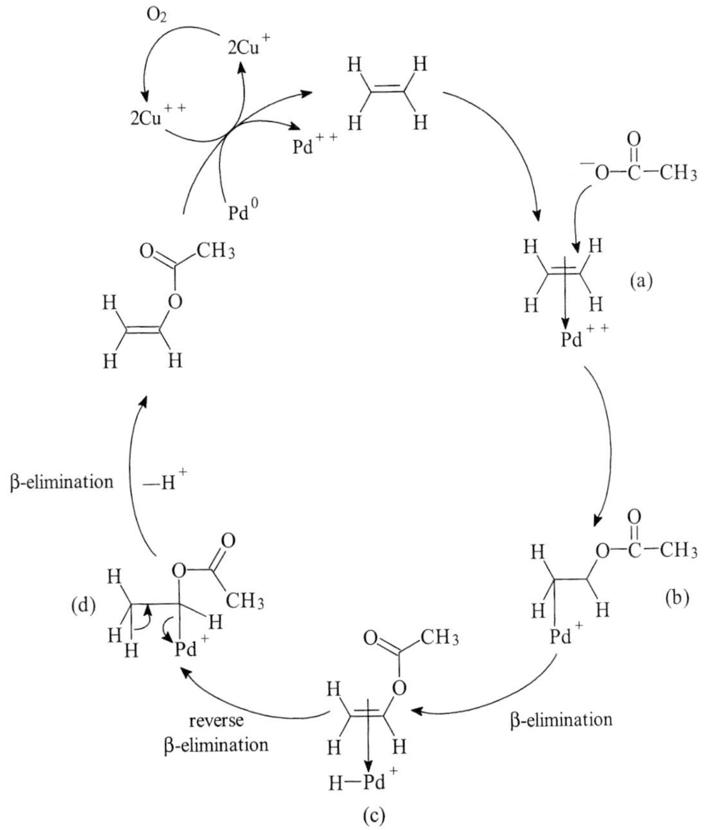

Fig. 10.8 The Wacker process for the production of vinyl acetate.

work so well that two large chemical corporations, ICI and Celanese, built large-scale plants while a third large company, Hoechst, went to a semi-commercial stage based on the same chemistry. It looked like a winner!

However, unforeseen problems of corrosion developed from the combined effect of acetic acid and the palladium and copper chlorides. Solutions were sought, but the corrosion problems were so severe that expensive materials such as titanium components were required for the plant. And mass transfer problems associated with foaming prevented sufficient ethylene from dissolving in the solutions. Although in principle, the essential chemistry is the same, there can be a big difference in the details of a process, which are not apparent by simply writing equations. After only a short time both ICI and Celanese had to shut down their plants, and Hoechst's process never became fully commercial. It was a nightmare!

What could be done? Vinyl acetate is a necessary industry intermediate and it was not desirable to return to acetylene, however simple the chemistry was. In spite of the problems with this Wacker chemistry, industry had to find another solution to the problem.

The essential chemistry of the Wacker process was used (Fig. 10.8) but the liquid phase was replaced by a heterogeneous vapor phase process so that the corroding molecules contacted the vessel walls to a lesser extent. The chloride salts were also removed, making impossible the last step of the process ($2\ CuCl + 0.5\ O_2 + 2\ HCl \rightarrow 2\ CuCl_2 + H_2O$). In the heterogeneous process for vinyl acetate, then, the cuprous ion is converted to the cupric ion by air oxidation. Corrosion problems were lessened, allowing the use of conventional stainless-steel reaction vessels while gaining other engineering advantages.

10.9
What is Valuable about Vinyl Acetate?

Vinyl acetate (Fig. 10.7), as we have noted earlier in Chapter 9 (Section 9.1), is an important product of the chemical industry. In the early 2000s about 720 million kilograms per year were produced in the USA. Conversion to poly(vinyl acetate) (Fig. 10.9), which is then dispersed in water, is the prerequisite for commercial use. The solid polymer particles suspended in water form a vehicle to be used as adhesives and as binders for paint. The adhesive is marketed in small, low-density polyethylene squeeze bottles for consumer use. A common trade name is "Elmer's Glue." Important uses for the adhesive are in bookbinding and woodworking, an example of which is furniture building.

The use of poly(vinyl acetate) water-borne paints was developed after World War II. At that time there were many styrene-butadiene rubber plants (see Chapter 7, Section 7.9) that had been built during the war but no longer needed. What to do with them? One solution was to add pigment to the synthetic rubber water dispersions. When this was done, water-based latex paints were born. These paints had the inherent advantages of replacing smelly and toxic organic solvents with water, and the additional advantage of easy clean-up. Subsequently, poly(vinyl acetate) provided a vehicle that gave a more easily washed surface to the cured paint and so it replaced styrene-butadiene. Today, water-borne paints for houses and building, so-called architectural paints, are based on polymers synthesized from vinyl acetate and esters of both acrylic (see Chapter 8, Section 8.9) and methacrylic acids (see Chapter 6). These paints allow more sophisticated properties to be formulated such as resistance to yellowing on exposure to light. But the basic idea of dispersing the polymer in water is still the same.

Poly(vinyl acetate) has a venerable history in the development of polymer science. It was one of the first known addition polymers, a class of polymers focused on in both Chapters 2 and 6. Thus, poly(vinyl acetate) has the same formula as vinyl acetate. The π-bonds are simply converted to the σ-bonds that hold the polymer chain together. This explanation is obvious now, but when poly(vinyl acetate) was first synthesized in Germany

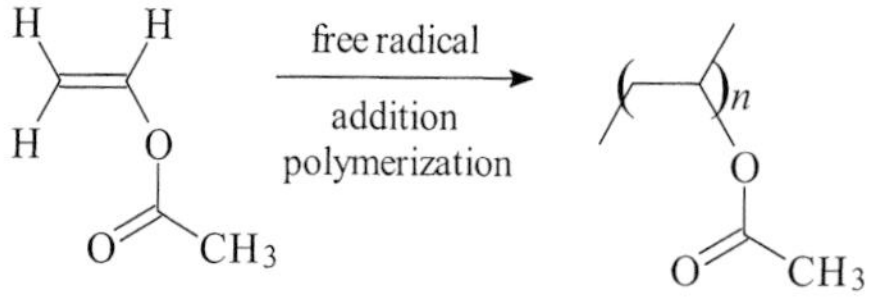

Fig. 10.9 Polymerization of vinyl acetate to poly(vinyl acetate).

in the early twentieth century, the method available for synthesis of the polymer from the monomer led to considerable controversy about its structure (Chapter 5, Section 5).

The only way the polymer could be made was by exposure of the vinyl acetate to sunlight. As reported by Morawetz, one can find in the archives of Farbwerke Hoechst a report by F. Klatte in 1929 suggesting that poor light intensity in the winter months "may be taken care of by increased production from March to October." The German chemists at Hoechst found poly(vinyl acetate) to be exceptionally useful as an adhesive compared to animal and vegetable glues, but did not understand how it was formed from its monomer.

This early work in Germany led poly(vinyl acetate) to play an important role in the argument about the nature of polymers. Were they aggregates of small molecules, or were they macromolecules bonded together in the same way as familiar small molecules? The experiment is shown in Fig. 10.10. Hydrolysis of poly(vinyl acetate) gave rise to a stable hydroxyl-containing polymer, poly(vinyl alcohol). But no carbonyl groups appeared. This result was a strong argument against the idea that polymers were "aggregates" of small molecules. The absence of the tautomeric equilibrium in poly(vinyl alcohol) (Fig. 10.10) suggested a profoundly different structure for the polymer compared to the monomer from which it is named. As we have seen, vinyl alcohol is rapidly tautomerized to acetaldehyde (Chapter 9, Fig. 9.13). This result reinforced the views of Herman Staudinger that polymers were very large molecules connected by the same kinds of bonds found in small molecules. Polymers were not aggregates or "colloids." The conflict concerning the nature of polymers, however, continued for many years after this observation on poly(vinyl alcohol) and, as we pointed out in Chapter 5 (Section 5.1), this conflict was one of the reasons that motivated Carothers to undertake the experiments at DuPont that led to the invention of nylon.

Poly(vinyl alcohol) is derived in modern industry by reaction of methanol with poly(vinyl acetate) to yield also methyl acetate, a transesterification (Fig. 10.11) (see Chapter 4, Section 4.10), which is a subject to be discussed in detail later in this chapter (Section 10.13).

Poly(vinyl alcohol) has many uses including reaction with butyraldehyde (the product of hydroformylation of propylene; Chapter 9, Section 9.9) to form the acetal (Fig. 10.12). This is the polymer that gave us safety glass. Originally, cellulose acetate held two sheets

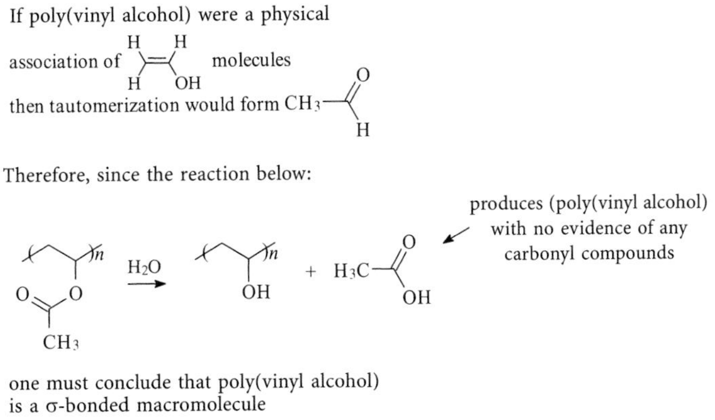

Fig. 10.10 Hydrolysis of poly(vinyl acetate) yields poly(vinyl alcohol).

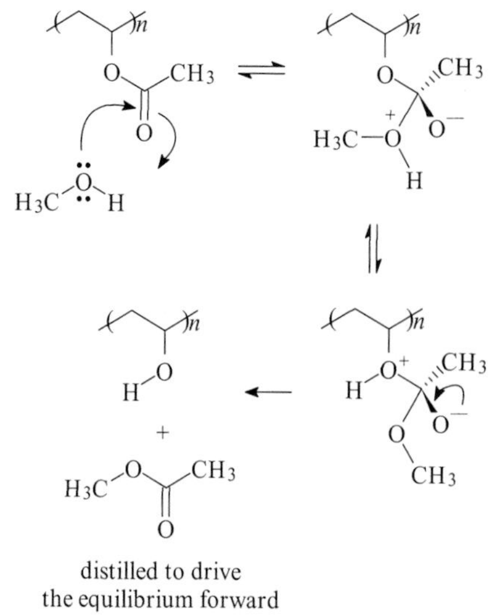

Fig. 10.11 Synthesis of poly(vinyl alcohol) via transesterification of poly(vinyl acetate) with methanol.

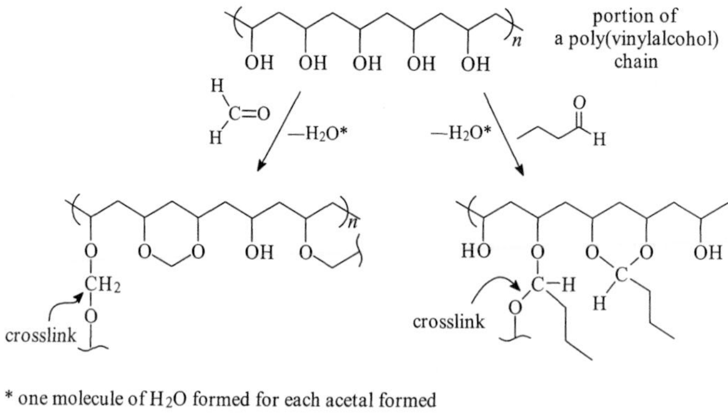

* one molecule of H_2O formed for each acetal formed

Fig. 10.12 The butyraldehyde and formaldehyde acetals of poly(vinyl alcohol).

of glass together to provide a primitive form of safety glass. Poly(vinyl butyral) (Fig. 10.12) was far more successful. With the proper ratio of acetal groups and free hydroxyl groups as shown in Fig. 10.12, a polymer can be made with the same refractive index as glass so that the binding polymer is invisible.

How marvelous, since this characteristic is combined with a tenacious adherence to the glass, which will keep the glass from breaking into flying fragments even on shattering. As early as 1941, 98 percent of safety glass depended on poly(vinyl butyral). Finally, we cannot leave this subject without mentioning that Japanese innovations led to use of the crosslinked formaldehyde acetal of poly(vinyl alcohol) as a fiber. This invention quickly jumped across the sea to China to be used for "Mao Uniforms."

10.10
Replacing Acetylene with Ethylene for the Production of Vinyl Chloride

F. Klatte, of Farbwerke Hoechst, the same man who played a key role in the discovery of poly(vinyl acetate) was also responsible for the development of poly(vinyl chloride), the subject of one of the first ever patents suggesting a commercial use for a polymer. As reported by Morawetz, in 1913 Klatte was issued a German patent for "the preparation of horn substitute, artificial fibers, lacquers, etc." But how was poly(vinyl chloride) synthesized in those early days? As for vinyl acetate discussed above (Section 10.9), sunlight again was the answer. Simply leaving the vinyl chloride in the sun for long periods of time converted the easily flowing liquid into the polymerized viscous mass.

Although Hoechst never used the patent, vinyl chloride has grown today to be one of the most important intermediates in the chemical industry. More than 8 billion kilograms are produced per year in the USA in the early 2000s, and all of it is used as a monomer for the synthesis of poly(vinyl chloride) and its copolymers. Poly(vinyl chloride) is used for plastics, molded products, films and sheets, and coatings. The major use is for pipe for plumbing. Klatte certainly had the right idea, even if he was ahead of his time by about twenty years. It was not until the 1930s that the commercial use of poly(vinyl chloride) began to boom. Today, more than one-third of the chlorine used industrially finds its way into vinyl chloride.

Although we have no direct information in hand, it is reasonable that Klatte obtained vinyl chloride by addition of HCl to acetylene, which would have foreshadowed the first large-scale process for manufacture of this important industrial intermediate. In this now obsolete process, Hg^{+2} in the form of $HgCl_2$ (see Fig. 10.7) is adsorbed onto activated charcoal, and acetylene and HCl is passed over it at the moderate temperature of 140–200 °C. Vinyl chloride is produced in very high yield with virtually no byproducts. The process is simple and involves low investment and operating cost, but has nevertheless been shut down all over the world – providing another example that the explosive and expensive nature of acetylene as a feedstock motivates its replacement, even if new processes are more complex.

Just as ethylene has replaced acetylene for the production of vinyl acetate (Section 10.8), ethylene has also universally replaced acetylene in modern processes for the production of vinyl chloride. In principle, the chemistry is straightforward – the addition of chlorine to a double bond and elimination of a hydrogen chloride to form a double bond. Thus, in oxychlorination, a process devised to save chlorine, a catalytic amount of $Cu(II)Cl_2$ is reacted with ethylene to produce ethylene dichloride followed by cracking at elevated temperatures to eliminate HCl and produce the vinyl chloride (Fig. 10.13).

Heating gaseous ethylene dichloride to high temperatures, 500–600 °C, yields enough energy to break the relatively weak bond between carbon and chlorine. As shown in Fig. 10.13, the chlorine radicals produced then initiate a free radical chain mechanism, examples of which we have seen before (Chapter 1, Section 1.11) and vinyl chloride is ultimately produced.

Previous discussions about the cost of chlorine (see Chapter 4, Section 4.22) clearly apply to vinyl chloride. In the production method described above (Fig. 10.13), two chlorine atoms are required but only one ends up in the vinyl chloride. Replacing acetylene with

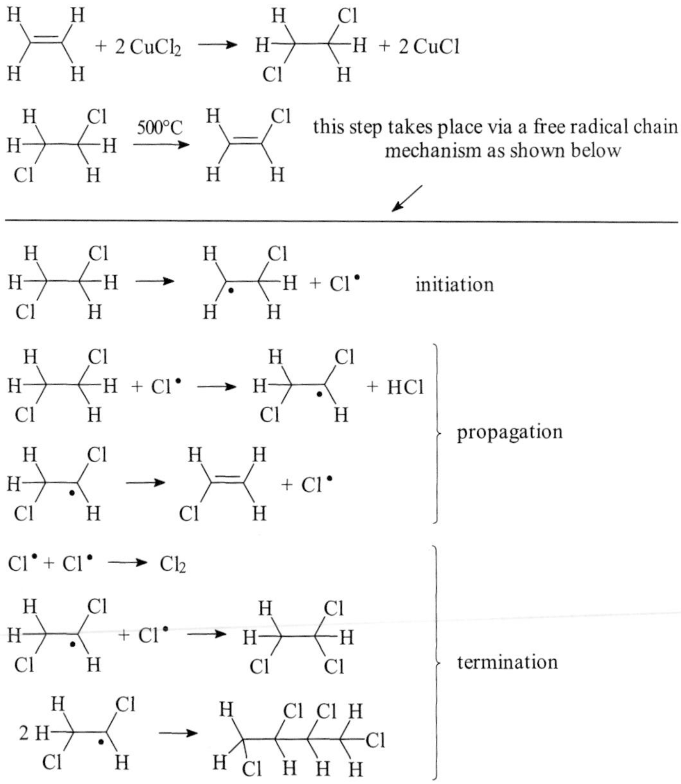

Fig. 10.13 Formation of 1,2-dichloroethane, and the free radical chain mechanism for elimination of HCl to form vinyl chloride.

ethylene therefore not only adds the complexity of two steps instead of one to produce vinyl chloride (Fig. 10.13 versus Fig. 10.7) but a chlorine atom, in the form of HCl, is now wasted for each molecule of vinyl chloride produced. Nevertheless, the expense and danger of acetylene in combination with the difficulty of obtaining large volumes of HCl turned industry away from acetylene and toward an ethylene route to vinyl chloride. Moreover, it stimulated an innovation to solve the wasted chlorine problem.

One of the first approaches was to develop a combined process using both ethylene and acetylene. Ethylene dichloride was first produced from ethylene, which was then heated to eliminate HCl to produce vinyl chloride. The HCl then, instead of being wasted, was reacted with acetylene to produce more vinyl chloride. This is an excellent idea, but at the cost of keeping the dangerous and expensive acetylene in the picture. Can this be improved? Can we produce vinyl chloride without wasting any chlorine, and without using acetylene?

The answer to these questions, although hardly addressed to vinyl chloride in those early days of chemistry in 1858 in England, can be found in a process credited to Henry Deacon (1822–1876). The Deacon reaction is the basis for oxychlorination.

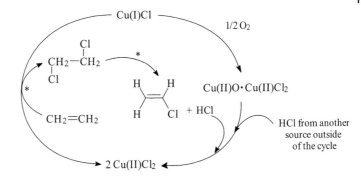

* as in Figure 10.13.

Fig. 10.14 The catalytic cycle for vinyl chloride production via oxychlorination.

Deacon played an early role in the development of the alkali industry in England, which accounts for his interest in chlorine derived from HCl. The alkali industry in England has serious environmental problems including the release of HCl fumes into the air. This problem was so severe that a law was passed, the Alkali Act of 1863, mandating that most of the HCl be recovered. Deacon played an important role in helping to solve this problem by creating a method for the oxidation of HCl to Cl_2, which was then used in reaction with lime to produce "bleaching powder," which in turn was used in the paper industry. Deacon found that HCl could be oxidized to Cl_2 by bricks soaked in cupric chloride. This redox reaction in which cupric ion is reduced to cuprous ion while chloride ion is oxidized to chlorine can be part of a catalytic cycle (Fig. 10.14), in analogy to the redox couple in the Wacker process where oxygen reoxidizes Cu^I to Cu^{II} (Fig. 10.8, and Chapter 9, Fig. 9.14). Deacon became a wealthy man from his inventions for the chemical industry.

Deacon, and also his mentor, one of the giants of nineteenth century experimental science, Michael Faraday (1791–1867), belonged to the Sandemanians (Glasites), a nonconformist protestant sect that believed in complete separation of church and civil matters – a point of view that dominates the American attitude toward religion. Faraday, whose accomplishments spanned the discovery of benzene and other organic molecules to laying the foundations of modern electrochemistry, was known in his day as "a model for scientific men." A famous piece of advice given traditionally to research students of chemistry by their mentors, "work, finish, publish," is ascribed originally to Faraday, and we might imagine that Henry Deacon received this advice. Faraday, in fact, was responsible for Deacon finding employment in the glass and alkali industry, which led Deacon to design the process that bears his name.

In the modern industrial process for the production of vinyl chloride from ethylene the Deacon process has been adapted to modern technology so that the chlorination of ethylene and the oxidation of the HCl cracked out of the ethylene dichloride (Fig. 10.13) are integrated. This process is known, as noted above, as oxychlorination, and is the way that 50% of all vinyl chloride is produced today. The other 50% is made by cracking the ethylene dichloride made by the addition of chlorine to ethylene. This is the source of the second mole of HCl shown in Fig. 10.14. Now, instead of the Cl⁻ in the HCl being wasted, this chloride anion finds its way back into the vinyl chloride. An overview of the process is shown in Fig. 10.14. The chlorination of ethylene is accomplished by direct re-action with cupric chloride and yields ethylene dichloride and cuprous chloride. The next step is the same thermolysis as we have seen in Fig. 10.13, producing the vinyl chloride and HCl.

The critical feature of the catalytic cycle of oxychlorination is the reformation of cupric chloride from the cuprous chloride produced in the first step (Fig. 10.14). Cuprous to cupric is an oxidation and the oxidizing agent is oxygen, which is then reduced to form water. The chemistry involves the intermediate double cupric salt, $(CuO)(CuCl_2)$. This double salt then reacts with the HCl produced in the thermolysis of the ethylene dichloride to regenerate the chlorinating agent cupric chloride, which completes the catalytic cycle (Fig. 10.14).

As discussed above stoichiometry is usually not seen as a problem but simply a descrip-tion of the moles of each substance necessary to describe a chemical reaction. As seen in Fig. 10.14, however, the stoichiometry of the production of cupric chloride reveals the neces-sity for 2 moles of HCl. But only 1 mole of HCl is produced in the thermolysis step for each ethylene dichloride conversion to vinyl chloride. This means that half of the molecules of vinyl chloride are produced via cupric chloride generated from HCl released in the thermo-lysis of ethylene dichloride, while half of the vinyl chloride has to find its chlorine from the addition of HCl from another source. Nevertheless, whatever the source of the chlorine, none is wasted. Every atom of chlorine ends up in a molecule of vinyl chloride.

In a recent adaptation of the oxychlorination process, HCl is produced in high purity in the production of isocyanates via reaction of amines with phosgene. This HCl is then piped to a vinyl chloride plant to provide the second mole of HCl needed to satisfy the stoichiometry of the oxychlorination, as discussed just above. When BASF built a chemi-cal complex in Antwerp, Belgium a major objective was to make isocyanates. In order to obtain value from the HCl resulting from the interaction of an amine with phosgene, they built a vinyl chloride unit in an adjoining plant so the HCl from one process could be piped to the other. That's efficiency!

10.11
The Production of 1,4-Butynediol shows an Entirely Different Face of Acetylene Reactivity

In the discussion about the formation of potassium and calcium carbides that first led to the discovery of acetylene, we described the enhanced Brønstead acidity of hydrogens at-tached to triply bonded carbon atoms (Fig. 10.4). The stability of a negative charge at sp-hybridized carbon, which is responsible for the relatively low pK_a of terminal alkynes, drives the reactivity of acetylene in a manner that is entirely different from the π-reactiv-ity of the triple bond. While the π-reactivity leads to addition to the triple bond as seen in the formation of vinyl acetate and vinyl chloride (Fig. 10.7), the relative ease of formation of the acetylide anion leads to formation of carbon–carbon bonds.

This kind of carbon–carbon bond formation as seen in Fig. 10.3 plays an important role in syntheses on a relatively small scale, but there is a related carbon–carbon bond-

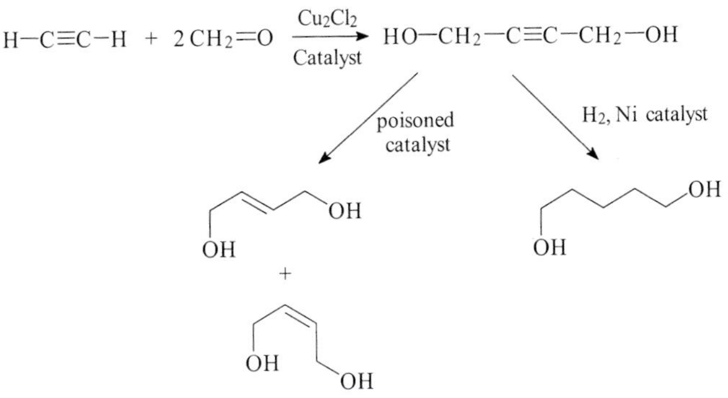

$$H-C\equiv C-H \ + \ 2\,CH_2{=}O \ \xrightarrow[\text{Catalyst}]{Cu_2Cl_2} \ HO-CH_2-C\equiv C-CH_2-OH$$

poisoned catalyst

H_2, Ni catalyst

OH / OH

+

OH / OH

Fig. 10.15 Synthesis of 1,4-butanediol from acetylene via the formation of 2-butyne-1,4-diol.

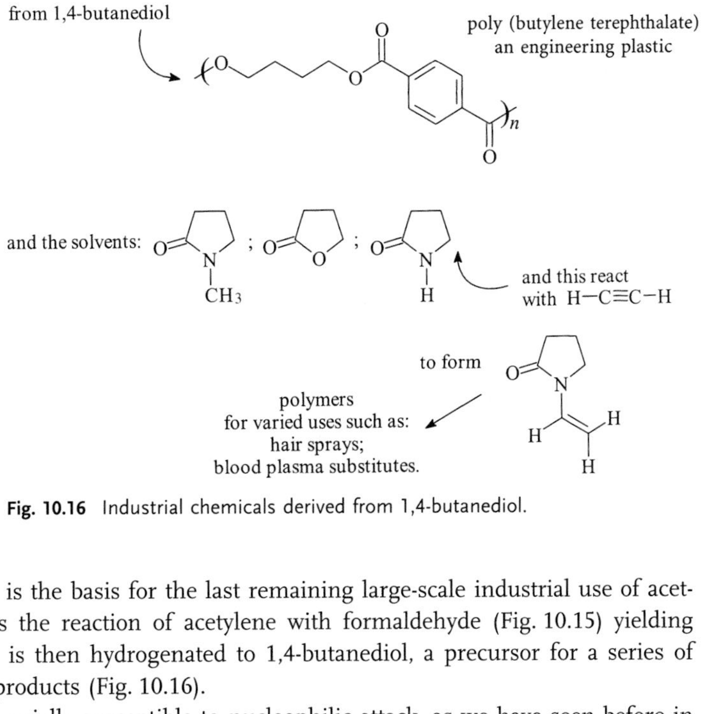

from 1,4-butanediol

poly (butylene terephthalate) an engineering plastic

and the solvents: ; ;

CH_3

H

and this react with $H-C\equiv C-H$

to form

polymers for varied uses such as: hair sprays; blood plasma substitutes.

Fig. 10.16 Industrial chemicals derived from 1,4-butanediol.

forming reaction that is the basis for the last remaining large-scale industrial use of acetylene. This process is the reaction of acetylene with formaldehyde (Fig. 10.15) yielding 1,4-butynediol, which is then hydrogenated to 1,4-butanediol, a precursor for a series of important industrial products (Fig. 10.16).

Formaldehyde is especially susceptible to nucleophilic attack, as we have seen before in the formation of the polyhydric alcohols (Chapter 9, Section 9.0). Although in the aldol reaction leading to the polyhydric alcohols, the carbanion acting as the nucleophile resided on a carbon adjacent to a carbonyl group, the susceptibility of formaldehyde to nucleophilic attack is no less when the nucleophilic agent is an acetylenic carbanion (Fig. 10.15).

Why has acetylene remained competitive here? The answer is that the plants producing 1,4-butynediol are already built, functioning well and are depreciated – which means that from an economic view the plant has no cost except maintenance and operating cost. Any new process must not only be more economical but must take into account the cost of a new plant. This identical factor was emphasized in competitive processes for producing methyl methacrylate in Chapter 6 (Section 6.10).

For acetaldehyde, vinyl chloride, vinyl acetate and acrylonitrile, chemistry based on ethylene has proved more economical than processes based on acetylene. But this economic advantage has not been apparent in the use of acetylene for production of 1,4-butynediol in depreciated plants – plants that have been fully paid for. The process is based on the work of Walter Reppe, who was responsible for the development, during the period between the two world wars, of a large number of acetylene-based processes. We have already seen one example of Reppe's ingenuity in the production of acrylic acid (Chapter 8, Section 8.13).

Although almost all other acetylene-based Reppe chemistry is now obsolete (for reasons we now well understand from this chapter), what is amazing about the production of 1,4-butynediol is that the process has been essentially unchanged since it was introduced in 1929! This is a remarkable longevity in the chemical industry, and even more remarkable because the process uses acetylene – a chemical that, as we know, has been widely abandoned as an industrial intermediate. This longevity stimulates us to look at the details of the formation of 1,4-butynediol from acetylene (Fig. 10.17).

In laboratory procedures in which acetylene is used for synthesis, the counterion for the acetylide anion is usually sodium, lithium or potassium (see Fig. 10.3). But in the industrial procedure the counterion is cuprous ion. The reason for this depends on the way the acetylide anion is formed. In a small-scale process, this anion can be formed by reaction of acetylene with bases that are commonly used in research laboratories, such as sodium hydride, or lithium N-diisopropyl amide. But these bases are expensive and would never do for a large-scale industrial process because they could not be used catalytically.

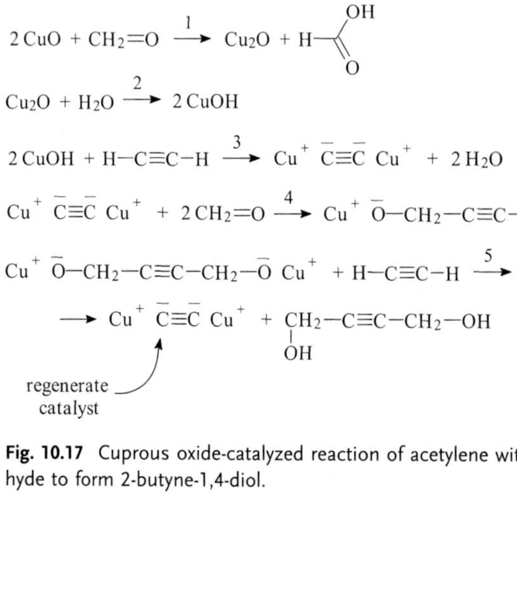

Fig. 10.17 Cuprous oxide-catalyzed reaction of acetylene with formaldehyde to form 2-butyne-1,4-diol.

In the Reppe process a transition metal, cuprous ion, Cu^{+1}, is the counterion and the active catalytic species is Cu_2C_2, cuprous carbide.

Cupric oxide, CuO, which can be converted to the active catalyst, initiates the process in an oxidation–reduction reaction with formaldehyde to form cuprous oxide, Cu_2O (step 1, Fig. 10.17). Formaldehyde, the reducing agent, is therefore oxidized to formic acid. Cuprous oxide is unstable in water and is rapidly converted to cuprous hydroxide, CuOH, which is the base that acts on the acetylene in an acid–base reaction to produce the active catalyst, Cu_2C_2 (steps 2 and 3, Fig. 10.17). The acetylide dianion, that is, carbide (C_2^{-2}), is exceptionally reactive with formaldehyde, forming the carbon–carbon bond (step 4, Fig. 10.17) and in so doing transferring the released cuprous ion to play the counterion role to the newly formed alkoxide anion. This sequence sets up the catalytic cycle. Let's see how.

The formation of the dialkoxide dianion is the key step in this catalytic reaction because this base abstracts two protons from another acetylene molecule to form another acetylide dianion (step 5, Fig. 10.17). A consequence of this reaction is that the cuprous ion is transferred back to its original role as counterion to an acetylide carbanion. Reaction of the newly formed acetylide dicarbanion with formaldehyde (repeat step 4, Fig. 10.17) then follows so that the cuprous counterion is transferred again to the position of counterion to the alkoxide anion, which is then set-up to react with another acetylene molecule. And so this catalytic cycle is repeated over and over and in this manner (steps 4 and 5) a relatively few cuprous ions act to chaperone many acetylide anions to their fate, to the formation of 1,4-butynediol.

Lovely, but were you disturbed by anything in the picture painted in the last two paragraphs? You rightfully could be. There are two problems with the mechanism outlined above (Fig. 10.17), two things that seem not to make sense.

The acetylide dianion formed in step 5 with its two cuprous counterions (Fig. 10.17) is surrounded by two reactants, water and formaldehyde. Reaction with the water would re-form acetylene, which would exactly parallel formation of acetylene by reaction of calcium carbide with water (see Fig. 10.2). But reaction with formaldehyde has to be more than competitive or the Reppe process could not work, and no 1,4-butynediol would be produced. This first problem with the mechanism outlined in Fig. 10.17 parallels a similar quandary we have seen before in the formation of polycarbonates from phosgene and bisphenol A, a reaction that also takes place in water.

Phosgene, although readily hydrolyzed by water to carbonic acid reacts competitively with bisphenol A to form links in the polymer chain. The same general explanation given in the last paragraph of Section 4.20 in Chapter 4 for the competition between water and the sodium salt of bisphenol A for reaction with phosgene applies just as well to the competition between water and formaldehyde for acetylide anion (Fig. 10.18).

Simply put, the rate constant for reaction of acetylide anion with formaldehyde is far larger than the rate constant for reaction with water. Therefore, although the water is in far higher concentration than the formaldehyde, most of the acetylide anion will react with the formaldehyde instead of the water. You may not be satisfied because we have offered no structural explanation for the higher rate constant for the reaction with formaldehyde. In other words, why is the reaction with formaldehyde so much faster than that with water? This is a separate question, which was not addressed in the water–bisphenol A competition for reaction with phosgene (Section 4.20 in Chapter 4) and will similarly

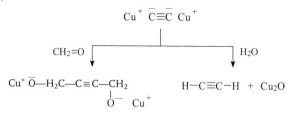

Fig. 10.18 Competition between water and formaldehyde for the acetylide anion.

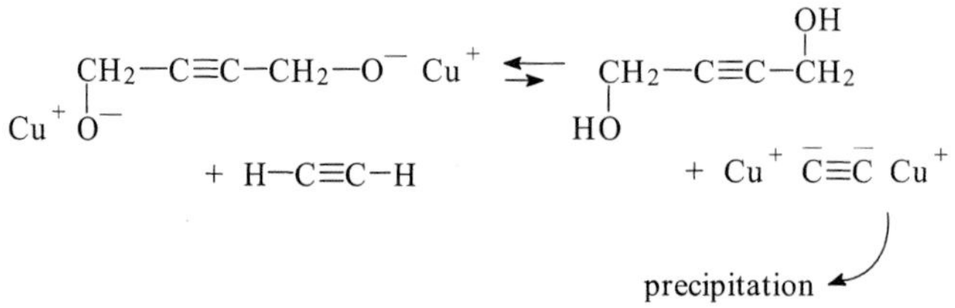

$$CH_2-C\equiv C-CH_2-O^- \; Cu^+ \rightleftharpoons \underset{HO}{CH_2-C\equiv C-\overset{OH}{CH_2}}$$

$$Cu^+ \, O^-$$

$$+ \; H-C\equiv C-H \qquad\qquad + \; Cu^+ \; \bar{C}\equiv\bar{C} \; Cu^+$$

precipitation

Fig. 10.19 Acid–base reaction of cuprous alkoxide with acetylene and precipitation of the Cu_2C_2.

not be addressed here, although there are answers that could be hypothesized. Can you think of good reasons; can you make a hypothesis?

What is the second apparent problem with the mechanism in Fig. 10.17? The disturbing aspect is that the alkoxide anion that arises in step 4 of Fig. 10.17 after the reaction of the acetylide dianion with formaldehyde is the conjugate base of an acid with a pK_a of about 20. A base with a pK_a of 20 would not form significant amounts of the acetylide anion by proton abstraction from acetylene, which is an "acid" with a pK_a of 25. The equilibrium between these species would greatly favor formation of the alkoxide anion since it is the conjugate base of the stronger acid. But transfer of the proton from acetylene to the alkoxide anion is the essential step in the catalysis (step 5, Fig. 10.17). It does take place, and has to occur to a large extent for the overall catalyzed process to proceed with a reasonable rate.

A straightforward explanation can be found in Parshall's book. The cuprous salt of the alkoxide is soluble while Cu_2C_2 is not soluble. So that once formed, the Cu_2C_2 will precipitate and therefore will no longer be available to participate in the equilibrium. This insolubility drives the alkoxide to produce even more Cu_2C_2, which then forms part of the slurry in the reactor. The catalyst is in the slurry (Fig. 10.19).

In the original Reppe process, thick-walled stainless-steel towers packed with the cupric oxide and other activating inorganic compounds were used with acetylene under high pressure. The towers, however thick the steel walls, were not adequate to contain the potential explosive decomposition of the acetylene. Modern plants have replaced the towers with suspended catalysts and low pressures of acetylene. But the original chemistry has not changed (Figs 10.15, 10.17, 10.18 and 10.19).

10.12
Phosgene and Chlorine – the Poison Gases of World War I. Can their Replacement for Industrial Processes by Safer Chemicals also be an Example of "Doing Well by Doing Good?"

Phosgene has many minor uses in the chemical industry and two major uses. One is re-action with bisphenol A to produce polycarbonates (see Chapter 4, Section 4.20), and the other is for synthesis of isocyanate intermediates for preparation of urethanes (Chapter 4, Sections 4.15 and 4.16). The use of phosgene for the synthesis of isocyanates was noted earlier (Section 10.10) as a source of HCl in discussing the oxychlorination production of vinyl chloride from ethylene. Phosgene has been difficult to replace in the formation of the two most important isocyanates, toluene diisocyanate (TDI) (Fig. 10.20) and methy-lene diphenyldiisocyanate (MDI), but more success has been encountered in finding ways to produce polycarbonate without phosgene. Moreover, emphasizing one of the themes of this chapter, the replacement process that does not use phosgene for polycar-bonate production has been found to be more economical. But before we look at the in-novative chemistry for the new process let's first go back and find out how phosgene and chlorine were first produced, because they are certainly not found naturally on earth in more than trace amounts.

By the mid-1770s, HCl was known as *acid of salt* (makes sense doesn't it?) and many experiments were conducted involving treatment of various substances with this acid. Among these experiments were those of the great Swedish-German chemist Carl Wil-helm Scheele, who although he only lived for 44 years from 1742 to 1786, was a prodi-gious experimentalist who made an astonishing range of discoveries working under very poor circumstances. His first laboratory was described by Partington as "a cold and draughty wooden shed." As late as 1775 – a short time after Scheele discovered the ele-ments chlorine and oxygen, and two years before he discovered the action of light on sil-ver salts (the foundation, more than a century later, of photographic film) – we read the following text by Partington: "In his position at Uppsala Scheele again had a considerate master, Lokk, who gave him one day a week for research."

Things certainly seem to have improved greatly for those interested in research. And Scheele was certainly one of the best during his time and the best even compared to the many who followed him to the present day. The discovery of oxygen and chlorine are only two of Scheele's contributions to chemistry. He also made several critical early dis-coveries in organic chemistry, a subject that interested him all his life. Among these were the discovery of tartaric, prussic, malic, lactic, uric and citric acids and also several neutral molecules, including notably glycerol (see Chapter 4, Sections 4.10–4.14). What a chemist!

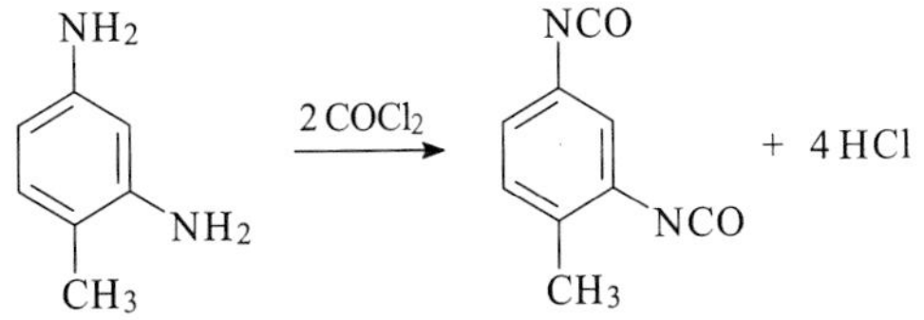

Fig. 10.20 Synthesis of toluene diisocyanate (TDI) using phosgene.

The experimental means that Scheele developed to produce chlorine has an interesting relationship to the Deacon process discovered one century after Scheele. Accordingly, it also relates to the foundation on which oxychlorination was developed two centuries after Scheele (Section 10.10). In 1773, Scheele announced that pyrolusite (the name then given to manganese dioxide), when dissolved in cold "acid of salt" (hydrochloric acid), forms a dark solution that on heating yields a greenish-yellow gas. This gas, which Scheele named dephlogisticated acid of salt, in modern terms would be acid of salt minus hydrogen or therefore $2HCl-H_2$, that is, Cl_2. This reaction of MnO_2 with HCl is clearly an oxidation of the HCl, which is directly parallel to the oxidation of HCl by $CuCl_2$, which is the basis of the chemistry of the Deacon process and modern oxychlorination (Section 10.13).

It was many years later in 1810 that Humphrey Davy, a scientific giant of the nineteenth century, took an interest in dephlogisticated acid of salt and gave this gas its modern name based on the Greek word for green. Following on the interest of his older brother, John Davy began his investigations into the properties of chlorine and discovered phosgene by exposing a mixture of carbon monoxide and chlorine to light. This is how phosgene got its name from the Greek as a contraction for the words, "light" and "I produce." Davy family members, all born in Penzance, in England, were amazing chemical innovators. Remember that it was Humphrey Davy and John Davy's cousin Edmund who, as we noted earlier in this chapter, first produced acetylene from potassium carbide and water (Section 10.3).

10.13
Is there a Way to Eliminate Phosgene in the Industrial Synthesis of Polycarbonate?

The synthesis of polycarbonates reaches back into the nineteenth century and in fact two competitive methods, which are still competitive as we shall see, were discovered within a few years of each other near the turn of the century, in 1898 and 1902. We have already investigated the phosgene route to polycarbonate, the older of these methods, in Chapter 4 (Section 4.20). The 1898 work was that of a distinguished German chemist, Alfred Einhorn, who was further distinguished as the teacher of one of the great synthetic chemists of the twentieth century, Richard Willstätter. Willstätter wrote his doctoral thesis on cocaine and later in his career was famous for his work on natural products, eventually winning a Nobel Prize. Many students know his name today as the chemist who first synthesized cyclooctatetraene which, like benzene, contains all conjugated double bonds in a ring. But cyclooctatetraene differs greatly from benzene in undergoing the usual reactions of double bonds, addition, rather than the chemistry benzene undergoes, substitution, as we have studied in Chapter 3. While the differences between benzene and cyclooctatetraene contributed to one of the great theoretical advances of the twentieth century, finally resolved by Hückel and his $4n+2$ rule, the discovery of polycarbonates by Willstätter's teacher led to a great industrial enterprise as we have discussed in Chapter 4 (Section 4.19). But this did not occur until more than half a century later.

Although we are unaware of any famous students of C. A. Bischoff, his work reported in Berichte in 1902 set the stage for our discussion below. Bischoff discovered a route to polycarbonate that does not require the use of phosgene. This transesterification

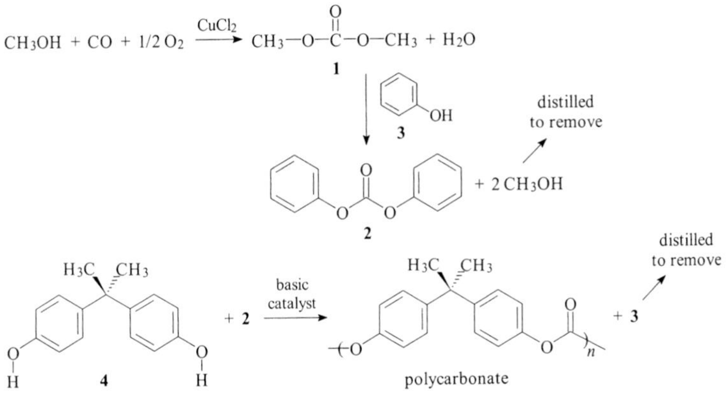

Fig. 10.21 Overall process for production of polycarbonate without using phosgene.

approach, discussed below, appears to be greatly attractive to industry as we move into the twenty first century and it may displace the phosgene route to bisphenol A polycarbonate. Let's find out how it works in the modern chemical industry.

Fig. 10.21 is an overall picture of a non-phosgene process developed by General Electric and also by Bayer and described in several patents, which is now gradually replacing the phosgene-based process for production of polycarbonate discussed in Chapter 4 (Section 4.20). We see three essential steps for the formation of polycarbonate by this route (Fig. 10.21). First, there is the production of dimethyl carbonate (**1**). Second, there is the formation of diphenyl carbonate (**2**) from dimethyl carbonate by reaction with phenol (**3**), which is driven to completion by distillation to remove the methanol produced. And finally there is the synthesis of polycarbonate from the reaction of diphenyl carbonate (**2**) with bisphenol A (**4**) using a basic catalyst and producing phenol (**3**) as the byproduct.

Fig. 10.22 is crowded, as are the complex tangle of reactions it is intended to represent. This figure shows some of the series of reactions resulting from the reaction of bisphenol A (**4**) with diphenyl carbonate (**2**), leading to high molecular-weight polycarbonate. Phenol (**3**) has to be driven off by heat and vacuum in order to drive the process to high molecular-weight polymer for reasons we will discuss in detail below. This is accomplished by conducting the polymerization under increasingly high vacuum, starting at 100 mmHg and ending at 0.5 mmHg, with the temperature rising by the end of the process to the range of 300 °C. As the molecular weight of the polycarbonate grows, the viscosity greatly increases; this gives rise to difficult engineering problems, which must be overcome to maintain stirring and the necessary evaporation of the phenol byproduct. This involves specialized equipment such as "wiped film evaporators, helicone reactors, or multiply vacuum-vented extruders" as reported in the Kirk-Othmer Encyclopedia.

All the reactions shown in Fig. 10.22 are ester exchange reactions. Every reaction in Fig. 10.22 proceeds via addition of a nucleophile to a carbonyl group. This is the primary mechanistic step linking the units together via formation of the carbonate bonds. Let's look further into the details of these ester exchange reactions, and discover how a familiar intermediate we have seen before plays a critical role.

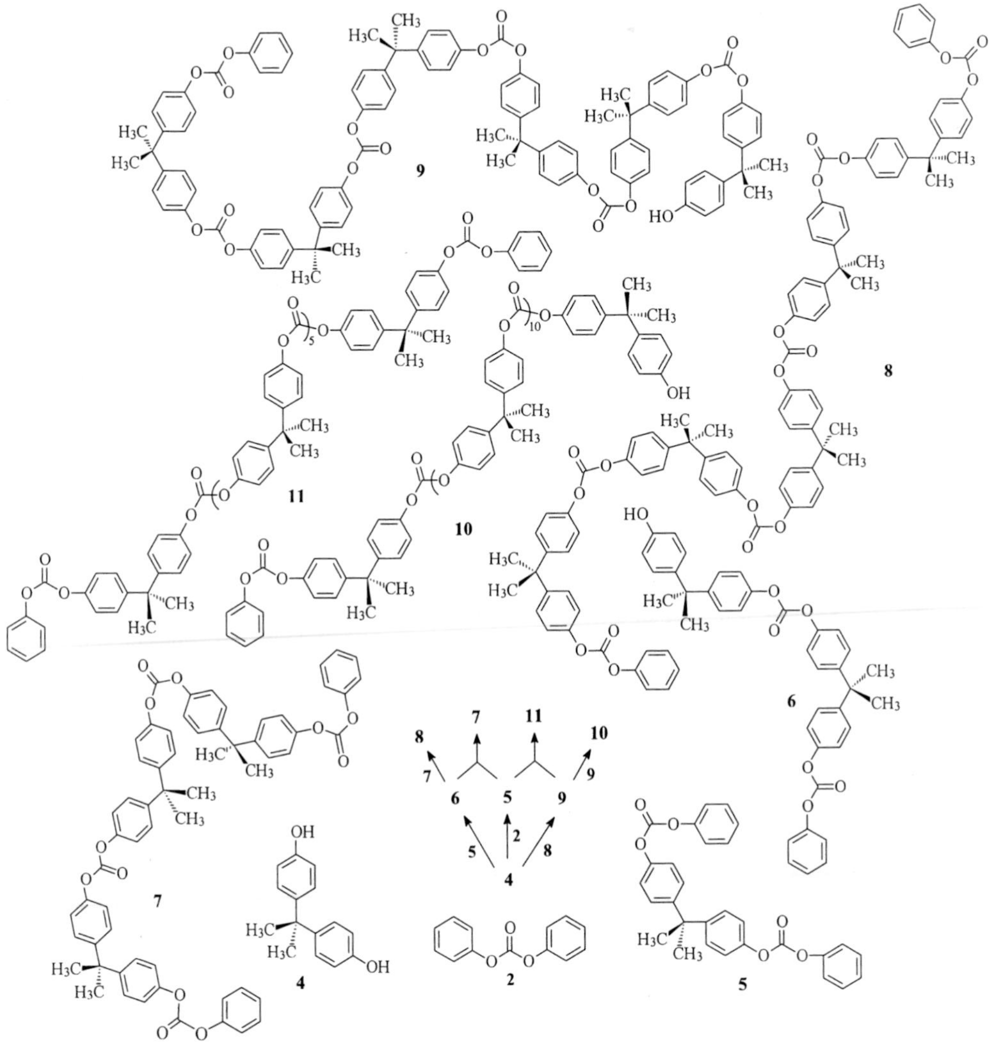

In every reaction shown in the figure, phenol is formed as the byproduct.

Fig. 10.22 A series of reactions of the type that will eventually lead to high molecular weight polycarbonate.

All reactions that exchange groups at an acyl carbon, such as reactions at the carbonyl groups of carboxylic acid derivatives, take place by addition of a nucleophile to the carbonyl group forming a four-coordinate intermediate We have seen this before, for example, in the discussion in Chapter 4 (Section 4.20), for the formation of polycarbonate from bisphenol A and phosgene. In the latter reaction the site of addition of the nucleophile is the carbonyl group of the phosgene. Chapter 4 (Section 4.10) offers another example in the reaction of glycerol with trigylcerides to form mono- and diglycerides useful

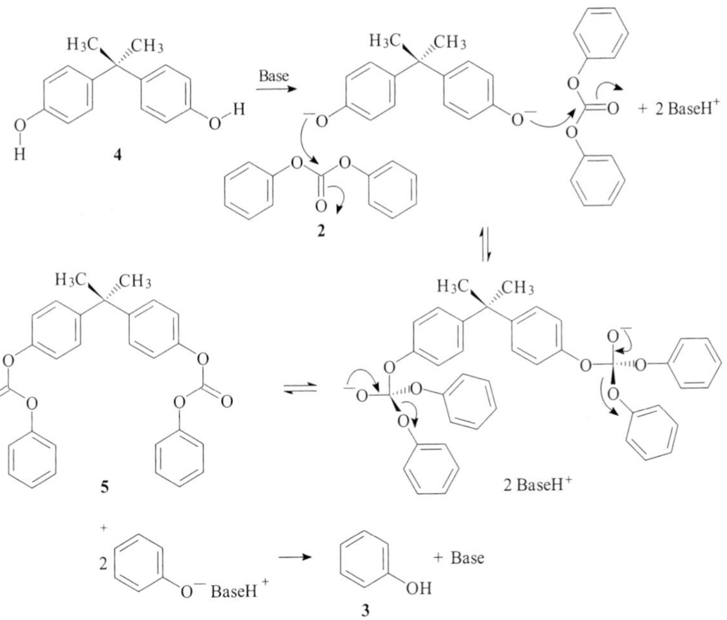

Although two molecules of **2** are shown reacting simultaneously with doubly ionized **5** to simplify the drawing of the figure, sequential reactions are likely.

Fig. 10.23 Mechanism for the base-catalyzed reaction of **4** with **2** to produce **5**.

for alkyd resin formation. An acyl transfer reaction is also found in Chapter 6 (Section 6.3) in the formation of the methyl ester in methyl methacrylate from the amide.

An important step however must take place before the nucleophilic additions to the carbonyl groups in Fig. 10.22 can occur. This step is the ionization of the phenolic hydroxyl group, and this is the reason for the basic catalyst that is used to drive the ester exchange polycarbonate synthesis to occur at reasonably fast rates. The mechanism for the first step on the path to high polymer, reaction of **4** with **2** (Fig. 10.22) is shown in Fig. 10.23, while Fig. 10.24 shows how the same basic mechanism is at work in the chain extending reactions that follow.

The four-coordinate intermediates shown in Fig. 10.23 and 10.24, which result from the nucleophilic addition to the carbonyl group, portray a critical aspect of the role of equilibrium in the acyl substitution chemistry that yields polycarbonate. The carbonyl group will re-form from the four-coordinate intermediate by pushing out one of the three groups around the former carbonyl carbon. Let's focus on the four-coordinate intermediate **A** in Fig. 10.24. Which leaving group will go? The answer is that the group that can best tolerate a negative charge will be the best leaving group. And this brings us immediately to acid–base chemistry since the best leaving group must then be the conjugate base of the strongest acid. This is a recurring theme in substitution chemistry of all kinds in organic chemistry and we have seen this idea several times throughout this book.

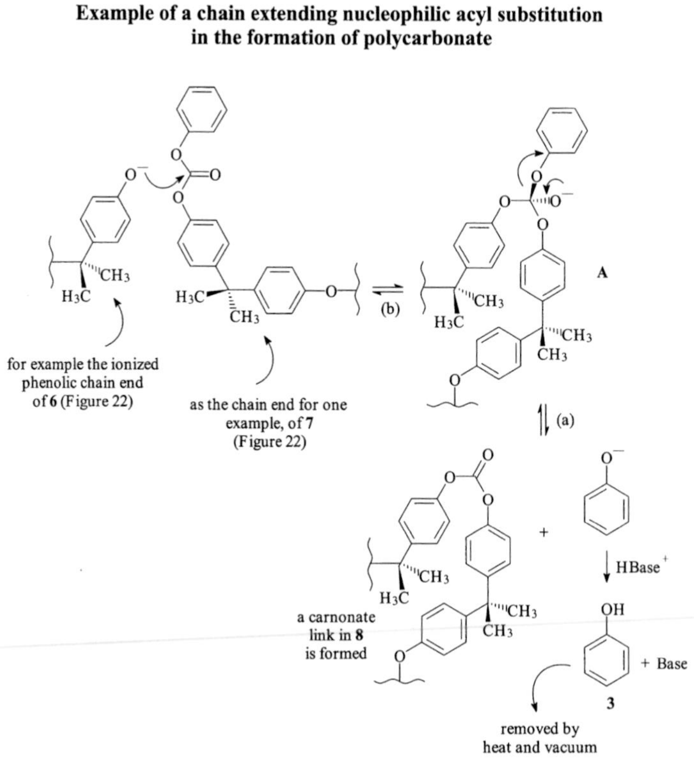

**Example of a chain extending nucleophilic acyl substitution
in the formation of polycarbonate**

for example the ionized
phenolic chain end
of **6** (Figure 22)

as the chain end for one
example, of **7**
(Figure 22)

a carbonate
link in **8**
is formed

removed by
heat and vacuum

Fig. 10.24 Mechanism for the base-catalyzed chain extension steps such as occur in
the reaction of: **5** with **4** leading to **6**; and **6** with **5** leading to **7**; and **7** with **6** leading
to **8**; and so on for the reactions shown in Figure 10.22.

In the four-coordinate intermediate **A** in Fig. 10.24, the struggle for leaving group supre-
macy is between similar entities, that is, phenolic groups. The competing groups are the
conjugate bases of acids of similar pK_a, about 10, so that both steps (a) and (b) of
Fig. 10.24 will occur competitively. But only step (a) leads to chain growth, to the formation
of a carbonate linkage in **8**. A certain fraction of the reforming of the carbonyl group, repre-
sented by step (b) in Fig. 10.24, will not advance the polymerization process. However, because step (b) is reversible, the chain-lengthening step (a) will be given another chance,
which is good news. And more good news is that the byproduct of step (a) is the conjugate
base of phenol, which on finding a proton forms phenol. The removal then of phenol from
the system, a consequence of the high temperature and vacuum, will drive the equilibrium
toward path (a) and therefore chain growth. Here we understand the great importance of the
engineering aspects of the process – the necessity to effectively remove the phenol byproduct.
Let's now look even more closely at nucleophilic attack at the acyl carbon of the chain
and the consequent leaving group competition and discover other complexities of the nature of the polymerization process. This is discussed below.

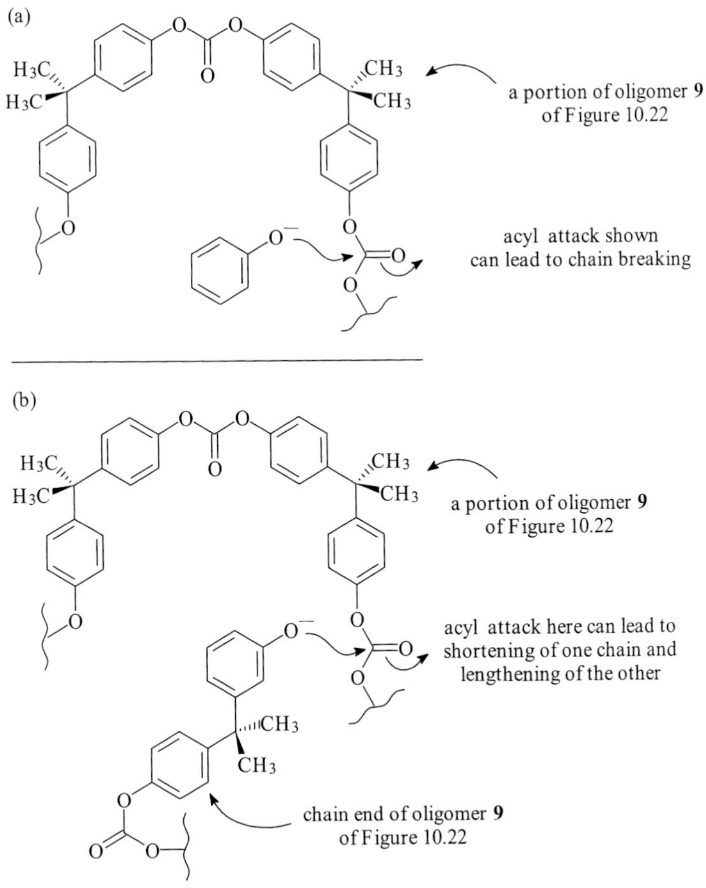

(a)

a portion of oligomer **9**
of Figure 10.22

acyl attack shown
can lead to chain breaking

(b)

a portion of oligomer **9**
of Figure 10.22

acyl attack here can lead to
shortening of one chain and
lengthening of the other

chain end of oligomer **9**
of Figure 10.22

Fig. 10.25 Nucleophilic substitution reactions on polycarbonate chains can lead to lengthening or shortening of the chain.

In the polymerization process, the nucleophilic attack at the carbonyl group, the consequent formation of the four-coordinate intermediate **A** and the reformation of the carbonyl group as in Fig. 10.24, means that as the chain grows it can also be broken as long as phenoxide groups are available. This is shown in Fig. 10.25 (part a) for the attack at an internal linking carbonate group as for example in **9** (Fig. 10.22) by a phenoxide anion, which had not been converted to phenol and removed before it could undertake its chain destructive action. Phenoxide groups arising either at the end of a growing chain via ionization of the phenolic hydroxyl group (Fig. 10.23) or from a bisphenol A (**4**), or from the byproduct phenol (**3**) can add to any carbonyl group along the chain forming the four-coordinate intermediate. Depending on which kind of phenoxide group adds, and then depending on how the resulting four-coordinate intermediate reforms the carbonyl group, the chain either extends, shortens, or breaks. For another example, examine the consequence of the reaction between two molecules of **9**, which as shown in Fig. 10.22 yields

Tab. 10.2 Polycarbonate cost of production (dollars per kg) *

	Phosgene process	*Diphenyl carbonate process*
Cash cost	0.28	0.27
Finance cost	0.19	0.11
Total	0.47	0.37

* Modified from "Polycarbonates," a report published by NEXANT/Chem Systems, White Plains, NY, September, 1998.

the oligomer **10**. This occurs via the reaction of the phenoxide end of one **9** with the phenyl carbonate end of another **9**. However, it is also perfectly possible for the reactive phenoxide end of one **9** to attack a carbonyl group of a carbonate bond anywhere in the chain gaining far less chain length in the two products produced (Fig. 10.25, part (b)). Does it seem like a complex mess of reactions is going on? You're right.

The kind of complex equilibrium described above (Fig. 10.25) can occur in many condensation polymerizations (see Chapter 4, Section 4.10; Chapter 5, Section 5.8) and is therefore of great interest to polymer scientists. In fact, the first one to understand the consequences of such equilibrium on the lengths of the various chains was the same man who first investigated condensation polymerization, the man who created the first nylon, Wallace Carothers. Also contributing to this theory was Paul Flory, a young man whom Carothers hired to work with him at DuPont (Chapter 5). Flory, who left DuPont to go to Cornell, then to the Mellon Institute, and finally to a career at Stanford University, later won a Nobel Prize for essentially creating the field of polymer physical chemistry. Flory's book, which comprehensively portrayed this field, is still important fifty years after it was written. We have come across Carothers and Flory in Chapter 4 (Section 4.2) where these chemists addressed another fundamental problem in polymer science – the nature and consequences of crosslinking.

Another result of the competition between the leaving groups displaced on re-formation of the carbonyl group from the four-coordinate intermediate arising from nucleophilic acyl attack (Figs 10.23–10.25) has a great deal to do with the economics of the process developed by General Electric. Examining the General Electric process in Fig. 10.21, one can see an advantage if dimethyl carbonate (**1**) instead of diphenyl carbonate (**2**) could have been used to produce the polycarbonate. The process cost would have been reduced significantly since the step of conversion to diphenyl carbonate (Fig. 10.21) would have been eliminated. However, consideration of the nature of the leaving group competition discussed above (Figs 10.23–10.25) shows us the necessity of using diphenyl carbonate.

The leaving group competition if dimethyl carbonate (**1**) had been used instead of diphenyl carbonate (**2**) would be between loss of methoxide versus the phenoxide anion. Methoxide is the conjugate base of a far weaker acid, methanol, with an acidity nearly ten orders of magnitude less than phenol, the conjugate acid of phenoxide. This disparity means that the methoxide would leave far less readily, instead forcing the incoming phenoxide out and therefore greatly slowing chain growth.

No industrial process, however much money may be saved, can avoid the principles behind the competitive nature of leaving groups in nucleophilic acyl substitution.

The diphenyl carbonate process, as outlined in Fig. 10.21, despite requiring the synthesis of two carbonate intermediates, **1** and **2**, still has a great financial advantage over formation of polycarbonate using phosgene. It is interesting to look at the economic details shown in the table below. It can be seen that cash costs – raw material, energy, labor – and related factors, are similar. The second item is finance costs. What are they? Finance costs are the depreciation costs plus the interest on the debt incurred to build the plant. Depreciation is based on the amount of money set aside each year so that in a set number of years the money, appropriately invested, would be sufficient to rebuild the plant.

Now why are the finance costs so much higher for the phosgene process? The answer is that the process using phosgene requires expensive equipment to protect workers and to ensure safety for the area surrounding the plant. This makes for a much higher plant cost, which in turn reflects itself in much higher depreciation and cost of capital. Tab. 10.2 is a perfect demonstration of doing well by doing good.

10.14
Let's look at Another Competition, the Production of Methyl Methacrylate, in Terms of Cash and Finance Costs

There is a lesson in industrial economics in seeing how the financial advantages compare in the production of polycarbonate without phosgene versus the production of methyl methacrylate without using HCN. We have already focused on the several methods under development to produce methyl methacrylate to circumvent the classical process based on acetone and HCN. This subject was discussed in Chapter 6, where we learned that the most economical of the competitors to replace the classical acetone/HCN cyanohydrin method (Section 6.3 in Chapter 6), and therefore the method most likely to be commercialized, is based on the oxidation of isobutylene (Fig. 6.10 in Section 6.9 in Chapter 6).

Tab. 10.3 reveals the economics of the isobutylene versus the HCN-acetone route to methyl methacrylate. The savings of 5 cents per kilogram translates to about US$100 million for the 450 million kilograms of methyl methacrylate manufactured in the United States each year. For the polycarbonate processes, most of the cost savings seen in Tab. 10.2 for the non-phosgene route arose from simplicity of engineering the plant. But in the methyl methacrylate situation, the cost savings in the isobutylene route

Tab. 10.3 Methyl methacrylate cost of production (dollars per kg)*

	HCN-acetone process	*Isobutylene oxidation*
Cash cost	0.26	0.22
Finance cost	0.11	0.10
Total	0.37	0.32

* Modified from "Methyl Methacrylate," a report published by NEXANT/Chem Systems, White Plains, NY, September, 2001.

arise primarily from the more expensive raw materials needed for the HCN-acetone route with a smaller contribution from finance costs.

Moreover, the reduced cost of the isobutylene process results despite its lower yield as compared to the HCN-acetone process. The yield is about 50% as compared to over 90% for the older process. Thus improvement in the yield of the isobutylene process, which is certain to occur, will make this process even more competitive by reducing material costs even more than they are in the table below.

The savings in replacing phosgene in polycarbonate production is mostly in plant costs, while the savings in eliminating HCN in production of methyl methacrylate is mostly in material costs. Although different both demonstrate that one way or another new technology can be more economical and safer.

10.15
Reducing the Use of Chlorine in Industrial Processes

Phosgene is produced by combining chlorine with carbon monoxide (Section 10.12), which means that reducing the use of phosgene (as we have seen above in the section on polycarbonate formation) certainly reduces the consumption of chlorine. We have seen another drop in chlorine consumption in the synthesis of vinyl chloride from ethylene. Here, oxychlorination allowed the Cl in the HCl cracked out of the 1,2-dichloroethane to be incorporated into the vinyl chloride instead of wasted (Section 10.10).

Another process that has been changed to reduce chlorine consumption is the synthesis of epichlorohydrin discussed in detail in Chapter 4 (Section 4.8). In the new process developed by Showa-Denko (see Chapter 4, Fig. 4.7) and now increasingly used for production of epichlorohydrin, the starting material is allyl alcohol instead of the allyl chloride used in the original 1936 Shell process (see Fig. 4.6).

This change from allyl chloride to allyl alcohol, in addition to saving chlorine cost, is also another of the multiple demonstrations of the power of catalysis. Both the Shell route to epichlorohydrin (Chapter 4, Section 4.7) and the Showa-Denko processes (Chapter 4, Section 4.8) start with propylene. However, while the Shell processes uses a high temperature of 500 °C to form allyl chloride, the Showa-Denko process uses palladium catalysis under far milder conditions to produce ally alcohol.

There are some interesting economics associated with the competitive routes to epichlorohydrin, because the Shell process – in spite of wasting chlorine and therefore having higher raw material costs – is still more favorable economically. This occurs because of the less expensive plant costs for formation of allyl chloride compared to the formation of allyl alcohol in the Showa-Denko process. Although this situation is not an example of doing well by doing good, because the new process using less chlorine is not more economical, the chemical industry is turning away from the original Shell process for the building of new plants. The reason is to avoid the use of chlorine at very high temperatures, which is central to formation of allyl chloride and offers a great deal of potential for accidents and danger to workers. Here, the industry is doing well by doing good in a different way.

10.16
HCN is a Dangerous Chemical Hastening its Replacement in the Synthesis of Methyl Methacrylate, as we have seen. But its Exquisite Reactivity has Fostered its Use in Other Processes and Particularly in a Potential Process for Getting rid of Ammonium Sulfate as a Byproduct in the Synthesis of Nylon 6

Despite HCN's "unfriendly" nature, to say the least, it has found a very important use in DuPont's synthesis of hexamethylenediamine as we have examined extensively in Chapter 5. The reason that HCN is still considered a viable industrial intermediate in spite of its intrinsic dangerous nature is that DuPont handles HCN with the utmost of care. Indeed, DuPont is the world's largest producer of HCN with a capacity for about 450 million kilograms per year. This ability over many years to handle HCN safely encourages DuPont to think of HCN as a viable reactant for other large-scale processes, especially where HCN use might solve otherwise intractable problems.

One such problem is the production of ammonium sulfate in the synthesis of caprolactam. We have discussed the synthesis of caprolactam and nylon 6 derived from it in Chapter 5 (Section 5.9), where we found how a rearrangement discovered by Ernst Beckmann in 1886, allowing the transformation of oximes to amides, could be applied to the oxime of cyclohexanone to produce caprolactam (Fig. 5.16 in Chapter 5). This chemistry to produce caprolactam has a great deal to do with ammonium sulfate, $(NH_4)_2SO_4$, a byproduct of caprolactam production, which is not to be confused with ammonium bisulfate, NH_4HSO_4. The latter is the byproduct of a different process, the HCN-acetone route to methyl methacrylate (Chapter 6, Section 6.3) and the useless chemical that was one of the primary driving forces for developing improved routes for this monomer for Plexiglas^{TM}.

Ammonium sulfate is a fertilizer for soils that require a lowering of pH. Ammonium bisulfate is too acidic to be used as a fertilizer. However, there is not much agricultural soil in the United States of this kind. So when ammonium sulfate is produced as a byproduct in a chemical process in the United States it must be shipped to the Far East, where there is a great deal of this kind of soil. Shipping bulk chemicals over long distances is expensive, so this added cost is the problem – a big problem.

The classical process for producing caprolactam via the oxime rearrangement (Chapter 5, Section 5.9) yields 2.0 kg of ammonium sulfate for each kilogram of caprolactam!! The ammonium sulfate does not arise from one step in the process but rather surprisingly from each step as follows: preparation of hydroxylamine sulfate, 0.7 kg; production of cyclohexanone oxime, 0.5 kg; Beckmann rearrangement, 0.8 kg, for a total of 2.0 kg for each kilogram of caprolactam produced (see Chapter 5, Fig. 5.16).

The goal of reducing the amount of ammonium sulfate in the United States and Europe stimulated the industry to develop no less than ten processes to produce caprolactam with less or no ammonium sulfate byproduct. Paradoxically, from the general point of view of this chapter, it was found that the most economical of these new processes uses HCN. We agree to one bad player, HCN, which we can handle, to get rid of another, NH_4SO_4.

Let's examine the process using HCN to produce caprolactam and another, which does not use HCN, neither of which has yet been commercialized. DuPont and BASF pioneered the former, while DuPont and DSM devised the latter.

10.17

Routes to Caprolactam that Avoid Production of Ammonium Sulfate

Hydroformylation is a process we have looked at in considerable mechanistic detail in the formation of butyraldehyde from propylene (see Chapter 9, Section 9.9). In Fig. 10.26 we come across this transition metal-catalyzed chemistry again. Step 1 in the DuPont/ DSM method for synthesis of caprolactam (Fig. 10.26) is a variation on hydroformylation in which a methoxy group, –OCH$_3$, replaces the hydrogen, which forms the normal alde-hyde product (Chapter 9, Fig. 9.0). Step 3, on the other hand, is precisely a hydroformyla-tion exactly parallel to that occurring with propylene (Fig. 9.15, Chapter 9). These two steps, step 1 and step 3, with the critical intervention of the isomerizing step 2, are cata-lyzed by cobalt catalysts or other transition metals in the same group VIII of the periodic table, and yield 5-formyl methyl valerate (Fig. 10.26).

One of the peculiar aspects of step 1 in Fig. 10.26 is that the double bond shifts from the original terminal position in the butadiene starting material to the 2,3-position. These kinds of shifts of double bonds are common in the interaction of alkenes with transition metal hydrides in a process that is not related directly to the formation of the methyl ester group and can arise by a variety of mechanisms. These mechanisms can be complex. Shifting the double bond back to its terminal position in step 2 (Fig. 10.26) is accomplished catalytically – a reaction we cannot go into here because we, and perhaps others, do not clearly understand it.

With an aldehyde group at one terminus of four methylene groups and an ester group at the other, 5-formyl methyl valerate is two conventional chemical steps from caprolactam. Reaction of the aldehyde group with ammonia forms the imine, which is then catalytically reduced with hydrogen to methyl 6-aminocaproate (step 4). The latter compound is the methyl ester of 6-aminocaproic acid, the molecule that Carothers used in his failed attempt to synthesize nylon 6 but where he instead obtained caprolactam (Fig. 5.16 of Chapter 5). For the DuPont and DSM process however, caprolactam is the intended target because

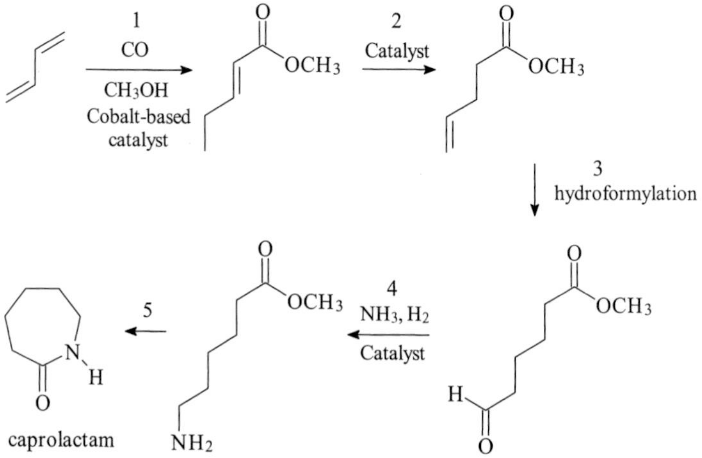

Fig. 10.26 The route to caprolactam developed by DuPont and DSM.

now it is understood how to convert this ring compound to nylon 6 (Chapter 5, Section 5.9). And the caprolactam is formed *without* ammonium sulfate, the goal of the industrial innovation outlined in Fig. 10.26.

Now let's turn our attention to the route developed by DuPont and BASF (Fig. 10.27).

When we analyze below the economic consequences of the classical route to caprolactam (Chapter 5, Figs 5.16 and 5.17) compared to the two new routes (Figs 10.26 and 10.27 discussed here), we shall discover the great advantage of the DuPont-BASF innovation, which is that both caprolactam and hexamethylene diamine are produced. A single process produces the key intermediates for both nylon 6 and nylon 6,6 (Fig. 10.27). The key to this industrial efficiency is adiponitrile, an intermediate produced directly from butadiene and HCN.

Although not created by Reppe – the master of acetylene chemistry (Chapter 8, Section 8.13) – the synthesis of adiponitrile from butadiene (Fig. 10.27) uses zero-valent nickel catalysis, as did Reppe in the synthesis of acrylic acid from acetylene. In both the production of adiponitrile (Fig. 10.27) and acrylic acid (Chapter 8, Figs 8.16 and 8.17) the catalyst has a nickel-to-hydrogen bond with the nickel attended by three ligands, that is, H–Ni(L)$_3$. In both situations, the Ni–H bond adds across a π-bond to form a nickel–carbon bond.

Two of the three ligands in the acrylic acid synthesis are CO, one of which will play an active role in inserting itself between the metal–carbon bond (Chapter 8, Fig. 8.17), while in the adiponitrile synthesis considered here (Fig. 10.27) this role is played by cyanide,

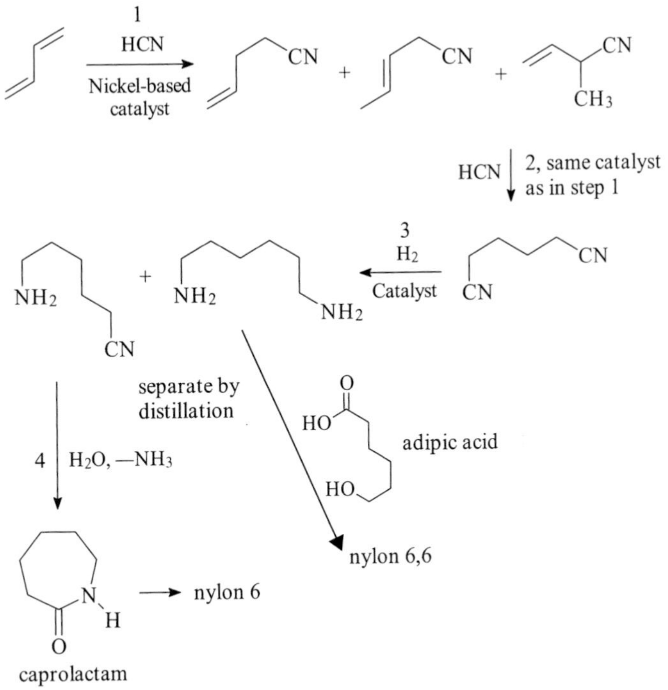

Fig. 10.27 The route to caprolactam developed by DuPont and BASF.

Tab. 10.4 Nylon production costs (dollars per kg) *

	Conventional process (Figure 5.16)	Butadiene carbonylation (Figure 10.26)	Butadiene hydrocyanation (Figure 10.27)
Cash costs	0.25	0.26	0.11
Finance costs	0.19	0.14	0.27
Total cost	0.45	0.41	0.38

* Modified from "Nylon 6/Nylon 6,6," a report published by NEXANT/Chem Systems, White Plains, NY, March, 2000.

CN^- so that the catalyst structure is $HNi(L)_2CN$. The most important steps, addition of Ni–H across a π-bond and insertion of a ligand in the Ni–carbon bond are related in both the formation of acrylic acid from acetylene (Chapter 8, Figs 8.16 and 8.17) and here in the addition of HCN across the double bond in 1,3-butadiene (Fig. 10.27).

Addition of HCN across the double bonds of 1,3-butadiene leads to unsaturated isomers as shown in step 1 of Fig. 10.27. The isomer mixture is the reason that the amount of HCN used in step 1 is limited so that the isomers do not go on to unwanted dinitrile products. The same catalyst as used for step 1 then causes the isomerization to the desired olefin precursor, which is reacted with more HCN to produce adiponitrile (step 2). The precise details of how this is controlled are not revealed. In industrial work, where yield and formation of specific isomers is key to profitability, these unrevealed details of how the process is conducted are critical to commercial success or failure. It makes sense therefore that competition for such an overall valuable route to both nylon 6,6 and nylon 6 would cause certain key details to be "closely held."

The hydrogenation step 3 (Fig. 10.27) produces a mixture of the hexamethylene diamine, by reducing both nitrile groups, while reduction of one of the nitrile groups produces the monoamine, the intermediate for production of caprolactam (Fig. 10.27, step 4). The monoamine and diamine differ widely in boiling point, allowing easy separation.

The process shown in Fig. 10.27 is quite profitable. Let's examine the economics in detail, that is, the cost per kilogram of nylon produced for the three processes (Tab. 10.4) and discover the source of the economic efficiency of the butadiene hydrocyanation process (Fig. 10.27). The butadiene carbonylation has a lower finance cost than the conventional process, primarily because no ammonium sulfate is produced and it is not necessary to pay for the part of the factory to arrange to ship the ammonium sulfate to fertilizer markets in the Far East. But why is the finance cost so high for the hydrocyanation process? The reason is the expensive equipment to protect workers and the surrounding environment from dangerous HCN. However, in spite of this high finance cost, the hydrocyanation process is the most economical with an exceptionally low cash cost. The reason is that this is a 2-for-1 process, producing both caprolactam, and hexamethylene diamine the intermediates for nylon 6 and nylon 6,6. Each intermediate therefore becomes an income credit for the other.

Simply put, while the conventional process produces an unwanted byproduct, ammonium sulfate, which adds costs, the hydrocyanation route to caprolactam produces a byproduct, hexamethylene diamine, which can be sold or used to produce nylon 6,6, producing income. This is a great advantage of the clever chemistry.

In a big surprise, however, DuPont in the early 2000s withdrew its participation in both of these processes. Subsequently there was a stunning announcement that DuPont is exiting from three fiber businesses, nylon, polyesters and spandex. DuPont feels that nylon has reached a degree of maturity that negates increased profitability. Polyesters have grown rapidly especially in the Far East. The resulting great deal of overproduction limited price for the polyesters. In spandex (see Chapter 7, Sections 7.12 and 7.13) profitability has been impeded by competition, also therefore limiting price. A major source of the profitability problems in all three fibers arises from the fact that patent protection has expired, which is the equivalent of maturity in these markets.

That's the chemical industry and that is the end of our story.

10.18
Summary

We started with the idea that solving problems about the kinds of chemicals used in the industry could lead to higher profit because innovation may lead to higher efficiency. We applied these ideas to the worst players in the chemical industry – acetylene, phosgene, chlorine, and hydrocyanic acid – and looked in detail at the problems associated with these molecules.

Acetylene is expensive to produce because its high reactivity is associated with its high energy, which means that large amounts of expensive energy must be used to produce acetylene. And this same high-energy content means that acetylene is dangerously explosive. We learned something about the history of acetylene and how much difference it makes if acetylene is to be used in large or small amounts.

But acetylene is such a great reactant. Its readily available π-bonds, allowing complexation with transition metals, open easy reaction paths to valuable industrial intermediates such as vinyl acetate and vinyl chloride. One can add to this other important industrial intermediates, which altogether have a value to the industry well in excess of US$12 billion. But the problems with acetylene have forced the chemical industry to turn away from it and to produce these intermediates in other ways. And we studied these other ways as well and discovered the advantages of the innovations. In the course of this effort we learned some history of both chemistry and the chemical industry.

But no matter how much innovation the industry applies, there is still one product that is difficult to divert from an acetylene source, and that is 1,4-butanediol. And the reaction path to this important intermediate demonstrates another important reactive characteristic of acetylene, the acidity of hydrogen bound to triple bonds. In the route to 1,4-butynediol from acetylene, which leads to 1,4-butanediol via hydrogenation, we found that copper plays a special role because the copper salt of acetylene is insoluble and this makes all the difference.

Acetylene is not the only bad actor. The fact that both phosgene and chlorine have been used as poison gases in war testifies to their danger. And similarly the use of HCN as the most effective chemical around to kill millions of people in the holocaust during World War II certainly puts this compound in a bad light.

First, we turned to chlorine and looked back in the eighteenth and nineteenth centuries to see its history and how discoveries of a disciple of a famous nineteenth century chemist invented a process that allows vinyl chloride to be made from ethylene without wasting any chlorine. Waste of chlorine has long been a driving force in industrial innovation and what could be worse than the use of phosgene to produce polycarbonate in which both chlorines are thrown away, as we have seen in Chapter 4. So we turned our attention to innovation at General Electric and Bayer, where new methods yielded commercially important ways to produce polycarbonate without phosgene. This led us into the world of ester exchange reactions, and we focused on the principles involved and the commercial importance of leaving group chemistry and the intervention of the ubiquitous four-coordinate intermediate encountered in all nucleophilic acyl substitution chemistry.

Finally, we turned our attention to HCN, and from what we already know it is certainly a bad neighbor in the chemical industry. But DuPont, after many years of experience, has learned how to tame HCN, allowing the use of this chemical to solve an important problem in the production of caprolactam, the intermediate necessary for the production of a key nylon, nylon 6. The problem is that large amounts of ammonium sulfate are produced in the formation of caprolactam via the time-honored Beckmann rearrangement route, in fact larger amounts than the desired product. What is one to do with this salt for which the markets as a fertilizer may be far away from the production site?

The answer is to find another route to caprolactam in which ammonium sulfate is not a byproduct, and it turns out that HCN plays a key role here. So we handle one bad player carefully and get rid of a big problem. And we have the extra benefit that this route to caprolactam also produces hexamethylene diamine necessary for production of nylon 6,6. The economics are outstanding in this process. As you might imagine, catalysis is essential in this innovation and of course it is those marvelous transition metals from Group VIII that do the trick in variations of Reppe chemistry and hydroformylation.

Some of the subjects treated in this chapter are listed below. These are key words and terms that act as reminders of the chapter's contents and should become a valuable part of your chemical vocabulary.

- Hybridization and electronegativity
- Acidity of terminal acetylene
- Carbide salts and their reaction with water
- Danger of acetylene
- History of acetylene
- π-Complexation
- Coordinate covalent bonds
- Vinyl chloride and poly(vinyl chloride)
- Vinyl acetate and poly(vinyl acetate) and poly(vinyl alcohol)
- Elmer's Glue
- Wacker reaction
- Tautomers and proving the concept of macromolecules
- Transesterification

- Nucleophilic acyl substitution
- Deacon chemistry
- Oxychlorination
- Formaldehyde
- 2-Butyne-1,4-diol
- 1,4-Butanediol
- Tetrahydrofuran
- Pyrrolidones
- Copper acetylide
- Carbonates
- Phosgene
- Hydrocyanic acid
- Chlorine
- Polycarbonate
- Leaving groups
- Caprolactam
- Nylon 6
- Nylon 6,6

Study Guide Problems for Chapter 10

1. Give one example each of a chemical reaction that exemplifies the reactive characteristics of each of the following chemicals: HCN, C_2H_2, $COCl_2$ and Cl_2.

2. Write a reasonable mechanism for the reaction of water with carbide salts such as K_2C_2 and CaC_2. What accounts for the large amount of heat released on reaction of these carbide salts with water? Would you expect these salts to react with methanol and if so what would be the products?

3. Propose a structure for $C_2^=$ and account for the fact that there are two negative charges within such a small molecule.

4. Propose a synthesis of sodium amide (Fig. 10.3).

5. The pK_a of ethanol is about 16. Could $C_2H_5O^-$ be used to form an acetylide anion? The pK_a of methane is about 60. Could CH_3^- be used to form an acetylide anion? Propose methods of synthesis of the two bases noted above.

6. Show how isoprene, which plays an important role in the synthesis of elastomers (Chapter 7), could be synthesized from the two industrial intermediates, acetone and acetylene. Write a mechanism for each step proposed.

7. Offer a reason for the experimentally observed connection between the chemical shift in the nuclear magnetic resonance spectrum of carbon and the hybridization of the carbon.

8. Using the molecular fragments suggested to arise in the formation of acetylene under the high temperature arc conditions described in the text (Section 10.6), predict the isotopic mixture of acetylene that could be detected by a mass spectrometer starting from a mixture of acetylene made from carbon-12 and acetylene made from carbon-13 and H_2 and D_2.

9. Propose a series of chemical reactions that could lead to benzene and graphite from the molecular fragments, C_2H, C_2, CH and H, proposed to be present in the electric

arc process for production of acetylene. How might the reactions you proposed be blocked by the presence of H_2 and H?

10. Propose syntheses that do not use transition metals for as many of the industrial intermediates in Fig. 10.6 as possible.

11. In what way do you see an analogy between the transition metal-catalyzed reactions in Fig. 10.7 and addition of bromine to acetylene?

12. Why are LDPE and not HDPE (Chapter 2) used for the plastic squeeze bottles for the marketing of Elmer's Glue?

13. Why were many physical chemists in the early part of the nineteenth century surprised when on hydrolysis of poly(vinyl acetate) no carbonyl groups could be detected?

14. Fig. 10.9 shows the free radical polymerization of vinyl acetate, but the structure of the polymer shown is oversimplified. Just as for low-density polyethylene, LDPE discussed in Chapter 2, the poly(vinyl acetate) is a branched structure. Draw a portion of the polymer chain showing the structure of these branches and write a mechanism for the free radical polymerization that can account for the branches.

15. The acetals shown in Fig. 10.12 tell only part of the story. What other acetal products could arise in the reaction of poly(vinyl alcohol) with the aldehydes shown in this figure? Could these products greatly change the macromolecular material properties?

16. We have come across the reactions between hydroxyl and carbonyl groups in Chapter 3 in the formation of phenol and acetone from cumene hydroperoxide. Outline the parallel mechanistic steps in the formation of the ketals and the acetals shown in Fig. 10.12.

17. Regarding the formation of hemiacetals, hemiketals, acetals and ketals, review Questions 14 and 15 in Chapter 3.

18. Propose termination steps for the free radical chain process shown in Fig. 10.13.

19. Trace all changes in oxidation state and account for all electrons for all the atoms in the conversion of HCl to cupric chloride. Do the same for the oxidation of HCl to Cl_2.

20. It's not obvious how $CuCl_2$ reacts with ethylene to produce ethylene dichloride, CH_2ClCH_2Cl (Fig. 10.13), although it is known that free Cl_2 is not involved. Can you propose a mechanism?

21. Amines, RNH_2, react with phosgene, $COCl_2$, to produce isocyanates, $R-N=C=O$. Write out all steps in a proposed mechanism.

22. Nucleophilic acyl substitution is involved in the conversion of the lactone in Fig. 10.16 to pyrrolidone and N-methylpyrrolidone. Propose mechanisms for these reactions.

23. Do you see a parallel in the conversion of pyrrolidone to N-vinylpyrrolidone compared to the conversion of acetylene to either vinyl chloride or vinyl acetate? Propose a catalyst for the formation of N-vinylpyrrolidone in Fig. 10.16.

24. In Question 6 above, acetone reacts with the acetylide anion. What factors control the relative rate of reaction of the acetylide anion with acetone versus formaldehyde and which reaction would be expected to be faster? Do you see any analogies to the aldol reaction discussed in Chapter 9?

25. Lithium diisopropyl amide, $Li\ N(CH(CH_3)_2)_2$ is a commonly used strong base for laboratory syntheses, and can be used to form the acetylide anion from terminal triple bond compounds. Show how this base can be synthesized from the following starting materials: isopropylamine, lithium metal, methane, and bromine.

26. What kinds of chemical reactions could you imagine might take place on inhaling phosgene or chlorine, two of the poison gases used in warfare in the twentieth century?
27. Fill in the detailed mechanism for the synthesis of toluene diisocyanate shown in Fig. 10.20. Propose a synthesis of the toluene diamine starting material to prepare the diisocyanate.
28. Write all steps in a detailed mechanism for the transformation of 4+8 to 9 in Fig. 10.22.
29. If the carbonate intermediate derived from dimethyl carbonate could be formed, could this intermediate lead to a polymer following the reaction sequence in Fig. 10.22? Answer the question in terms of the four-coordinate intermediate that would be formed in the linking of **9** with **9** to form **10** if a methyl group replaced the phenyl group at the chain end of **9**.
30. In the synthesis of diphenyl carbonate from dimethyl carbonate as shown in Fig. 10.21, why is it absolutely essential to adjust the reaction conditions to remove methanol as it is formed?
31. Explain the essential difference between finance costs and cash costs in the economics of the chemical industry.
32. The first step forming the unsaturated ester in Fig. 10.26 can be catalyzed by the same cobalt carbonyl catalyst used for hydroformylation studied in Chapter 9 to synthesize butyraldehyde from propylene. However, a variation is required to yield the ester instead of the aldehyde. What might this variation be and how might the whole process be described in mechanistic terms?
33. In the first step in Fig. 10.26, the product is an unsaturated ester with the remaining double bond shifted to a new position compared to the two double bonds in 1,3-butadiene. Catalysts allow shifting this double bond back to the terminal position, which then allows a conventional hydroformylation to produce the 5-formylmethyl valerate. Show all reagents and write a mechanism for this hydroformylation.
34. Show all intermediates and the mechanism for their formation in the conversion of 5-formyl valerate to 6-aminocaproic acid as shown in Fig. 10.26.
35. What byproducts(s) might be expected in the first step in Fig. 10.26?
36. What byproducts might be expected for the third step in Fig. 10.26 in which the unsaturated ester is converted to 5-formyl methyl valerate?
37. Show all mechanistic details in the conversion of 6-aminocaproic acid to caprolactam (Fig. 10.26).
38. The key intermediate in Fig. 10.27 is adiponitrile, which has been produced in an electrochemical process developed by Monsanto and discussed in Chapter 5. Show all steps and mechanistic details of the electrochemical process and offer a reason why the butadiene route in Fig. 10.27 is more economical than the electrochemical route.
39. Using the Reppe process for production of acrylic acid shown in Chapter 8 as a model, develop a series of mechanistic steps accounting for the nickel-catalyzed addition of HCN across one of the double bonds in 1,3-butadiene (Fig. 10.27).
40. There is a big difference between nickel-catalyzed Reppe addition to acetylene and the nickel-catalyzed addition of HCN to 1,3-butadiene, that is, the formation of regioisomers. Explain and support this statement with chemical details.

An Epilogue – The Future

This book has dealt with the here and the now–how the 1.5 trillion dollars worth of chemicals produced annually by the global chemical industry relates to the theory that organic chemistry students learn. But what about the future? The theory may not change much. But the chemistry will change a great deal. Let's devote a few pages of this book to the future.

Today, 95% of the 800 million tonnes of chemicals produced in the world are based on petroleum and, to a lesser extent, on gas. But we are running out of petroleum. In 1910 it was predicted that petroleum would be exhausted in 50 years, though obviously this was a wrong prediction. But predictions judged wrong are often correct, just off in their timing. The classic example was provided in 1798 by the English demographer, Malthus, who extrapolated growth of population and growth of food production. By 1900 the two curves crossed and, accordingly, Malthus predicted that there would be rampant starvation at that time. Feedback factors such as improvements in agriculture proved him wrong. But today there are those who ask if his prediction was wrong only in its timing.

Being wrong in timing was obviously so with the 1910 prediction about oil supplies. For today in the early 2000s it is pretty well agreed that petroleum supplies will not last beyond a generation or two. This prediction takes into account the fact that a large amount of oil is left in the rock formation simply because the pumping process used today can extract only a relatively small percentage of it. So-called enhanced recovery techniques can produce considerably more. However, these techniques are expensive and will not be used extensively if cheaper alternatives are available.

BP, the world's fourth largest petroleum company, has announced that BP no longer stands for British Petroleum. Rather, it designates "beyond petroleum." And BP was the world's biggest supplier of solar energy equipment in the early 2000s. Similarly, the Chief Executive Officer of Atlantic Richfield Petroleum Company is said to have stated, "We have embarked on the beginning of the last days of the age of oil."

So what is next? A large part of the answer lies in gas. In 2002, petroleum provided 50% of our global energy, with 25% each coming from coal and natural gas. It is predicted that petroleum use will decrease markedly during the next two decades in favor of gas. Coal, apparently, will hold its own.

Gas has much to commend it, particularly from the point of view of pollution reduction. Moreover, in the 1980s huge amounts of gas were discovered in the world. One estimate indicates that gas reserves, expressed as oil equivalents, are four times as great as oil reserves! Much of the gas is in remote locations and is called "stranded" gas. How is this gas to be valorized – a term used in this industry which means how is the gas to be brought to commercial use? Gas is difficult and expensive to transport. The best way is by pipeline, but economics determine how long a pipeline can be. And the stranded gas is often so far away from where it would be used that a pipeline is simply not feasible. Another approach is to transform the gas at its source to a transportable material. One reasonable possibility is to convert the gas to synthesis gas, CO and H_2, which can then be transformed to liquid materials that can be shipped without using pipelines. This technology goes back to the industrial revolution when it was learned how to convert coal to synthesis gas for use as cooking gas as well as gas for energy. The major reaction was

$$C + H_2O \Rightarrow CO + H_2$$

Today, the major uses for synthesis gas from natural gas ($CH_4 + H_2O \rightarrow 3H_2 + CO$) are for the production of methanol and ammonia. However, there are two major developments based on synthesis gas that could influence markedly the future of the chemical industry. Both are being developed aggressively as this book goes to press. One is methanol cracking to olefins for which zeolites play a critical role, and the other is conversion of CO and H_2, via the so-called Fischer-Tropsch reaction, to a petroleum-like product. This is termed gas-to-liquids, or GTL.

There are several signals of change in the early 2000s that could seriously affect the global chemical industry, as we know it today. Most are beyond the scope of this book, but let's briefly examine three of them: the functionalization of alkanes; the use of immobilized enzymes; and fermentation with engineered organisms.

The functionalization of alkanes – that is, the use of ethane or propane to make functional compounds instead of using alkenes, ethylene and propylene – can make us less dependent on expensive, energy-consuming steam crackers. Success depends on catalyst development, and in the early 2000s two processes appeared ready for commercialization. The first is the conversion of ethane to vinyl chloride, and the second is the conversion of propane to acrylonitrile.

The ability to immobilize enzymes, essentially converting them to heterogeneous catalysts, makes enzymes much more practical. An Israeli scientist, E. Katzir, who later became the president of Israel, devised the immobilization process. Thousands of enzymes are in existence, and these carry out numerous chemical processes with high selectivity and low energy input. But their isolation is expensive and they are destroyed in a batch reaction. The advent of immobilized enzymes made possible the development of high-fructose corn syrup from glucose. Today, this sweeter-then-sucrose product is used in all soft drinks and has replaced at least 25% of sugar use in the USA.

The advantage of enzyme technology was demonstrated by the Japanese company, Nitto, who immobilized an enzyme that can convert a nitrile to an amide, and applied this to the production of acrylamide from acrylonitrile $CH_2 = CH - CN \Rightarrow CH_2 = CH - CONH_2$. Acrylamide is polymerized to polyacrylamide useful for the coagulation of very finely divided particles or "fines" from process water such as the water that results from papermaking or from processes for mineral production. The water can then be reused.

This enzyme chemistry caught on rapidly, and by 2002 there were at least six plants using this technology. It was also applied to the synthesis of the B-vitamin, nicotinamide, from cyanopyridine. There are numerous chemicals whose processing could be improved by use of immobilized enzymes.

Fermentation has been known since biblical times. Most primitive people knew how to ferment grain to make an alcoholic beverage resembling beer. Until the 1960s fermentation was used to make acetone and *n*-butanol. Lactic acid and citric acid are still made by fermentation since these processes are more economical than chemical processes. Fermentation is used to make antibiotics, the synthesis of which involves expensive multi-step processes. And one step in the classical vitamin C process is carried out using fermentation. But fermentation is a very slow process and frequently not very selective, providing byproducts as exemplified by the production of butyl and higher alcohols in fermentation to produce ethanol. Modern biotechnology makes possible the engineering of enzymes that cause the reaction they catalyze to proceed more quickly and with greater selectivity to the desired product.

Two examples demonstrate this new approach to chemical synthesis in the early 2000s. One was the development of a biodegradable polymer, poly(lactic acid) by Dow and Cargill. This polymer has been known for decades, but its economics depends on the cost of the monomer, lactic acid. This was made by a slow fermentation process, satisfactory for a relatively expensive food-grade product. But genetic engineering led to the production of an organism that improved the economics of the fermentation markedly so that it can be considered as a monomer for a semi-commodity polymer.

The second example is the production of 1,3-propanediol. Shell devised a unique chemical synthesis based on the hydroformylation of ethylene oxide, $C_2H_4O + 2H_2 + CO \Rightarrow HO(CH_2)_3OH$. This allowed them to make economically poly(trimethylene terephthalate), PTT, an analogue of poly(ethylene terephthalate), PET, made from ethylene glycol and terephthalic acid.

PTT has superior properties for carpet manufacture. Since most carpeting is made from nylon and DuPont enjoys much of this business, DuPont decided that it too would make PTT. But where to get the 1,3-propanediol? Shell would not license their process. The answer, DuPont hopes, is by fermentation with an engineered organism using glucose as a substrate. Their success or failure should be evident by the mid-2000s. If successful, their work will stimulate research aimed at producing many other specialty and semi-commodity chemicals by fermentation with engineered organisms.

We have discussed only a tiny part of a future that, at the very least, promises to be an exciting one. And it is particularly important to you because you have to spend all the rest of your life in it! Bon Voyage!!

Books for Further Study and Reference

K. Weissermel and H.-J. Arpe, *Industrial Organic Chemistry*, 4th edition, VCH Publisher, 2003.

H. A. Wittcoff and B G. Reuben, *Industrial Organic Chemicals*, Wiley-Interscience Publisher, 1996.

J. M. Tedder, A. Nechvatal and A. H. Jubb, *Basic Organic Chemistry, Part 5, Industrial Products*, John Wiley & Sons, 1975.

An organic chemistry textbook directed to the use of sophomore students.

A. Streitwieser, Jr. and C. H. Heathcock, *Introduction to Organic Chemistry*, 3rd edition, Macmillan Publisher, 1985.

C. R. Noller, *Chemistry of Organic Compounds*, W.B. Saunders Publisher, 1965.

J.K. Stille, *Industrial Organic Chemistry*, Prentice-Hall, Inc., 1968.

F. A. Carey and R. J. Sundberg, *Advanced Organic Chemistry, Parts A and B*, 3rd edition, Plenum Press, 1990.

F. A. Cotton and G. Wilkinson, *Advanced Inorganic Chemistry, A Comprehensive Text*, 3rd edition, Interscience Publishers, 1972.

H. Morawetz, *Polymers, the Origins and Growth of a Science*, Dover Publishers, 1995.

F. Aftalion (translated by O.T. Benfey), *A History of the International Chemical Industry*, University of Pennsylvania Press, 1991.

A.J. Ihde, *The Development of Modern Chemistry*, Harper & Row Publishers, 1964.

J. R. Partington, *A History of Chemistry*, Volume 3 (1962) and Volume 4 (1964), Macmillan Publishers.

I. Asimov, *A Short History of Chemistry*, by Doubleday & Company Publishers, 1965.

H. A. Liebhafsky, with S. S. Liebhafsky and G. Wise, *Silicones Under the Monogram*, John Wiley & Sons, 1978.

S. Hochheiser, *Rohm and Haas, History of a Chemical Company*, University of Pennsylvania Press, 1986.

D. Yergin, *The Prize, The Epic Quest for Oil, Money and Power*, Touchstone Press, 1991.

P.H. Spitz, *Petrochemicals, The Rise of an Industry*, John Wiley & Sons, 1988.

M.E. Hermes, *Enough for One Lifetime, Wallace Carothers, Inventor of Nylon*, American Chemical Society and Chemical Heritage Society, 1996.

S. Fenichell, *Plastic, The Making of a Synthetic Century*, HarperCollins Publisher, 1996.

W. Brushwell, *Coatings Update*, American Paint Journal, 1974.

Kirk-Othmer Encyclopedia of Chemical Technology, 4th edition, Wiley Publishers.

G. Odian, *Principles of Polymerization*, 3rd edition, Wiley-Interscience Publishers, 1991.

P.J. Flory, *Principles of Polymer Chemistry*, Cornell University Press, 1953.

J.E. Mark and B. Erman, *Rubberlike Elasticity, A Molecular Primer*, Wiley-Interscience Publishers, 1988.

G. W. Parshall, *Homogeneous Catalysis, The Applications and Chemistry of Catalysis by Soluble Metal Complexes*, John Wiley & Sons, 1980.

Index